Muon Spin Spectroscopy

Muon Spin Spectroscopy

Methods and Applications in Chemistry and Materials Science

Donald G. Fleming
Iain McKenzie
Paul W. Percival

WILEY VCH

Authors

Prof. Donald G. Fleming
TRIUMF and Department of Chemistry
University of British Columbia
Vancouver, BC
Canada V6T 1Z1

Dr. Iain McKenzie
TRIUMF, Centre for Molecular and
Materials Science
Vancouver, BC
Canada V6T 2A3

Department of Chemistry
Simon Fraser University
Burnaby, BC
Canada V5A 1S6

Department of Physics and Astronomy
University of Waterloo
Waterloo, ON
Canada N2L 3G1

Prof. Paul W. Percival
TRIUMF and Department of Chemistry
Simon Fraser University
Burnaby, BC
Canada V5A 1S6

Cover Image: © Prof. Paul W. Percival,
Simon Fraser University, Burnaby,
Canada

Library of Congress Card No.: applied for

British Library Cataloguing-in-Publication Data
A catalogue record for this book is available from the British Library.

Bibliographic information published by the Deutsche Nationalbibliothek
The Deutsche Nationalbibliothek lists this publication in the Deutsche Nationalbibliografie; detailed bibliographic data are available on the Internet at <http://dnb.d-nb.de>.

Print ISBN: 978-3-527-34236-5
ePDF ISBN: 978-3-527-80608-9
ePub ISBN: 978-3-527-80606-5
oBook ISBN: 978-3-527-80605-8

Typesetting: Straive, Chennai, India
Printing and Binding: CPI Group (UK) Ltd, Croydon, CR0 4YY

C9783527342365_290124

Contents

Preface

The theme of our book is the application of muons in chemistry and materials science. There are a few introductory chapters to provide the background of muonium chemistry, including an introduction to the techniques of muon spin spectroscopy. We then continue with specific theme areas of varying degrees of complexity, from chemical reactivity in the gas phase to materials science, even biological systems. A final chapter serves to summarize those aspects of the muon which make it such a valuable probe of chemical systems and suggests how the field may develop in the future.

Although muon spin spectroscopy has its origins in particle physics (the discovery of parity violation in the nuclear weak interaction in 1957), it has gradually evolved into a practical tool which can be applied over a wide range of topics in physics and chemistry, now also with potential applications in the biological sciences. Essential to this evolution was the production of high-intensity spin-polarized muon beams and the development of various spectroscopic techniques. The physicists responsible for this work naturally applied their new tools in familiar areas, e.g. atomic spectroscopy in the gas phase or muon diffusion in simple solids. Although studies in chemistry were included from the early days of muon spin spectroscopy, the advancement of muon and muonium chemistry over the past 50 years has largely been due to the efforts of a small number of specialists.

That situation has changed in recent years. Muon sources are inevitably sited at particle accelerators, but those institutions now welcome users from a wide range of scientific disciplines. Indeed, the accessibility of muon beams to non-expert users has now become reality, with all the major muon facilities in the world operating in a 'user-friendly' mode, implemented by facility scientists and technical support personnel. The modus operandi of 'suitcase physicists' is now increasingly common to chemists, who frequently travel to neutron, X-ray and synchrotron light sources. But how will new users with a chemistry background learn about the many facets of muon spin spectroscopy? Since the focus of their research will be on chemistry, they cannot be expected to be familiar with the variety of experimental tools which might benefit their investigations. Our book aims to introduce researchers in the chemical sciences to the merits of muon spin spectroscopy, in a reference text that includes the key methods and typical applications, written at a graduate student level.

To date there has been only one book dedicated to muonium chemistry, that of David Walker, titled *Muon and Muonium Chemistry* (Cambridge University Press, 1983; reprinted but not revised in 2009). Although this text provided a reasonably complete presentation of the subject at the time, it predates many important advances, such as the development of avoided level-crossing resonance, a muon spectroscopic technique that is key to the study of muoniated radicals. A later book filled this gap and indeed concentrated on muoniated radicals: *The Positive Muon as a Probe in Free Radical Chemistry* (Springer-Verlag, 1988) by Emil Roduner. Nevertheless, there has been tremendous growth in the applications of muon spin spectroscopy in the three decades since. Our book aims to bring the subject up to date. Some readers will use it as a comprehensive guide to the current state of the art in the utility of muons in the chemical sciences, while others will read it more selectively, in order to judge the potential merit of applying the techniques of muon spin spectroscopy in their own fields of research.

While the major focus of our work is on the utility of positive muons, the capture of negative muons by helium leads to the formation of the charge-neutral $^{4}He\mu$ atom, which behaves in matter as the heaviest isotope (4.11 u) of the H-atom, making it possible to probe isotope effects via muon spin spectroscopy over a remarkable factor of 36 in atomic mass.

In addition to developing experimental techniques, much of the early work in muonium chemistry focused on understanding how the introduction of a positive muon inherently perturbs the molecular environment. Over the years the emphasis has shifted from studying the effects of muon irradiation to utilizing the muon as a probe of chemical systems not easily studied by other means.

Our accumulated experience in the field of muonium chemistry spans well over a century, but our understanding and appreciation of muon spin spectroscopy and its applications was influenced by sage advice from many individuals. In particular we would like to thank Jess Brewer, Jonathan Connor, Steve Cottrell, Steve Cox, Dave Garner, Hanns Fischer, Fredy Gygax, Jörn Manz, Emil Roduner, Alex Schenck, George Schatz, Robert Scheuermann and Don Truhlar. In addition, our own research would not have been possible without the help and expertise of Donald Arseneau and Jean-Claude Brodovitch.

Vancouver
January 2024

Donald G. Fleming
Iain McKenzie
Paul W. Percival

1

Perspective and Introductory Remarks

1.1 What Do Muons Bring to Chemistry?

Like their leptonic cousins, the electron and positron, muons come in two charge states, μ^+ and μ^-. They were first discovered in cosmic ray showers [1] and have been actively studied in accelerator-based experiments ever since. The discovery of parity violation in muon decay [2, 3] soon led to studies of the interactions of muons in matter and the development of various experimental techniques commonly referred to as 'μSR', for muon spin rotation/relaxation/resonance, and here collectively known as 'Muon Spin Spectroscopy'.

The fundamentals and applications of μSR to solid-state physics are well covered in a variety of texts [4-8]. In contrast, the current book focuses on the importance of muons in chemistry. There has been only one previous book dedicated to this subject, but this was published in 1983 [9] and there have been major developments in the field since then.

Particle physicists view the negative muon as a heavy electron, and indeed it plays this role in muonic atom chemistry. However most chemical studies of muons make use of the antiparticle, the positive muon. The muon rest mass is 105.66 MeV, which is 206.8 times heavier than the electron, and 0.1126 (roughly 1/9th) the mass of a proton. The single-electron atom with μ^+ as nucleus is known as muonium ($Mu = \mu^+e^-$), and from a chemical point of view this can be viewed as the lightest isotope of hydrogen.

At the other end of the mass scale, the interaction of an energetic negative muon with helium can result in the muonic helium atom, i.e. a helium atom in which one electron has been replaced with a μ^-. Given the large mass of the muon (relative to the electron) the μ^- resides in a tight atomic orbital close to the nucleus, where it effectively screens half of the nuclear charge. Thus Heμ is a single-electron atom with an effective nuclear charge of +1, just like the H atom [10].

The properties of a single-electron atom of nuclear charge Ze are readily calculated by either the Bohr atomic model or standard quantum mechanics. The allowed electron energies are given by

$$E_n = -\frac{m_r Z^2 e^4}{2n^2 \hbar^2 (4\pi\varepsilon_0)^2} \tag{1.1}$$

Muon Spin Spectroscopy: Methods and Applications in Chemistry and Materials Science, First Edition.
Donald G. Fleming, Iain McKenzie, and Paul W. Percival.

and the Bohr radius (charge separation for the lowest energy state of the atom) is

$$a_0 = \frac{\hbar^2}{m_r Z e^2}(4\pi\varepsilon_0), \tag{1.2}$$

where e is the elementary charge, ε_0 is the electric constant (vacuum permittivity), $\hbar$ is the Planck constant divided by 2π, and m_r is the reduced mass for the two-body system:

$$m_r = \frac{m_e m_N}{m_e + m_N}, \tag{1.3}$$

where m_e and m_N are the electron mass and the mass of the nucleus, respectively. As long as m_N is large compared with m_e, the reduced mass is approximately equal to m_e. Thus the fundamental atomic properties given by Eqs. (1.1) and (1.2) depend only weakly on nuclear mass. This is why isotopes are considered to have the same chemistry – ionization energies and charge separation are the key *chemical* properties of atoms.

Isotopes are normally thought of as atoms with the same number of protons (same Z) but differing numbers of neutrons in their nuclei. However, application of Eqs. (1.1)–(1.3) shows that the series of single-electron atoms Mu, H, D, T, Heμ (Figure 1.1) should have the same chemistry. Their atomic properties are summarized in Table 1.1.

Thus muonium and Heμ can be expected to react in the same manner as the other atomic hydrogen isotopes, e.g.

Abstraction	$Mu + H_2 \rightarrow MuH + H$
Addition to unsaturated molecules	$Mu + CH_2{=}CH_2 \rightarrow MuCH_2{-}CH_2^{\cdot}$
Oxidation–reduction	$Mu + Ag^+ \rightarrow Mu^+ + Ag^0$
Acid–base	$Mu + OH^- \rightarrow MuOH + e^-$
Electron spin exchange	$Mu(\uparrow) + NO(\downarrow) \rightarrow Mu(\downarrow) + NO(\uparrow)$

Of course, the rate constants can vary with isotopic mass, and this is the basis of the kinetic isotope effect discussed in later chapters. At this stage it is sufficient to point out that muons greatly extend the range of isotope effect studies, providing a remarkable mass range of 36 from Mu to Heμ.

Another facet of isotopes that finds great utility in chemistry is their application as tracers. Reaction mechanisms are often deduced or tested by following the fate

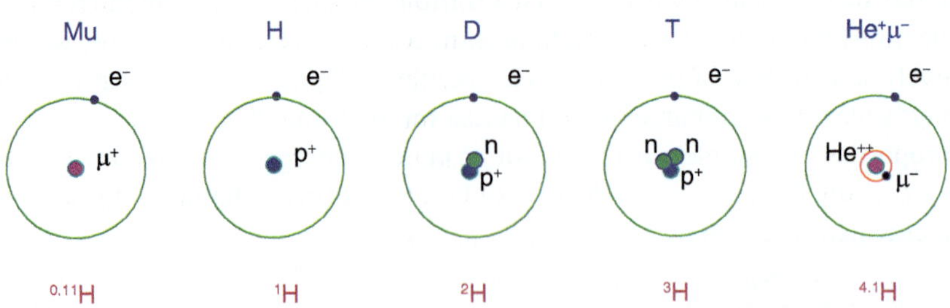

Figure 1.1 The series of single-electron atoms which behave as isotopes of hydrogen.

Table 1.1 Mass, reduced mass, ionization energy and atomic radius of hydrogen isotopes.

	Mu = μ^+e^-	H = p^+e^-	D = np^+e^-	T = nnp^+e^-	Heμ = $[^4He\mu^-]^+e^-$
Mass (u)	0.114	1.01	2.01	3.01	4.11
m_r/m_e	0.9952	0.9995	0.9997	0.9998	0.9999
I.E. = $-E_1$ (eV)	13.548	13.606	13.609	13.611	13.611
r_1 (Å)	0.5314	0.5292	0.5290	0.5290	0.5290

of 'labelled' molecules. The labelling may be radioactive (e.g. tritium) or spin (e.g. deuterium). Muonium qualifies in both categories: the positive muon spontaneously decays with a mean lifetime of 2.197 μs, irrespective of medium, and it has spin ½, just like a proton but with a larger magnetic moment (by a factor of 3.183). Both of these properties are utilized in muon spin spectroscopy. Furthermore, the relatively low intensity of muon beams means that each Mu atom is effectively isolated, so cross-reactions are avoided.

The addition of Mu to unsaturated bonds results in the formation of muonium-labelled molecules (formally referred to as 'muoniated' radicals [11]). Radicals are atoms or molecules that have one or more unpaired electrons, and are therefore typically highly reactive. They play an important role in chemistry, often as intermediates in reactions, but their transient nature makes them challenging to study with conventional spectroscopic techniques. Muoniated radicals are studied for similar reasons that Mu is studied in place of H, either to explore isotope effects, or to use Mu as a tracer. In the latter case, it could be to investigate radicals that would be difficult to produce or study with other spectroscopic techniques. Alternatively, the aim could be to label specific parts of a complex system to learn about the dynamics and local environment. Chapters 6–9 contain many examples of these varied uses.

Some Comments on Nomenclature

The common isotopes of hydrogen are named protium (H), deuterium (D) and tritium (T), and if a similar convention were followed for Mu it would be named muium. Indeed, according to the nomenclature of particle physics, the 'onium' ending implies the bound state of a particle with its antiparticle (e.g. positronium, Ps = e^+e^-). Nevertheless, the term muonium for μ^+e^- has been in use since 1957 [3] and is so well entrenched that it is endorsed by IUPAC [11].

Older literature used the term *muonated* radical instead of muoniated radical. This practice has been discontinued, as 'muonation' is now defined [11] to be the equivalent of protonation, i.e. the addition of Mu^+ rather than neutral Mu. Even older literature refers to *muonic* radicals. The adjective 'muonic' is now reserved for negative muon entities.

1.2 Muon Facilities and Background to Experimental Muon Techniques

There are currently four nuclear accelerators in the world that produce intense beams of spin-polarized muons: the TRIUMF cyclotron in Vancouver, Canada; the ISIS Facility at the Rutherford Appleton Laboratory in the UK; the Paul Scherrer Institute (PSI) in Switzerland; and the Japan Proton Accelerator Research Complex (JPARC) at Tokai in Japan. These accelerators have different features, but all generate muons from the decay of charged pions, $\pi^{\pm}$, which are themselves produced from the nuclear reactions of energetic protons. Pion production targets typically feed several muon beamlines, and at some accelerators there are two targets per proton beam. Nevertheless, the total number of muon beamlines suitable for muon spin spectroscopy is less than twenty worldwide.

TRIUMF and PSI produce quasi-continuous (CW) beams, while ISIS and J-PARC are pulsed sources. The CW facilities produce muons one at a time, with an even but stochastic time distribution. Such beams permit the use of single-particle counting techniques and the high time resolution necessary for studying fast processes and high precession frequencies. Pulsed machines produce bursts of muons at a relatively low repetition rate. The width of the muon burst ($\sim$100 ns) precludes high time resolution. On the other hand, the long interval between pulses (20 ms at ISIS; 40 ms at J-PARC) provides a very clean background for studying slow processes. Furthermore, pulsed machines are inherently suitable for experiments that require intermittent stimulation, such as pulsed RF or laser excitation.

Of particular importance to muon spin spectroscopy is that muons produced from pion decay are 100% spin-polarized (in the rest frame of the pion), with their spins aligned anti-parallel to their momentum (for μ^+; parallel for μ^-). In general, this polarization is retained in μ^+ which stop in matter, although negative muons can lose as much as 80% polarization in their capture and thermalization process. This is a key factor in the sensitivity of muon spin spectroscopy, compared to conventional magnetic resonance that relies on thermal population of spin states.

Another important feature of muon spin spectroscopy stems from parity violation in muon decay. This is a three-body process:

$$\mu^+ \rightarrow e^+ + \nu_e + \overline{\nu}_\mu$$
$$\mu^- \rightarrow e^- + \overline{\nu}_e + \nu_\mu \quad (1.4)$$

which produces a positron (or electron from μ^-) and two neutrinos ($\overline{\nu}$ is an anti-neutrino). Of these only the positron (or electron) is detected, but most importantly there is angular correlation between the positron momentum and the muon spin direction at the moment of its decay. The maximum of the probability distribution is aligned with the muon spin, so that a commonly stated description is that the positron is emitted preferentially in the direction of the muon spin. Further details of muon production and decay are given in Chapter 2. Suffice it to say here that muon spin polarization can be monitored by counting decay positrons in specific directions. There is no need for a stimulating field (radio-frequency in NMR or microwave in ESR) to interrogate the spin system.

1.3 The Development of Muonium Chemistry

The advent of muonium chemistry can be traced right back to the first demonstration of parity violation in muon decay in 1957, when it was remarked that the muon asymmetry depends on material [2], and that this might be due to the possibility of muonium formation [3]. A systematic survey of simple materials was published in the following year [12], although muonium itself was not detected directly until 1960, in a target of argon gas at ~50 bar pressure [13]. By 1966 understanding had developed to the extent that Hughes could write in a review article that 'Muonium can be regarded as a light isotope of hydrogen' [14].

The earliest experiments were conducted in the United States, at the cyclotron laboratories located at the University of Chicago and Columbia University. Within a few years, however, investigations of muonium chemistry were undertaken at the JINR synchrocyclotron at Dubna, Russia. Various attempts were made to deduce the rate constants for reactions of muonium from muon asymmetry values, and it was even suggested that the muonium data could be used to calculate rate constants for atomic hydrogen [15]. Remarkably, both kinetic isotope effects and the formation of muoniated radicals were discussed as early as 1963 [16]. Another advance from Russia was the first detection of Mu precession in solids – in quartz, in solid carbon dioxide and in ice [17]. This was followed by studies of two-frequency Mu precession [18], from which the muonium hyperfine frequency can be calculated (see Section 2.6.1).

At Columbia University there were early attempts to study gas-phase muonium reactivity by monitoring the muon polarization in longitudinal magnetic fields [19]. However, the first convincing measurements of muonium reaction rates were not until 1976, when Fleming et al. studied the decay of the characteristic transverse-field precession signal of muonium at the Lawrence Berkeley Laboratory [20]. Their ability to stop muons in gases at low pressures was due to a major advance in muon beam technology: the development of the surface muon beam (Section 2.1).

Despite many attempts in those early days, Mu precession had never been detected in liquid samples. Instead, the Berkeley researchers determined rate constants by fitting the concentration dependence of the magnitude and phase of the 'residual polarization' – transverse polarization in the diamagnetic products of muonium reactions [21] (see Section 5.4). They were also able to deduce the involvement of muoniated free radicals, again without direct detection [22]. A review article from that period gives a comprehensive account of the 'state of the art' of muonium chemistry in 1974 [23]. It also suggested the acronym μSR for 'muon (or muonium) spin relaxation, rotation, resonance, etc.', to suggest the analogy with NMR and ESR.

The next major advance came with the emergence of the 'meson factories' SIN (1968), TRIUMF (1968) and LAMPF (1972), accelerators designed to produce beams of protons in the energy range of 400 to 1000 MeV and capable of generating muon beams 100–1000 times more intense than hitherto.

It was at SIN (Swiss Institute for Nuclear Research, now part of the Paul Scherrer Institute) that muonium precession signals were first detected in liquids, originally in water [24] and then in other common solvents [25, 26]. A major factor in this advance was the removal of dissolved oxygen from the samples. Being paramagnetic, oxygen can both react chemically with Mu and engage in non-reactive spin-exchange encounters (Section 7.4). Both of these interactions lead to loss of coherent Mu precession.

Another reason that Mu signals were so elusive in liquids is that they have small amplitudes, corresponding to about 20% of the original muon spin polarization. Since typically 60% of the muon spin polarization precesses at the diamagnetic frequency in liquids, this left 20% unaccounted – the so-called 'missing fraction' (Section 3.2). In contrast to liquid water, ice has a larger Mu fraction, smaller diamagnetic signal, and no missing fraction. There are also much larger Mu fractions in gases, and the work of Fleming et al. at TRIUMF showed that these depend largely on the muon thermalization process, in particular on charge-exchange cross sections [27]. Small deviations from expectations based on proton thermalization data were ascribed to competing hot-atom reactions of the Mu atom. The need to explain this puzzling phase dependence of the Mu fractions prompted the proposal of radiolysis effects, including the 'spur model' of muonium formation [28]. Until this time it was thought that positive muons would thermalize as muonium sufficiently far from the last charge exchange that it would escape any effects of radiation damage [23]. The partition of muons between muonium and diamagnetic compounds in the last 30 eV or so of thermalization was thought to arise from hot-atom reactions [29], modelled after the then current understanding of tritium chemistry. The spur model met considerable opposition for many years, and radiolysis effects were not generally accepted until the electric field studies of Storchak et al. [30]. Further details can be found in Chapter 3.

Whatever the mechanism of Mu formation in condensed matter, being able to detect its signal on a microsecond timescale made it possible to determine rate constants by measuring Mu decay rates in reactant solutions of known concentration [31]. Comparison of Mu rate constants with those of H soon revealed a wide range of kinetic isotope effects. Gas-phase reactivity is discussed in Chapter 4 and solution kinetics in Chapter 5.

Of course, the direct study of muonium decay kinetics requires that the liquid solvent (or buffer gas) be non-reactive. By analogy with H it was expected that reaction of Mu with an unsaturated molecule should result in a free radical. However direct detection of muoniated radicals did not occur until 1978, when Emil Roduner, then a graduate student, proposed searching for the precession signals at high magnetic field [32]. These first experiments utilized pure organic liquids as targets, and even though hundreds of muoniated radicals have now been studied in liquids, including in solutions, the requirement for fast transfer of muon polarization from Mu to radical limits detection of radical precession signals to highly concentrated samples (Section 2.7.3). Nevertheless, muoniated radicals have also been studied in the gas phase, initially at high pressure [33], but later at pressures as low as 1 atm [34, 35].

As will be explained in Chapter 6, the isotropic muon hyperfine coupling constant can be readily determined from the precession frequencies of a muoniated radical. However, this is often not enough to identify the radical, and full characterization should include determination of the hyperfine constants of other spin-active nuclei, typically protons in organic radicals. Abragam [36] suggested the means of doing this – avoided level-crossing resonance (ALCR) – and teams at TRIUMF [37] and SIN [38] raced to demonstrate the feasibility of the method. At SIN the experiment was limited by the available magnet, so the $(CH_3)_2\dot{C}OMu$ radical was chosen because its six equivalent protons give rise to a single resonance at only 1.5 kG. The TRIUMF group had the advantage of a superconducting magnet, and were able to scan a wide field range and detect all four resonances due to the distinct fluorines in muoniated hexafluorocyclohexadienyl, C_6F_6Mu. Interestingly, this radical was also chosen to limit the scan range (3–12 kG), it having a muon hyperfine constant less than half that of C_6H_6Mu, whose avoided level-crossing resonances extend to almost 30 kG [39].

The ability to determine hyperfine constants (hfcs) for other nuclei in addition to the muon permits a more thorough investigation of intramolecular dynamics. Thus Ramos et al. [40] interpreted the temperature dependence of the muon hfc in the muoniated ethyl radical in terms of CH_2Mu rotation, but it was only with the advent of ALCR that the corresponding effect on the proton hfc could be followed, and even contrasted with the unsubstituted CH_3 groups in muoniated tert-butyl, $(CH_3)_2CH_2Mu$ [41]. These data eventually led to a deeper understanding of the origins of hyperfine isotope effects, i.e. why it is that muon hfcs are typically 10–30% larger than the equivalent proton hfcs scaled by the ratio of magnetic moments. This topic is discussed in Section 6.2.

Once the basic techniques of muon spin spectroscopy had been developed it became possible to extend muon studies to chemical problems difficult to solve by conventional means. An early example is the mapping of unpaired spin density in the $C_{60}H$ radical [42]. At that time theoretical and computational papers expressed divergent opinions ranging from localization of unpaired spin on a single carbon atom to complete delocalization over the C_{60} 'globe'. The first μSR paper on fullerenes [43] is interesting for two reasons. First, it predates ESR detection of $C_{60}H$ [44] – in fact the search for narrow ESR lines was guided by reports of the muon hfc in $C_{60}Mu$. Secondly, the μSR spectrum exhibited signals from encapsulated muonium, $Mu@C_{60}$, as well as from the exohedral muonium adduct $C_{60}Mu$. This was the first time that muonium and a muoniated radical had been detected simultaneously in the same sample.

The development of muonium chemistry was initially driven by advances in technology, resulting in both more intense beams and in new experimental techniques. These in turn enabled the study of more difficult problems. For example, extremely high data statistics were necessary to study the decay kinetics of the weak precession signals of muonic helium [45]. Similarly, an intense muon beam and novel experimental setup enabled the first observation of light emitted from a muonium-containing molecule, namely the chemiluminescence resulting from excimer transitions in the Rydberg molecule NeMu* [46]. However, development of the field in recent decades has largely involved applications in ever more complex systems and extreme environments: muonium and muoniated radicals on surfaces, in porous materials, in supercritical fluids and soft matter. Examples of all these will be given in subsequent chapters.

References

1 Neddermeyer, S.H. and Anderson, C.D. (1937). *Phys. Rev.* 51: 884–886.
2 Garwin, R.L., Lederman, L.M., and Weinrich, M. (1957). *Phys. Rev.* 105: 1415–1417.
3 Friedman, J.I. and Telegdi, V.L. (1957). *Phys. Rev.* 105: 1681–1682.
4 Schenck, A. (1985). *Muon Spin Rotation Spectroscopy. Principles and Applications in Solid State Physics*. Adam Hilger.
5 Lee, S.L., Kilcoyne, S.H., and Cywinski, R. (ed.) (1999). *Muon Science. Muons in Physics. Chemistry and Materials*. IOP Publishing.
6 Nagamine, K. (2003). *Introductory Muon Science*. Cambridge University Press.
7 Yaouanc, A. and Delmas de Reotier, P. (2011). *Muon Spin Rotation, Relaxation and Resonance. Applications to Condensed Matter*. Oxford University Press.
8 Blundell, S.J., De Renzi, R., Lancaster, T., and Pratt, F.L. (2021). *Muon Spectroscopy: An Introduction*. Oxford University Press.
9 Walker, D.C. (1983). *Muon and Muonium Chemistry*. Cambridge University Press.
10 Fleming, D.G., Arseneau, D.J., Sukhorukov, O. et al. (2011). *Science* 331: 448–450.
11 Koppenol, W.H. (2001). Names for muonium and hydrogen atoms and their ions (IUPAC Recommendations 2001). *Pure Appl. Chem.* 73: 377–380.
12 Swanson, R.A. (1958). *Phys. Rev.* 112: 580–586.
13 Hughes, V.W., McColm, D.W., Ziock, K., and Prepost, R. (1960). *Phys. Rev. Lett.* 5: 63–65.
14 Hughes, V.W. (1966). *Ann. Rev. Nucl. Sci.* 16: 445–470.
15 Babaev, A.I., Balats, M.Y., Myasishcheva, G.G. et al. (1966). *Sov. Phys. JETP* 23: 583–591.
16 Brodskii, A.M. (1963). *Sov. Phys. JETP* 17: 1085–1088.
17 Myasishcheva, G.G., Obukhov, Y.V., Roganov, V.S., and Firsov, V.G. (1968). *Sov. Phys. JETP* 26: 298–301.
18 Gurevich, I.I., Ivanter, I.G., Meleshko, E.A. et al. (1971). *Sov. Phys. JETP* 33: 253–259.
19 Mobley, R.M., Amato, J.J., Hughes, V.W. et al. (1967). *J. Chem. Phys.* 47: 3074–3075.
20 Fleming, D.G., Brewer, J.H., Garner, D.M. et al. (1976). *J. Chem. Phys.* 64: 1281–1287.
21 Brewer, J.H., Crowe, K.M., Johnson, R.F. et al. (1971). *Phys. Rev. Lett.* 27: 297–300.
22 Brewer, J.H., Crowe, K.M., Gygax, F.N. et al. (1974). *Phys. Rev. A* 9: 495–507.
23 Brewer, J.H., Crowe, K.M., Gygax, F.N., and Schenck, A. (1975). *Muon Physics, Chemistry and Solids*, vol. 3 (ed. V.W. Hughes and C.S. Wu). New York: Academic Press.
24 Percival, P.W., Fischer, H., Camani, M. et al. (1976). *Chem. Phys. Lett.* 39: 333–335.
25 Percival, P.W., Roduner, E., and Fischer, H. (1979). *Positronium and Muonium Chemistry*, vol. 175 (ed. H.J. Ache), 335–355. American Chemical Society Advances in Chemistry.

26 Ito, Y., Ng, B.W., Jean, Y.C., and Walker, D.C. (1980). *Can. J. Chem.* 58: 2395–2401.

27 Fleming, D.G., Mikula, R.J., and Garner, D.M. (1982). *Phys. Rev. A* 26: 2527–2544.

28 Percival, P.W., Roduner, E., and Fischer, H. (1978). *Chem. Phys.* 32: 353–367.

29 Fleming, D.G., Senba, M., Arseneau, D.J. et al. (1986). *Can. J. Chem.* 64: 57–66.

30 Storchak, V., Brewer, J.H., Morris, G.D. et al. (1999). *Phys. Rev. B* 59: 10559–10572.

31 Percival, P.W., Roduner, E., Fischer, H. et al. (1977). *Chem. Phys. Lett.* 47: 11–14.

32 Roduner, E., Percival, P.W., Fleming, D.G. et al. (1978). *Chem. Phys. Lett.* 57: 37–40.

33 Roduner, E. and Garner, D.M. (1986). *Hyperfine Interact.* 32: 733–739.

34 Fleming, D.G., Kiefl, R.F., Garner, D.M. et al. (1990). *Hyperfine Interact.* 65: 767–771.

35 Fleming, D.G., Pan, J.J., Senba, M. et al. (1996). *J. Chem. Phys.* 105: 7517–7535.

36 Abragam, A. (1984). *Comptes Rendus Acad. Sci. II* 299: 95–99.

37 Kiefl, R.F., Kreitzman, S., Celio, M. et al. (1986). *Phys. Rev. A* 34: 681–684.

38 Heming, M., Roduner, E., Patterson, B.D. et al. (1986). *Chem. Phys. Lett.* 128: 100–106.

39 Percival, P.W., Kiefl, R.F., Kreitzman, S.R. et al. (1987). *Chem. Phys. Lett.* 133: 465–470.

40 Ramos, M.J., McKenna, D., Webster, B.C., and Roduner, E. (1984). *J. Chem. Soc., Faraday Trans. I* 80: 267–274.

41 Percival, P.W., Brodovitch, J.-C., Leung, S.K. et al. (1988). *Chem. Phys.* 127: 137–147.

42 Percival, P.W., Addison-Jones, B., Brodovitch, J.-C. et al. (1995). *Chem. Phys. Lett.* 245: 90–94.

43 Ansaldo, E.J., Boyle, J., Niedermayer, C. et al. (1992). *Z. Phys. B-Condens. Mat.* 86: 317–318.

44 Morton, J.R., Preston, K.F., Krusic, P.J., and Knight, L.B. (1993). *Chem. Phys. Lett.* 204: 481–485.

45 Fleming, D.G., Arseneau, D.J., Sukhorukov, O. et al. (2011). *J. Chem. Phys.* 135: 184310.

46 Baer, S., Fleming, D.G., Sloan, J.J. et al. (1994). *J. Chem. Phys.* 101: 1202–1218.

2

Muon Beams and Spin Spectroscopy

2.1 Spin-Polarized Muon Beams

In conventional magnetic resonance techniques, the signal is proportional to the polarization of the spins of interest, which arises from the Boltzmann distribution over quantized spin states. This polarization is typically very small; for protons in a magnetic field of 100 kG at 300 K, the population difference is 3.5×10^{-5}. The polarization decreases in smaller magnetic fields or at higher temperatures. One of the main factors that contributes to the sensitivity of the μSR technique is that high-intensity beams of muons can be produced with polarization close to 100%. The polarization of the muon beam is independent of the magnetic field or the temperature, being a consequence of how they are produced.

Muons are generated in the decay of pions, which are in turn produced by high-energy proton (p^+) bombardment of a low-Z material, such as graphite. Single-pion production reactions have an energy threshold of ~280 MeV in the laboratory frame, but the cross section for pion production increases dramatically with energy, peaking at proton energies between 500 MeV and 1 GeV, which is the energy range of the current μSR facilities. Typically, low-Z targets are used, because they are favourable for pion production and there is a reduced probability of multiple scattering events, which lead to downstream activation of the beamline. Positive pions (π^+) and negative pions (π^-) are produced in the reactions

$$p^+ + p^+ \rightarrow \pi^+ + p^+ + n, \tag{2.1}$$

$$p^+ + n \rightarrow \pi^+ + n + n, \tag{2.2}$$

$$p^+ + n \rightarrow \pi^- + p^+ + p^+. \tag{2.3}$$

Neutral pions (π^0) are also produced but are not discussed here since they do not generate muons.

Positive muons (μ^+) are produced from the decay of π^+ (Figure 2.1), while negative muons (μ^-) are produced from the decay of π^-:

$$\pi^+ \rightarrow \mu^+ + \nu_{\mu}, \tag{2.4}$$

$$\pi^- \rightarrow \mu^- + \overline{\nu}_{\mu}. \tag{2.5}$$

ν_μ and $\overline{\nu}_\mu$ are the muon neutrino and anti-muon neutrino, respectively. Both π^+ and π^- have a spin of 0, a mass of 2.49×10^{-28} kg, or 139.57 MeV, and lifetime of 26 ns.

Muon Spin Spectroscopy: Methods and Applications in Chemistry and Materials Science, First Edition.
Donald G. Fleming, Iain McKenzie, and Paul W. Percival.

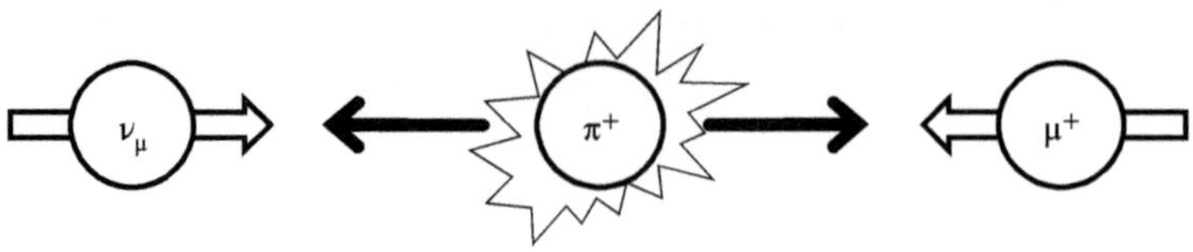

Figure 2.1 Decay of positive pions at rest. The black arrows depict the direction of motion of the decay products and the block arrows depict the spin of the particles.

Conservation of linear momentum requires that when a pion at rest decays the resulting muon and neutrino are emitted in opposite directions. Being leptons, the μ^+, μ^-, ν_μ and $\overline{\nu}_\mu$ all have spins of 1/2. The ν_μ has negative helicity (i.e. the spin is antiparallel to the linear momentum), while $\overline{\nu}_\mu$ has positive helicity (spin parallel to the momentum). Since the pion has a spin of zero, conservation of angular momentum results in the spin of μ^+ being polarized with its spin anti-parallel to its momentum. The degree of spin polarization is decreased if the pion decays in flight.

Muons are guided to the spectrometer down evacuated beamlines by magnetic steering elements, which are a series of dipole and quadrupole magnets. Muon beamlines are characterized by the momentum of the muons. *Surface muon beams* are produced from the decay of positive pions on or just below the surface of the production target. Since muons generated inside the production target lose kinetic energy on their way out, the maximum muon momentum corresponds to those at rest on the surface, i.e. $p_\mu^{\max} = 29.8\,\mathrm{MeV/c}$, or 4.12 MeV of kinetic energy in the pion's rest frame. Surface muon beamlines are tuned to transport muons in a narrow momentum range $\Delta p_\mu/p_\mu$ of a few percent, with p_μ usually chosen to be just below $p_\mu^{\max}$. They typically have high intensity, good luminosity and spin polarization close to 100%. Unlike μ^+, it is not possible to produce a surface μ^- beam because the π^- undergoes nuclear capture from low-lying orbitals of pionic atoms before it has a chance to decay [1].

Muons interact with matter mainly by the electromagnetic interaction and lose energy/momentum along their path, as discussed in Section 3.1. The stopping range of surface μ^+ beams in matter is approximately $0.15\,\mathrm{g\,cm^{-2}}/\rho$, where ρ is the density of the material. In water ($\rho \sim 1\,\mathrm{g\,cm^{-3}}$) the stopping range is ~1.5 mm, but in a typical gas at 1 bar pressure it is of order 100 cm, necessitating much larger target environments.

Modern surface muon beamlines incorporate separators (crossed electric and magnetic fields that act as velocity selectors and which remove positron contamination from the muon beam). Such devices can also be used to rotate the muon spin away from the beam direction [2].

Muons can also be selected from pions that decay in flight. These *decay beamlines* can produce μ^+ and μ^- beams with a wide momentum range. The pions decay to muons in a section of beamline several metres in length, where the particles are constrained by a high longitudinal magnetic field, typically produced by a superconducting magnet. If the muon is emitted in the opposite direction to the pion motion then the momentum of the muon is less than that of the pion and the spin is parallel to the momentum. These are known as *backward muon beams*. By appropriate setting of the bending magnets of the beamline one can select the momentum of the muons, which typically ranges from ~50 to 100 MeV/c. The intensity drops off dramatically outside this range. The disadvantages of a backward muon beam are the larger momentum spread (and therefore lower stopping density), poor optics (because the beamline is not monochromatic) and positron contamination. The polarization of a backward muon beam is ~70–80% due to the finite acceptance of the beamline, i.e. muons emitted at an angle to the beam axis.

If the muon is emitted in the direction of the pion motion then their momenta add and the muon spin points backwards. This is known as a *forward muon beam* and has the highest energy but the lowest muon polarization. Forward muon beams have higher intensity than backward beams but are usually highly contaminated with pions and positrons of the same momentum. Although early μSR experiments used such beams, they are rarely used nowadays.

Muon beams of energies down to the keV range can be produced by moderating a surface muon beam using a material such as thin layer of a van der Waals gas frozen on a substrate. The energy of the muon beam can then be tuned between ~1 and 30 keV by applying a bias voltage between the beamline and the sample, resulting in implantation depths in solids ranging from approximately a nanometre up to several hundred nanometres. These are called *low-energy muon beams* and are used for studies of thin films, multi-layers and near-surface effects. Currently only PSI has such a beamline [3]. The moderation efficiency is quite low (~1 in 10^4), so the rate is low compared with surface muon beams. A different type of low-energy muon beamline is under construction at J-PARC. The muons are created by the resonant ionization of thermal Mu atoms generated from the surface of a hot tungsten foil placed at the end of an intense surface muon beam line [4].

2.2 Muon Decay and Detection of Its Spin Polarization

The second key to the μSR technique is the manner in which the polarization is detected. Unlike conventional magnetic resonance, where an electric current is induced in a receiver coil, in μSR the signal is obtained from the asymmetry of the β decay of the muon. Muons are radioactive, and the decay of both positive and negative muons is a three-body process:

$$\mu^+ \rightarrow e^+ + \nu_e + \overline{\nu}_\mu, \tag{2.6}$$

$$\mu^- \rightarrow e^- + \overline{\nu}_e + \nu_\mu, \tag{2.7}$$

where ν_e and $\overline{\nu}_e$ are the electron neutrino and anti-electron neutrino, respectively. This is depicted for a positive muon in Figure 2.2. The mean lifetime of μ^+ is 2.197 μs, independent of its environment. The lifetime of μ^- is the same in vacuum but is less in matter (to be discussed in Section 3.4).

The kinetic energy of the decay positron (electron for μ^- decay) depends on how its momentum is distributed among the three particles and can range from 0 to $E_{max} = 52.3$ MeV (half of the muon mass). Most positrons are very energetic, which

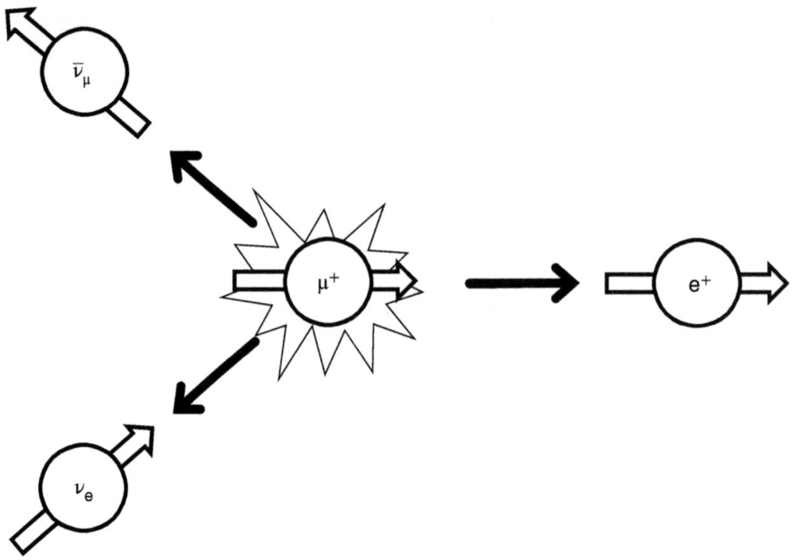

Figure 2.2 Decay of positive muons. The black arrows depict the direction of motion of the decay products and the block arrows depict the spin of the particles.

means that they can escape from the sample and pass through any target environment, such as a cryostat, furnace or pressure vessel.

Muon decay involves the weak nuclear force and thus, like muon production, does not conserve parity. The decay probability of the muon per unit time for positron emission at an angle θ to the muon spin is given by [1]

$$W(\theta) \propto [1 + a(\varepsilon)\cos\theta], \tag{2.8}$$

where $a(\varepsilon)$ is the asymmetry parameter, which depends on the positron energy ($\varepsilon = E/E_{max}$). Positrons with high energy are emitted preferentially along the direction of the muon spin, because $a(\varepsilon) \to 1$ as $\varepsilon \to 1$. In contrast, low-energy positrons ($\varepsilon < 0.5$) are emitted preferentially in the opposite direction because $a(\varepsilon)$ is then negative. Integrating over all positron energies and accounting for the positron energy distribution result in an average asymmetry, $\bar{a}$, of $1/3$. Representative angular distributions are shown in Figure 2.3. One can increase the asymmetry by putting a degrader between the sample and the positron detectors in order to remove the low-energy positrons, but this results in a decrease in the rate.

2.3 Continuous Versus Pulsed Muon Beams

Muon beams are also characterized by their time structures: continuous (CW) or pulsed. The time structure of the muon beam affects the type of experiments one can perform and each type of beam has advantages and disadvantages. There are two continuous (CW) muon facilities (TRIUMF and PSI) and two pulsed muon facilities (ISIS and J-PARC).

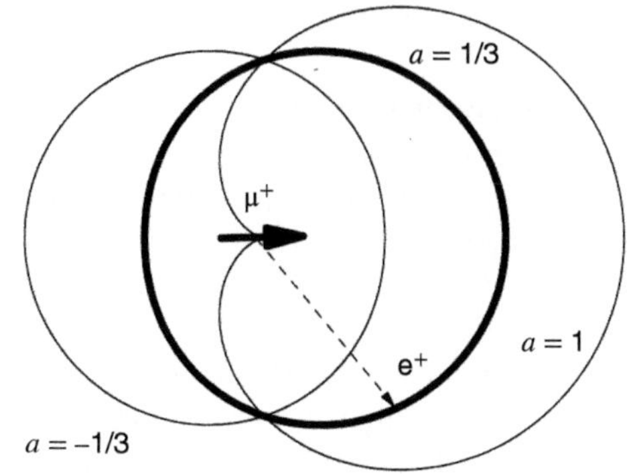

Figure 2.3 Angular distribution of positrons from muon decay for $a = -1/3$, 1/3 and 1. The bold circle corresponds to the average asymmetry, $\overline{a} = 1/3$, obtained from integrating over all positron energies and accounting for the positron energy distribution. The solid arrow indicates the orientation of the muon spin at the moment of decay.

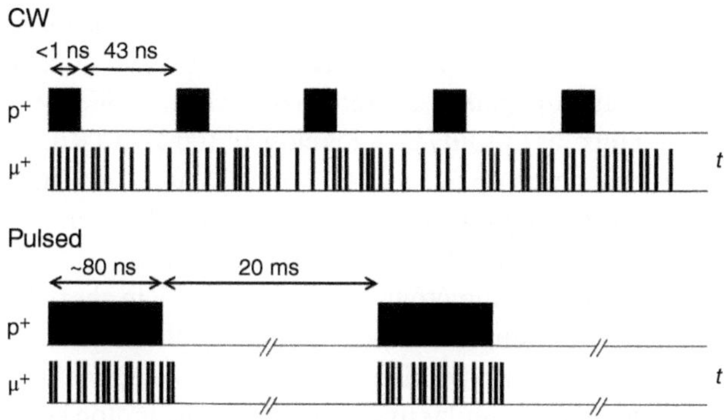

Figure 2.4 Time structures of proton and muon beams at a CW facility (TRIUMF) and a pulsed facility (ISIS). Not to scale.

Although nominally continuous, the beams at CW facilities have time structures dictated by their particle accelerators. For example, the 23 MHz RF frequency of the cyclotron at TRIUMF results in pulses of protons with sub-ns width that are 43 ns apart. However, this time structure is averaged out by the pion lifetime (26 ns) so that muons arrive at the sample in the μSR spectrometer essentially randomly. This is shown schematically in Figure 2.4.

Given the stochastic nature of the beam, time-resolved measurements at a CW facility need a detector before the sample to register the arrival of each muon. Furthermore, it is necessary to limit the muon implantation rate so that there is only one muon in the sample at a time. This is typically achieved with collimators and/or slits in the beamline. In addition, the data acquisition system is set up to reject events for which there is more than one muon in the sample within a set time window (typically $\lesssim 10\,\mu s$ long). This ensures an accurate correlation of decay positrons with their parent muons. If no positron is detected in a given period (the data gate), the system is reset to allow for the next muon. The good-event rate (correlated muon stop/positron emission) is much less than the incoming muon rate because the positron counters don't completely surround the sample. Thus a muon rate of $7 \times 10^4\,\mu^+/s$ typically results in a good-event rate of $\sim 6 \times 10^3\,e^+/s$ in a data gate of $6\,\mu s$. The use of veto detectors to distinguish between muons stopping in a sample and its surroundings can result in even lower data rates.

The advantage of the CW facility for time-resolved measurements is that one can measure high frequencies and fast spin relaxation rates. The observable asymmetry depends on the frequency of the signal, ν, and the time resolution, Δt_R, of the instrument [5]:

$$A(\nu) = A_0 e^{-(\pi\nu\Delta t_R)^2/(4\ln 2)} \tag{2.9}$$

The best current instruments have $\Delta t_R < 50$ ps, which means they can measure precession frequencies up to several GHz. For studies of muoniated free radicals (Chapter 6), spectrometers need $\Delta t_R \sim 1$ ns in order to measure frequencies up to 500 MHz. A limitation of CW facilities is that one can only measure data for a few μs before the random background begins to exceed the true decay events. This makes it difficult to measure slow (i.e. $\lesssim 0.1\ \mu s^{-1}$) muon spin relaxation rates. CW facilities are the preferred choice for time-integral measurements, such as avoided level crossing (ALC) muon spin resonance, as the duty factor is close to 100%. At TRIUMF the μ^+ rate in an ALC experiment can be up to $\sim 5 \times 10^5\ \mu^+/s$ with an e^+ rate up to $\sim 2 \times 10^5\ e^+/s$. The μ^+ and e^+ rates vary significantly with magnetic field in an ALC experiment, due to the focusing effect of the applied longitudinal field.

At pulsed muon sources protons hitting the production target arrive in short and widely separated pulses. At ISIS, for example, the pulses have a full width at half maximum (FWHM) of 80 ns, and the 50 Hz repetition rate results in the pulses being 20 ms apart. The muon beams then have a similar time structure to the proton beam, as shown schematically in Figure 2.4. The pulses are relatively intense with $\sim 300\ \mu^+$ per pulse at ISIS. An advantage of pulsed muon sources is that one can run at higher rates for time-resolved measurements compared to CW facilities, as long as the spectrometer has highly segmented detectors [6]. For example, the EMU spectrometer at ISIS has 96 detectors and can run at rates of $4 \times 10^4\ e^+/s$. The data rate is the same for time-differential and time-integral measurements. The positron detectors are active only after the incoming muon pulse, which strongly reduces the accidental background counts. The negligible background means that one can measure up to about ten times the muon mean lifetime. The time resolution is set by the width of the muon pulse. At ISIS one can measure precession frequencies only up to ~10 MHz, which corresponds to fields of ~600 G for diamagnetic muons. Pulsed facilities are best for studying very slow relaxation and for experiments involving intermittent stimulation, such as pulsed RF and laser excitation.

2.4 μSR Spectrometers

Muons are transported by the beamline to an end-station, where they arrive at a μSR spectrometer. At CW facilities the muons pass through a thin (~0.125 mm) plastic muon scintillation detector that provides a 'start' pulse to the data acquisition electronics. At pulsed facilities the trigger is set by the RF pulse of the proton accelerator. The muon then enters the sample, where it comes to a stop and eventually decays.

Positron detectors are arranged around the sample. Ideally, one would like to completely cover the 4π steradian solid angle around the sample, but in practice only a fraction of this is covered, to allow access for sample support and control (e.g. a cryostat).

When one of the positron[1] detectors registers an outgoing positron, a 'stop' pulse is generated and the elapsed time is determined by a time-to-digital converter (TDC) – essentially a stopwatch able to measure time intervals (Δt) with a resolution of about 150 ps to 1 ns at the CW facilities and 16 ns at the pulsed facilities. The histogram for that positron detector is then incremented in the bin corresponding to the measured time interval. Electronic logic units are used to discriminate against events where more than one muon or more than one positron are detected within the data gate. The positron detectors are usually mounted in pairs on opposite sides of the sample: Back–Front, Left–Right and Up–Down. Spectrometers are designed for specific tasks, such as measurements in high transverse field or to accommodate different sample environments, and this determines the positions of the positron detectors. A schematic drawing of a spectrometer for transverse-field measurements is shown in Figure 2.5, and one for longitudinal-field measurements is shown in Figure 2.6.

The number of decay positrons recorded per time bin in the ith counter is given by:

$$N_i(t) = N_0 \left\{ B_g + e^{-t/\tau_\mu}[1 + a_0 P_i(t)] \right\}, \tag{2.10}$$

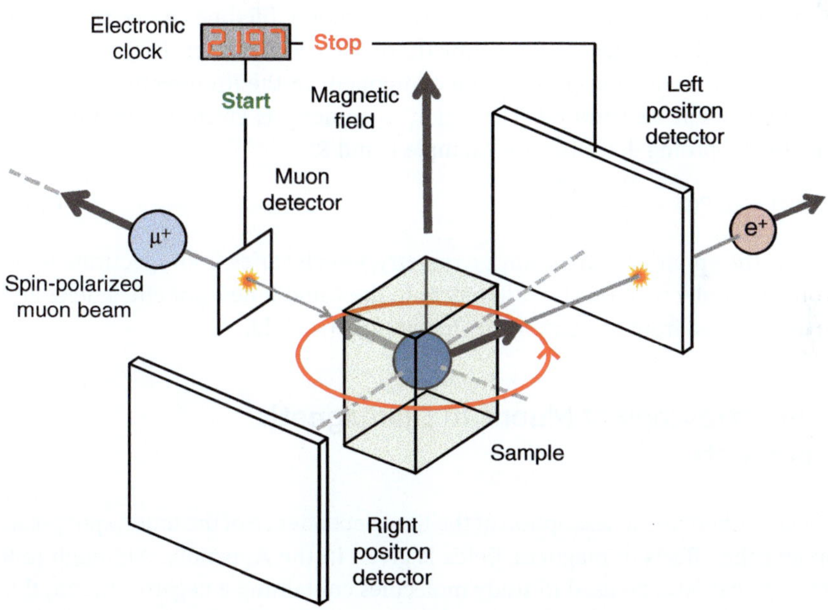

Figure 2.5 Schematic of a transverse-field geometry. At pulsed facilities there is no muon detector.

1 The same μSR spectrometers can be used for negative muons, but then it would be electrons which are detected. For simplicity the description here will refer only to positive muons and positrons.

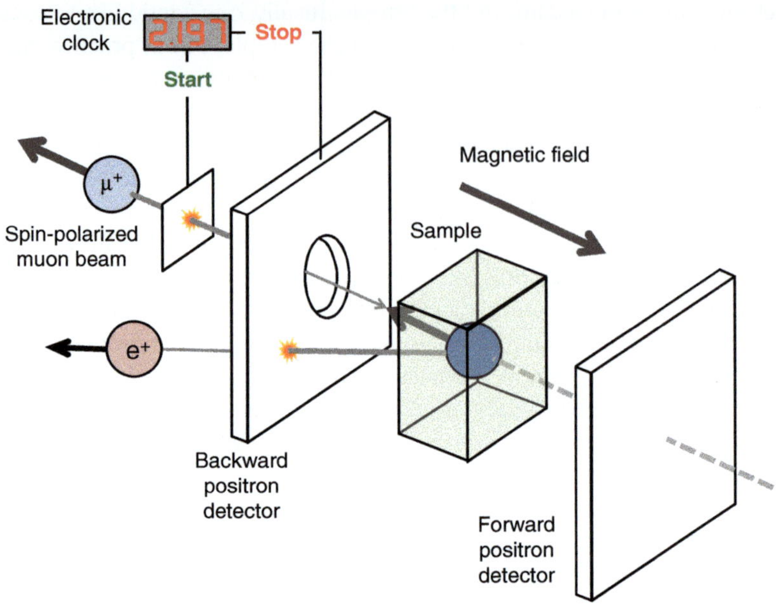

Figure 2.6 Schematic of a longitudinal-field geometry. At pulsed facilities there is no muon counter.

where N_0 is a normalization factor, τ_μ is the muon lifetime, a_0 is the initial asymmetry, which depends on the geometry of the spectrometer, $P_i(t)$ is the time evolution of the muon spin-polarization component in the ith direction and B_g is a time-independent background. The muon asymmetry is given by the difference between two opposing detectors, for example L and R:

$$A(t) = a_0 P(t) = \frac{N_L - \alpha N_R}{N_L + \alpha N_R}, \tag{2.11}$$

where a_0 is the empirical maximum asymmetry, which varies from spectrometer to spectrometer, and α accounts for differences in the L and R detector efficiencies and can be determined from measurements in a transverse field.

2.5 Spectroscopy of Muons in Diamagnetic Environments

A detailed mathematical description of the time dependence of the muon spin polarization and the effects of magnetic fields is given in the Appendix. Although μSR spectroscopy can also be used to study molecules containing a negative muon, this section deals only with muoniated (i.e. positive muon) species.

2.5.1 Diamagnetic Muons in a Longitudinal Field

This is conceptually analogous to a classical NMR experiment, except that the muon is 100% spin-polarized, while the spin polarization in NMR usually comes from a

Boltzmann population of the spin states. The muon polarization as a function of time is given by:

$$P_{\mathrm{D}}(t) = h_{\mathrm{D}} \mathrm{e}^{-\lambda_{\mathrm{D}} t}, \tag{2.12}$$

where h_{D} is the fraction of the muons that reside in a diamagnetic environment and λ_{D} characterizes their reaction or relaxation rate. An applied magnetic field locks the muon polarization, which only changes amplitude through its decay rate λ_{D}. This is the origin of the exponential term in Eq. (2.12). Spin relaxation is caused by a fluctuating magnetic field with a component at the muon Larmor frequency, which depends on the applied magnetic field. If there is no conversion to another state, λ_{D} corresponds to the muon spin-lattice relaxation rate, $1/T_1^{\mu}$. In most non-magnetic materials the relaxation rate of diamagnetic muons is negligible due to T_1 values being on the order of seconds. An example is the LF-μSR spectrum of muons in silver (Figure 2.7), where 100% of implanted muons reside in a diamagnetic environment and there is negligible spin relaxation [7].

2.5.2 Diamagnetic Muons in a Transverse Field

In a transverse-field experiment, a magnetic field is applied in one direction (usually defined as the z-axis) which is perpendicular to the initial spin polarization of the muon (defined here as the x-axis direction). The time dependence of the muon polarization is then given by:

$$P_x(t) = h_{\mathrm{D}} \mathrm{e}^{-\lambda_{\mathrm{D}} t} \cos(2\pi \nu_{\mu} t). \tag{2.13}$$

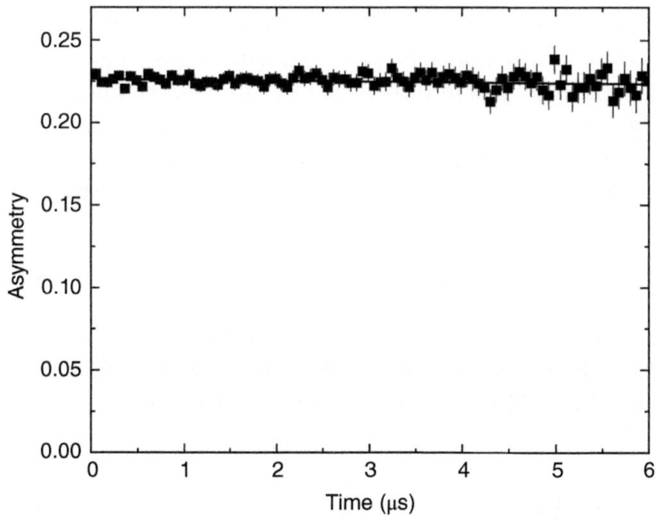

Figure 2.7 Time-domain μSR spectrum of the diamagnetic signal in silver in a longitudinal magnetic field of 368 G parallel to the muon momentum and initial spin orientation. The asymmetry a_0 for this spectrometer is 0.23.

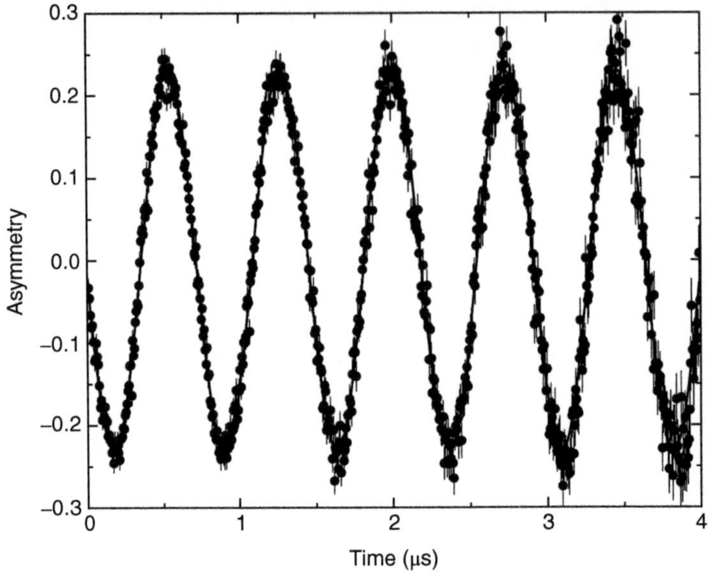

Figure 2.8 Time-domain TF-μSR spectrum of the diamagnetic signal in CCl_4 in a transverse magnetic field of 95 G. a_0 for this spectrometer is 0.23.

The muon polarization precesses at the Larmor frequency, ν_μ, matching the more intuitive classical picture. The TF-μSR spectrum of CCl_4 is shown in Figure 2.8; 100% of the implanted muons are in a diamagnetic environment, precessing at ν_μ.

Damping of the precession signal can be caused by transformation to a different chemical state, or transverse spin relaxation $\left(1/T_2^\mu\right)$ or a distribution of magnetic fields. In the latter case this can arise from inhomogeneities in the spectrometer magnetic field, which are invariably present to a greater or lesser degree, and from dipolar fields generated by neighbouring spins. The simplest case is when the applied magnetic field is much larger than the field distribution, so that Eq. (2.13) can simply be multiplied by a Gaussian envelope. In the case of a static magnetic field distribution this gives

$$P_x(t) = h_D e^{-\lambda_D t} e^{-\frac{1}{2}\gamma_\mu^2 \Delta_{TF}^2 t^2} \cos(2\pi\nu_\mu t), \tag{2.14}$$

where Δ_{TF}^2 is the variance of the magnetic field distribution along the z-axis.

If the magnetic field distribution is not static, which can be caused by motion of the muon and the surrounding spins, the Gaussian envelope is replaced by the Abragam function [8]:

$$\exp\left[-\frac{\gamma_\mu^2 \Delta_{TF}^2}{\nu_c^2}\left(e^{-\nu_c t} - 1 + \nu_c t\right)\right], \tag{2.15}$$

where ν_c is the magnetic fluctuation frequency. In the fast fluctuating limit, where $\nu_c t \gg 1$, this simplifies to an exponential function with a decay constant, λ_0, that

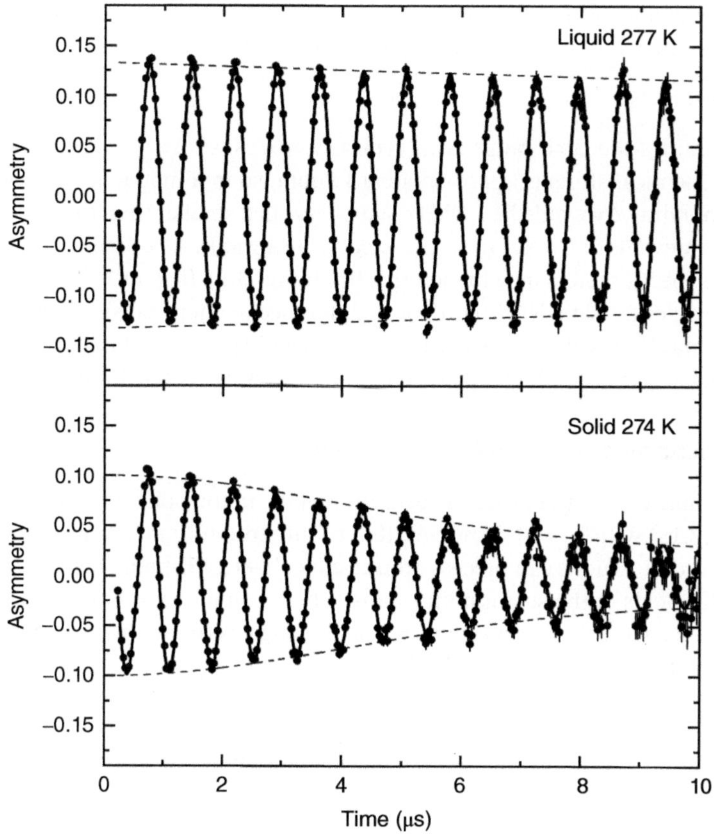

Figure 2.9 Time-domain TF-μSR spectra of the diamagnetic fraction in liquid (277 K) and solid (274 K) water in a transverse magnetic field of 100 G.

equals $\gamma_\mu^2 \Delta_{\mathrm{TF}}^2 / \nu_c$. This is typically the case for diamagnetic muon species in gases and liquids.

$$P_x(t) = h_D e^{-\lambda_0 t} \cos(2\pi \nu_\mu t). \tag{2.16}$$

An example of such motional effects is the difference between the diamagnetic signal in liquid water and ice, as shown in Figure 2.9. In ice the molecule MuOH is static, and there is a distribution of local fields that arise from the dipole moments of surrounding protons, leading to a Gaussian decay of the muon spin polarization. In the liquid the orientation of the spins and the distance between them are fluctuating rapidly, and this leads to a Lorentzian decay.

In every material a fraction of the implanted muons will thermalize in a diamagnetic environment (h_D). The reason for this will be discussed in detail in Chapter 3. In water, muons are incorporated in a variety of molecules such as MuOH, MuH_2O^+ and MuO^-. Unfortunately, little is known about the nature of diamagnetic muon states as we have been unable to distinguish them (so far) by measuring chemical shifts, as is routinely done in ^{1}H- and ^{2}H-NMR. The reason for this is the short lifetime of the muon and the limited time range over which a μSR measurement is made. The natural linewidth is determined by lifetime broadening:

$$\Delta\nu \sim \frac{1}{2\pi\Delta t}. \tag{2.17}$$

At CW μSR facilities the time range Δt is typically ~10 μs, so $\Delta\nu \sim 16$ kHz. The range of proton chemical shifts is approximately 13 ppm, so in a magnetic field of 10 kG the corresponding $\Delta\nu$ would be 1.8 kHz, which is much smaller than the natural linewidth and prevents one measuring chemical shifts by μSR. A new generation of very high-field spectrometers may breach this limitation. In a field of 100 kG the precession frequencies for signals with a 13 ppm difference in chemical shift would be 17.6 kHz, so it may be possible to resolve them. This would provide valuable information about the early events which determine muon states in matter (Chapter 3).

2.5.3 Diamagnetic Muons in Zero Magnetic Field

In zero magnetic field one would expect no evolution of the muon polarization, consistent with Eq. (2.13) with $\nu_\mu = 0$. However, this is only true if there are no spins surrounding the muon. Muons can precess around a local field, $|B|$, being the vector sum of the dipolar fields of the surrounding spins, which could be protons or electrons. The amplitude of the oscillations depends on the angle between the local field and the initial direction of the muon's spin, θ:[2]

$$P_z(t) = \cos^2\theta + \sin^2\theta \cos(2\pi\gamma_\mu|B|t). \tag{2.18}$$

For a polycrystalline material one has to average over all values of θ: $\langle\cos^2\theta\rangle = 1/3$ and $\langle\sin^2\theta\rangle = 2/3$. This gives

$$\overline{P}_z(t) = \frac{1}{3} + \frac{2}{3}\cos(2\pi\gamma_\mu|B|t), \tag{2.19}$$

which is widely used in condensed matter physics to study magnetic ordering, but it also has applications in materials science.

Another frequently encountered case is when there is a Gaussian distribution of local fields with width Δ/γ_μ due to concentrated dipole moments. An example of this is muons stopping in polyethene [9]. Each muon experiences a local field whose orientation is randomly oriented. Integrating over all orientations and magnitudes of the local magnetic field gives the Gaussian Kubo–Toyabe relaxation function [10, 11]:

$$P_z(t) = \frac{1}{3} + \frac{2}{3}e^{-\frac{1}{2}\Delta^2t^2}(1 - \Delta^2t^2). \tag{2.20}$$

This relaxation function looks like a Gaussian at early times, goes to a minimum at $t = \sqrt{3}/\Delta$ and has a 1/3 tail at long times. Fluctuations of the local field at a rate ν_c will affect the shape of this function, as shown in Figure 2.10. When the fluctuation rate is small (i.e. $\nu_c < 0.5\gamma_\mu\Delta$) the main effect is a relaxation of the 1/3 tail. In the case of fast fluctuations ($\nu_c \geq 0.5\gamma_\mu\Delta$) one observes an exponential decay of the polarization. An example of a static system with Kubo–Toyabe relaxation function is shown in Figure 2.11.

---

2 By convention, the initial direction of the muon spin defines the z axis.

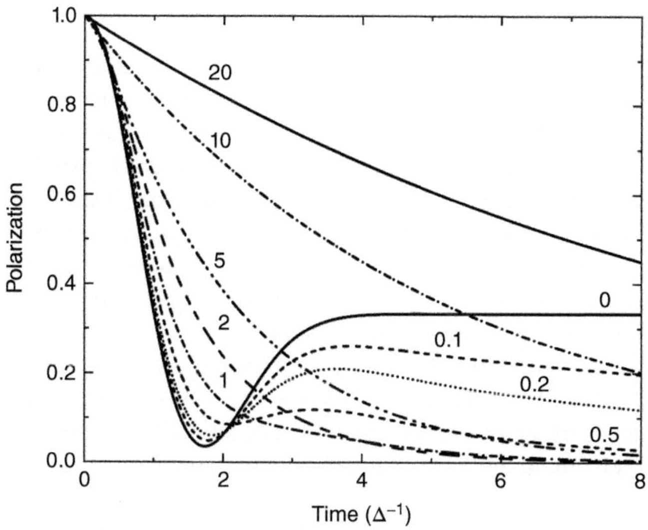

Figure 2.10 Fluctuation-rate dependence of the zero-field depolarization function derived for an isotropic Gaussian field distribution. The values of the fluctuation rate are given in units of Δ.

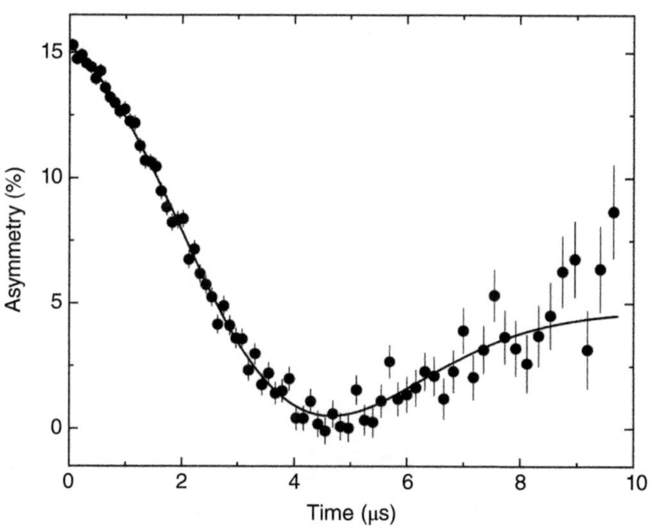

Figure 2.11 ZF-μSR spectrum of polyethene at 50 K. Source: The data are taken from Pratt et al. [9].

2.6 Spectroscopy of Muonium

A detailed mathematical description of the spin states, transitions, the time dependence of the muon spin polarization and the effects of magnetic fields is given in Sections A.3 and A.4.

2.6.1 Muonium in a Transverse Field

In muonium the muon and electron spins interact with each other and with the magnetic field. There are four spin states, labelled $|1\rangle$ to $|4\rangle$, that are linear combinations of the electron and muon Zeeman states. The energies of the spin states depend on the applied magnetic field and are given in Section A.3 and shown (Figure 2.12) in a Breit–Rabi diagram, so named after Gregory Breit and Isidor Isaac Rabi (Nobel Prize in Physics 1944). Rabi developed the magnetic resonance method on beams of atomic hydrogen and deuterium, and measured the magnetic moments of protons, deuterons and, by inference, neutrons [12]. He is reputed to have said 'Who ordered that?' when informed about the discovery of the muon [13].

In a transverse field one observes precessions that correspond to allowed transitions between the spin states. The selection rule is $\Delta M = \pm 1$, where M is the sum of the spin quantum numbers of the muon and electron. The allowed transitions are labelled according to their initial and final states, so the ν_{12} transition is between spin states $|1\rangle$ and $|2\rangle$.

In zero field the spin states of Mu are grouped into a triplet of degenerate states, $|\alpha^e\alpha^\mu\rangle$, $|\beta^e\beta^\mu\rangle$ and $\frac{1}{\sqrt{2}}(|\alpha^e\beta^\mu\rangle + |\beta^e\alpha^\mu\rangle)$, and a lower energy singlet, $\frac{1}{\sqrt{2}}(|\alpha^e\beta^\mu\rangle - |\beta^e\alpha^\mu\rangle)$. The energy gap corresponds to the hyperfine frequency, A_μ, which is 4463 MHz for Mu in vacuum. Detecting such high frequencies requires special effort but is the most direct and precise way to measure A_μ. The timing resolution must be at least 140 ps, and this requires special detectors and

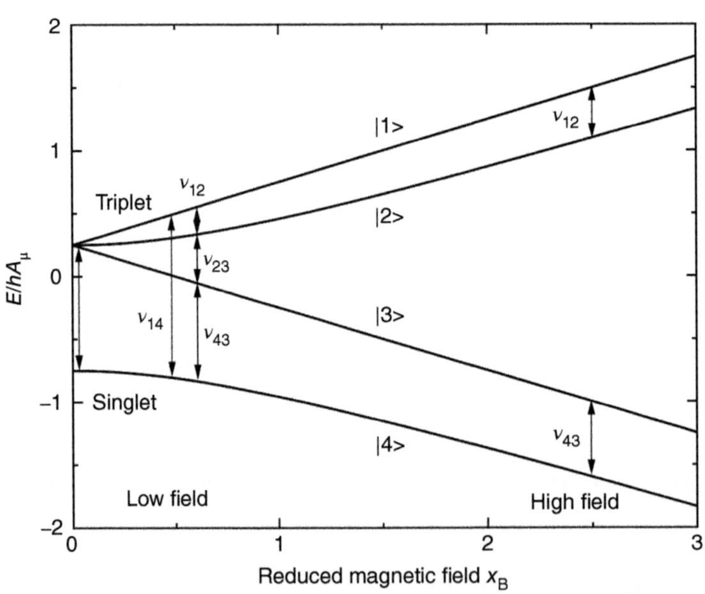

Figure 2.12 Breit–Rabi diagram showing the energies of the four muonium spin states as a function of magnetic field (in reduced units of $x_B = (\gamma_e + \gamma_\mu)B/A_\mu$). The use of reduced units means that the same diagram applies to the H atom if the field and energy units use A_H instead of A_μ.

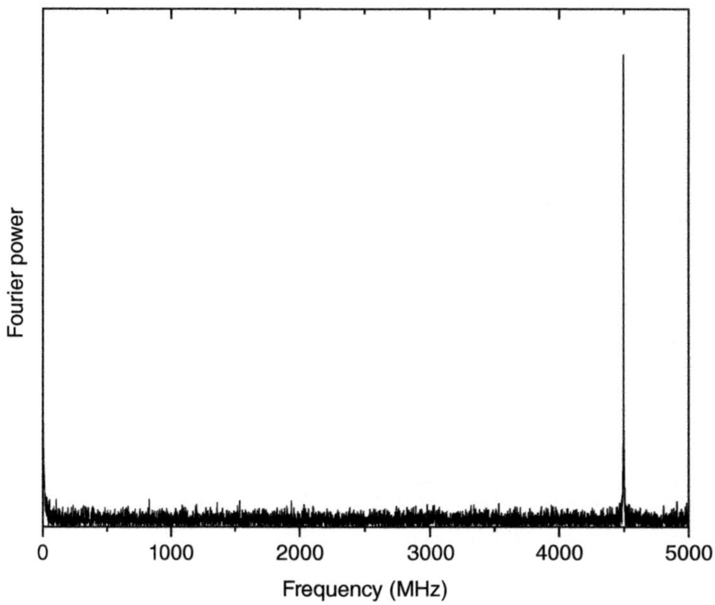

Figure 2.13 Fourier spectrum of the muon precession signal in fused quartz at room temperature in zero magnetic field. More than 90% of the implanted muons form muonium. The peak at 4500 MHz corresponds to the singlet–triplet transition in muonium. The spectrum was obtained on HAL-9500 at PSI by R. Scheuermann.

electronics [14]. An example of a high-frequency spectrum is shown in Figure 2.13 for Mu in quartz. This method was also used to measure A_μ of Mu trapped in silsesquioxane cages [15].

In low transverse field (i.e. a few gauss, where $A_\mu \gg \nu_e + \nu_\mu$) there are four allowed transitions: ν_{12}, ν_{23}, ν_{14} and ν_{43}. The intra-triplet ν_{12} and ν_{23} transitions are essentially degenerate in low field, and one observes a rapid precession with a frequency given by:

$$\nu_{\mathrm{Mu}} = \frac{1}{2}(\nu_e - \nu_\mu) = \frac{1}{2}(\gamma_e - \gamma_\mu)B. \tag{2.21}$$

The factor $\frac{1}{2}(\gamma_e - \gamma_\mu)$ can be considered an effective gyromagnetic ratio for muonium and has a value of 1.394 MHz/G [16], which is 102.8 times that of diamagnetic muons (0.01355 MHz/G). The ν_{14} and ν_{43} frequencies are much higher (of order A_μ) and are usually unobserved, given typical detector resolutions of ~1 ns. The muon polarization as a function of time in such weak fields is given by

$$P_x(t) = h_{\mathrm{D}} e^{-\lambda_{\mathrm{D}} t} \cos(2\pi\nu_\mu t + \phi_{\mathrm{D}}) + \frac{1}{2} h_{\mathrm{Mu}} e^{-\lambda_{\mathrm{Mu}} t} \cos(2\pi\nu_{\mathrm{Mu}} t + \phi_{\mathrm{Mu}}). \tag{2.22}$$

The first term in Eq. (2.22) is due to diamagnetic muons and the second term is due to muonium. The factor of ½ in the Mu term is because only two of the four Mu frequencies precess at ν_{Mu}. An example of a TF-μSR spectrum is shown in Figure 2.14. With an external field of 8 G there is a fast oscillation at 11.2 MHz (Mu) superimposed on the slow 0.108 MHz oscillation due to the diamagnetic muon precession,

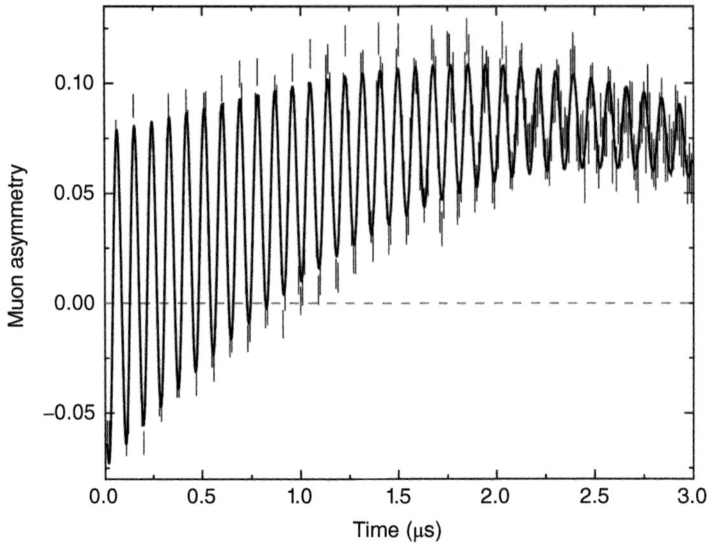

Figure 2.14 Time-domain TF-μSR spectrum of Mu in supercritical water in a field of 8 G [17]. The curve is a fit of Eq. (2.22) to the data.

of which only approximately a quarter of a cycle is within the time window of the μSR experiment.

This type of Mu signal is the preferred measurement for the study of Mu kinetics, which will be discussed in detail in Chapters 4 and 5. The Mu reaction rates can be determined from the decay of the Mu signal.

A small field lifts the degeneracy of ν_{12} and ν_{23}, and ~20 G is sufficient to observe the individual frequencies. The splitting increases with field and is large enough at a few hundreds of gauss to allow the determination of A_μ, as indicated in Eq. (2.23). This method has the advantage that it does not require specialized high time resolution apparatus. Examples of its application can be seen in studies of Mu in supercritical water [18] (see Figure 2.15) and Mu in aqueous methanol and salt solutions [19].

$$A_\mu = 2\frac{(\nu_{23} + \nu_\mu)(\nu_{12} + \nu_\mu)}{(\nu_{23} - \nu_{12})}. \tag{2.23}$$

2.6.2 Muonium in a Longitudinal Field

The initial polarization (i.e. at $t = 0$) for Mu in a longitudinal field is given by (Section A.4)

$$P_z^{\mathrm{Mu}}(t = 0) = h_{\mathrm{Mu}} \left[\frac{1}{2} + \frac{1}{2}\left(\frac{x_{\mathrm{B}}^2}{1 + x_{\mathrm{B}}^2}\right)\right], \tag{2.24}$$

where $x_{\mathrm{B}} = (\gamma_{\mathrm{e}} + \gamma_\mu)B/A_\mu$ and is equal to 1585 G for Mu in vacuum. $P_z^{\mathrm{Mu}}(t = 0)$ is thus $\frac{1}{2}h_{\mathrm{Mu}}$ at zero magnetic field and h_{Mu} at high fields, when $\nu_{\mathrm{e}} + \nu_\mu \gg A_\mu$.

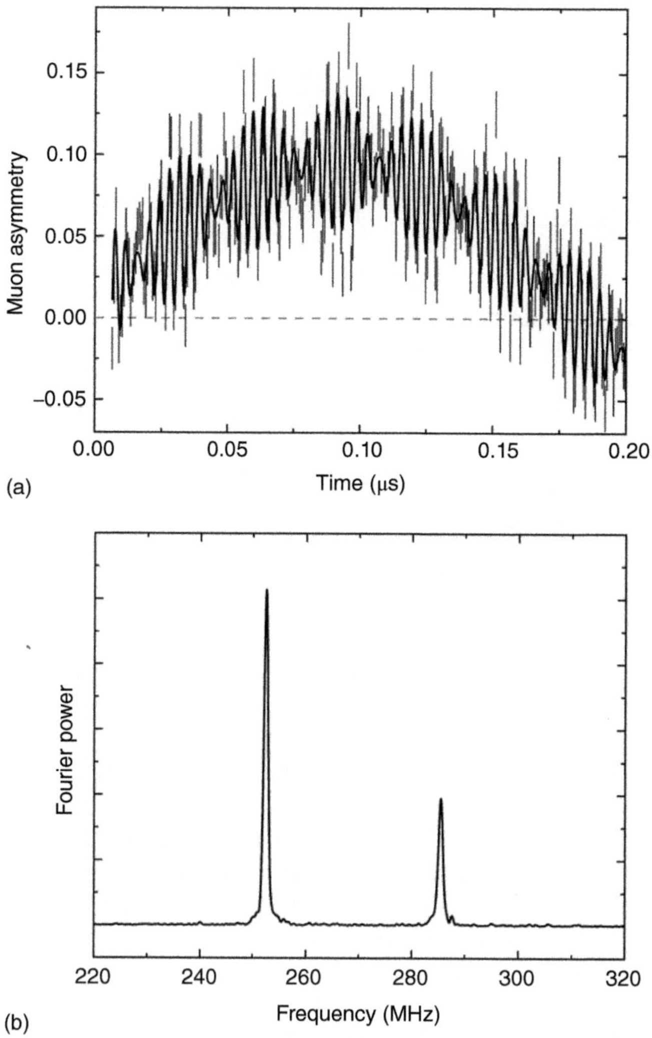

Figure 2.15 (a) Time-domain TF-μSR spectrum; and (b) Fourier transform TF-μSR spectrum of Mu in supercritical water in a field of 194 G [17]. The amplitude of the ν_{23} frequency is less than that of ν_{12} because of the field dependence of the amplitude factors ($c^2 > s^2$ for $B > 0$) and also the time resolution of the spectrometer.

In the 'repolarization' technique the initial polarization is measured as a function of the applied longitudinal magnetic field. For isotropic Mu at low field, half of the initial muon polarization is lost through averaging of the ν_{24} frequency, but this is gradually recovered in sufficiently large fields. At high fields the Zeeman coupling of the spins to the external field exceeds their hyperfine interaction. This is often incorrectly described as decoupling of the muon and electron spins.

The inflection point, $x_B = 1$ ($P_z = 0.75$), occurs at ~1585 G for vacuum Mu and usually at smaller values for Mu in condensed matter, where some spin density is transferred to the surroundings. In principle the shape of the curve can be used to

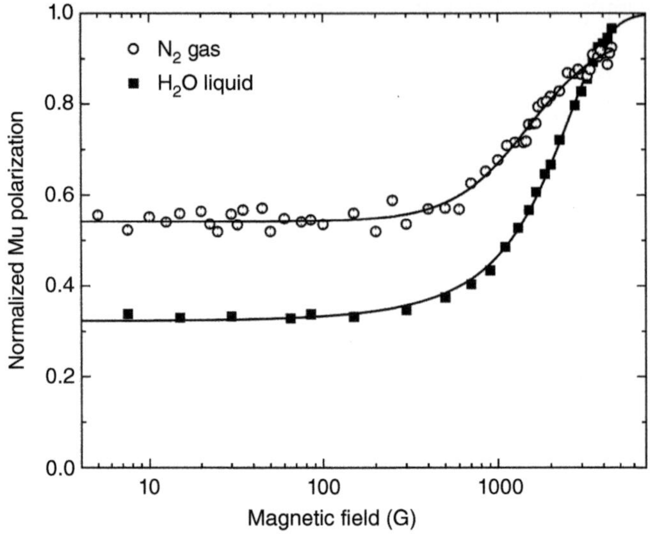

Figure 2.16 Repolarization curve of Mu in nitrogen gas [20] and liquid water at 300 K (McKenzie and Cottrell, ISIS). The low-field polarization is less than 0.5 in water due to hyperfine coupling with protons in the surrounding solvation shell.

estimate the value of A_μ by fitting with Eq. (2.24). An example of a repolarization curve is shown in Figure 2.16 for Mu in N_2 gas. The initial polarization is ~0.5 at low fields and approaches 1 at high fields.

The repolarization technique is critically dependent on the accuracy with which the diamagnetic component can be determined over the full longitudinal magnetic field range of the spectrometer (a non-trivial exercise) and the concomitant effect of the field on both the incident μ^+ and decay e^+ trajectories, affecting both the parameter α (Eq. (2.11)) and the intrinsic asymmetry of the spectrometer. Thus, the 'repolarization' method is the least reliable method to determine A_μ, although it is popular at pulsed facilities, where direct measurement of high frequency (>10 MHz) is impossible.

There is often hyperfine coupling of the Mu electron to other nuclei ('superhyperfine coupling' in the ESR literature), and this can result in reduction of the low-field polarization below the expected 0.5. An example is shown in Figure 2.16, for the case of Mu in water. Muonium is a hydrophobic solute and resides in a transient clathrate-like cage or 'bubble' in water. The 'leakage' of spin density from Mu to the surrounding solvent molecules also reduces the hyperfine constant below the vacuum value [21]. Other examples are discussed in Chapter 8.

2.7 TF-μSR of Muoniated Radicals

2.7.1 Muoniated Radicals in Isotropic Environments

The magnetic moment of the unpaired electron couples to the magnetic moments of the muon and any other nuclei, which results in a multitude of spin states. There

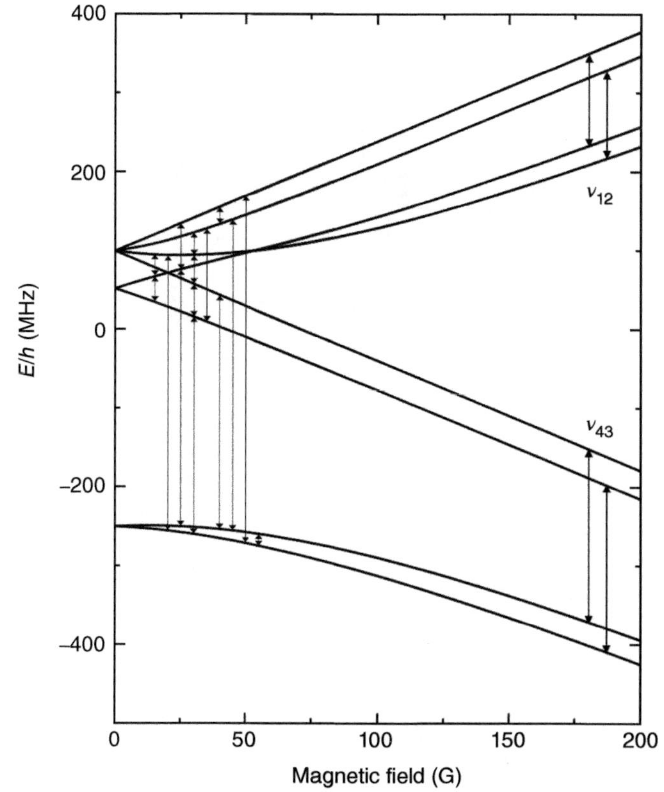

Figure 2.17 Energy levels versus magnetic field for a three-spin system of muon, electron and proton with isotropic hyperfine coupling constants A_μ = 330 MHz and A_p = 67.1 MHz. There are 15 allowed transitions in low magnetic field. The two sets of degenerate transitions labelled ν_{12} and ν_{43} correspond to the allowed radical transitions in high magnetic field.

are $4\Pi(2I^k+1)$ spins states, where the product is over all nuclei with non-zero spin. The Breit–Rabi diagram, i.e. the variation of the energy levels with applied magnetic field, is shown for the $e^- - \mu^+ - p$. system in Figure 2.17. There are eight spin states due to muonium-like spin states being split by the interaction with the proton.

In low magnetic fields the muon polarization is distributed over many transition frequencies; there are 15 allowed transitions in the $e^- - \mu^+ - p^+$ system and 792 allowed transitions in the complete spin system of the muoniated ethyl radical ($MuCH_2CH_2$) radical. Distributing the polarization over so many transitions reduces the intensity of the μSR signal, which makes detection of muoniated radicals difficult or impossible in low fields. Portions of the TF-μSR spectra of C_6H_6Mu and C_6D_6Mu in various transverse fields are shown in Figure 2.18. The splitting is larger in C_6H_6Mu and a higher magnetic field is required for the system to be in the high-field limit, due to the larger magnetic moment, and hence hfc, of H compared with D.

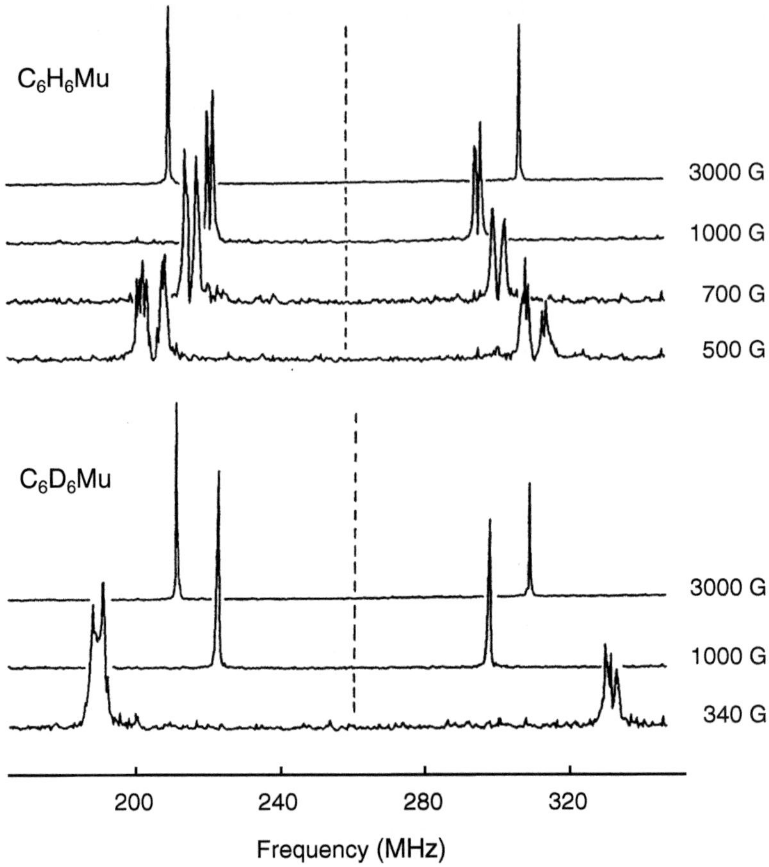

Figure 2.18 TF-μSR spectra of C_6H_6Mu and C_6D_6Mu in low transverse fields. The diamagnetic signal is at lower frequencies and is not shown. Source: Roduner [22], figure 4.1.

In high magnetic fields, where the Zeeman energy is much larger than the hyperfine interactions, one can treat the hyperfine interactions as perturbations. Applying the selection rules in high transverse fields, $\Delta m_\mu = \pm 1$, $\Delta m_e = 0$ and $\Delta m_k = 0$, one finds that the frequency spectrum is considerably simplified, with two radical frequencies (ν_{12} and ν_{43}) observed, in similar fashion to electron-nuclear double resonance (ENDOR) spectroscopy. This is true regardless of the number of nuclear spins coupled to the electron. The observable radical frequencies, which correspond to pure muon spin flips, are given by

$$\nu_{12} = \nu_{\text{mid}} - \frac{1}{2}A_\mu, \tag{2.25}$$

$$\nu_{43} = \nu_{\text{mid}} + \frac{1}{2}A_\mu, \tag{2.26}$$

where ν_{mid} is close to, but not exactly equal to, the muon Larmor frequency:

$$\nu_{\text{mid}} = \frac{1}{2}\left\{ \left[A_\mu^2 + (\nu_e + \nu_\mu)^2\right]^{1/2} - \nu_e + \nu_\mu \right\}. \tag{2.27}$$

The hyperfine constant, A_μ, is simply given by the difference in the two precession frequencies. In some cases, the higher radical precession frequency is not observed due to the time resolution of the μSR spectrometer, but A_μ can be determined from the lower radical precession frequency and ν_μ. Combining Eqs. (2.25) and (2.27) leads to

$$A_\mu = 2\frac{(\nu_\mu - \nu_{12})(\nu_e + \nu_{12})}{\nu_e - \nu_\mu + 2\nu_{12}}. \tag{2.28}$$

A TF-μSR spectrum obtained from liquid benzene at 280 K and 22 kG is shown in Figure 2.19. There are three precession frequencies, at 43.3, 300.1 and 559.1 MHz. The peak at 300.1 MHz, labelled ν_D, is due to the diamagnetic signal, i.e. muons precessing at the muon Larmor frequency, ν_μ. This can be checked by multiplying γ_μ by the applied magnetic field. The peaks at 43.3 and 559.1 MHz are the ν_{12} and ν_{43} frequencies of the muoniated cyclohexadienyl radical C_6H_6Mu. A_μ is determined from the difference between these frequencies and is 515.8 MHz. The lower amplitude of ν_{43} compared with ν_{12} is primarily due to the time resolution of the spectrometer.

The splitting of the radical frequencies is independent of the applied magnetic field, as long as the system is in the high-field limit. It is good practice to perform TF-μSR measurements at more than one magnetic field and confirm that the radical frequencies shift appropriately (Figure 2.20). In situations where ν_μ is less than $A_\mu/2$ one of the radical frequencies will be negative, which can be distinguished from positive frequencies by applying a complex Fourier transform to the data from two sets of orthogonal positron detectors.

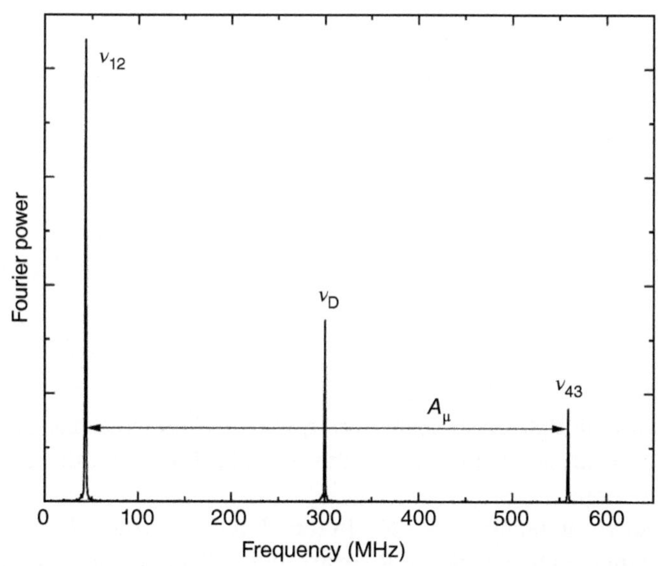

Figure 2.19 Fourier transform TF-μSR spectrum of the muoniated cyclohexadienyl radical (C_6H_6Mu) in liquid benzene at 280 K and 22 kG (peaks ν_{12} and ν_{43}). The peak at ~300 MHz labelled ν_D is due to muons in diamagnetic environments.

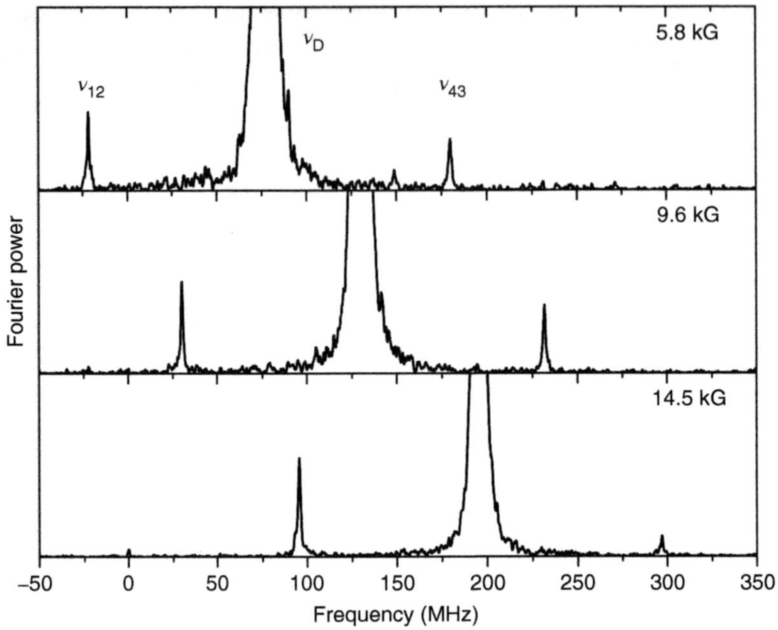

Figure 2.20 Fourier transform TF-μSR spectra obtained from liquid ketene at 165 K in high magnetic fields. The observed radical is the muoniated methyl radical [23]. Note that ν_{12} is *negative* in the top panel.

2.7.2 Muoniated Radicals in Anisotropic Environments

TF-μSR measurements can also be made of radicals in the solid state, but the ν_{12} and ν_{43} frequencies are typically broadened by the dipolar interactions and the generally more rapid spin relaxation. In a muoniated radical with an axially symmetric muon hyperfine tensor, which has three principal components, $D_\mu^\parallel$ and two $D_\mu^\perp$, with $D_\mu^\parallel + 2D_\mu^\perp = 0$, the frequencies in the high-field limit are given by

$$\nu_{12} = \nu_\mu - \frac{1}{2}\left[A_\mu + \frac{D_\mu^\parallel}{2}(3\cos^2\theta - 1)\right], \tag{2.29}$$

$$\nu_{43} = \nu_\mu + \frac{1}{2}\left[A_\mu + \frac{D_\mu^\parallel}{2}(3\cos^2\theta - 1)\right]. \tag{2.30}$$

In a single crystal with multiple crystallographically inequivalent molecules one can observe multiple radical signals for a single type of radical. In a powder one would observe a distribution.

Consider the Mu adducts of naphthalene [24]. In the TF-μSR spectrum of a concentrated solution of naphthalene in acetone there are four radical lines due to the two types of muoniated radical formed by Mu addition to the secondary carbons (Figure 2.21a). In the TF-μSR spectrum of the single crystal there are multiple lines due to radicals with different orientations to the magnetic field (Figure 2.21b).

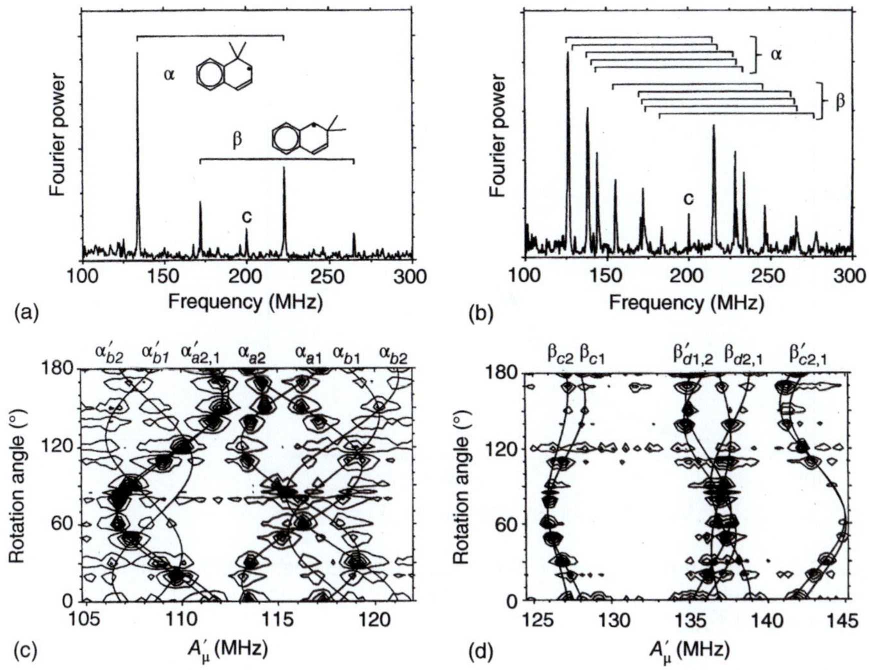

Figure 2.21 (a) TF-μSR spectrum of a concentrated solution of naphthalene in acetone in a magnetic field of ~3 kG. Assignments of the lines to the α- and β-type radicals are indicated. The feature 'c' at 200 MHz is an artefact of the digital clock. The lower frequency of the pair of lines is a negative frequency due to $\nu_\mu < A_\mu/2$. (b) TF-μSR spectrum of a single crystal of naphthalene under the same experimental conditions. Panels (c) and (d) show contour plots used to determine the angular dependence of the reduced muon hyperfine coupling constants. The angles refer to rotation about an axis perpendicular to the applied magnetic field. Source: Adapted from Reid and Roduner [24] with permission from Springer Nature.

In single crystals the angular dependence of ν_{12} and ν_{43} can be used to construct the muon hyperfine tensor. In polycrystalline materials the distribution of θ leads to powder patterns of asymmetric shape, and large $D_\mu^\parallel$ values can broaden the lines beyond detection. Figure 2.21c and d show contour plots for the angular variation of the radical signals.

2.7.3 Polarization Transfer from Precursor to Radical

A muoniated radical must be formed promptly in order for the muon spin polarization to be transferred from the Mu precursor to the radical. Slower formation reduces muon spin polarization because of dephasing of the precessing muons in the transverse field. The fraction of spin polarization transferred to a radical (P_R) in the high-field limit is given by

$$P_R \approx \frac{\lambda_{Mu}^2}{\lambda_{Mu}^2 + \delta\omega^2}, \tag{2.31}$$

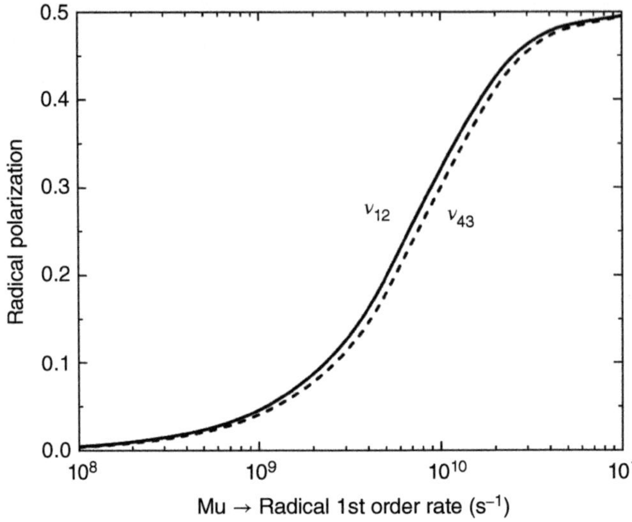

Figure 2.22 Amplitude of ν_{12} (solid line) and ν_{43} (dashed line) frequencies of the muoniated cyclohexadienyl radical (A_μ = 515 MHz) as a function of its first-order formation rate in a field of 15 kG.

where λ_{Mu} is the Mu reaction rate and $\delta\omega$ is the difference between the muon precession frequencies in Mu and in the radical. λ_{Mu} depends on both the addition rate constant, k_{Mu}, and the concentration of the parent compound. If the rate of addition is slow or if the concentration of the parent compound is low, then only a small amount of polarization is transferred to the muoniated radical and it will be difficult to observe. The minimum concentration at which a radical can be observed in a TF-μSR spectrum depends on the muonium reaction rate constant. As an example, k_{Mu} is 3.7×10^9 M^{-1} s^{-1} for the addition of Mu to benzene in *n*-hexane. The concentration of benzene in *n*-hexane needed for the amplitude of the ν_{12} line to be 0.01, which is ~2% of the value in neat benzene, is 0.07 M. If the addition rate constant was one hundred times slower, the minimum concentration would be one hundred times larger. The minimum concentration for detection of muoniated radicals by TF-μSR will in practice be larger if multiple types of radical are formed and if spin relaxation and chemical reactions broaden the radical lines and decrease their amplitude. Typical values are 0.1 to 0.5 M for muoniated cyclohexadienyl radicals. The variation of the amplitude of the ν_{12} and ν_{43} frequencies with the conversion rate from Mu to C_6H_6Mu, which is given by $k_{Mu}[C_6H_6]$, is shown in Figure 2.22.

2.8 Avoided Level-Crossing Resonance of Muoniated Radicals

The ALC-μSR technique involves measuring the time-integrated asymmetry of the muon decay as a function of a magnetic field applied parallel to the initial direction of the muon spin. In contrast to TF-μSR, there is no restriction on the number of muons in the sample at one time, so it is possible to run at a much higher incident muon rate. Also, there is no dephasing problem in longitudinal fields, so it is possible to observe species using ALC-μSR that are formed within ~1 μs of muon implantation, and this makes it possible to study samples with concentrations of the precursor molecule as low as 1 mM.

The setup for an ALC-μSR experiment is shown in Figure 2.6. The time-integrated asymmetry is measured as the magnetic field is scanned in a series of small steps. A challenge of ALC-μSR is dealing with a non-linear background. The baseline asymmetry varies with the magnetic field due to: (i) variation of the efficiencies of the detectors (gain variation of photomultiplier tubes); (ii) fluctuations of the muon beam spot; (iii) field-dependent positron trajectories. There are several methods to deal with the non-linear baseline. The baseline can be removed by subtracting the spectrum of a reference sample, but this only works if the reference has the same density and geometry as the sample of interest. Typically, this method accounts for most of the baseline, but some remains due to slight differences between the sample and the reference. A method that is frequently used is to fit the baseline with a polynomial while simultaneously fitting the resonances. This works well when the resonances have a large amplitude and are narrow compared with the curvature of the baseline, but is difficult for weak and broad resonances. Often, reference subtraction and polynomial fitting are used together. Another technique that is used is field modulation, which gives the resonances a differential lineshape. This involves applying a small square-wave modulation field (e.g. $B_{\mathrm{mod}} \sim 150$ G), measuring the asymmetry at $B + B_{\mathrm{mod}}$ and $B - B_{\mathrm{mod}}$ and taking the difference. This method should only be used when the width of the resonances is less than or comparable to B_{mod} and therefore is usually only used in measurements on radicals in liquids.

In high magnetic fields the eigenstates of the radical can be approximated by simple products of Zeeman states of the muon, electron and nuclei, so there is no evolution of the muon's spin with time and the asymmetry is independent of the magnetic field. At specific values of the applied magnetic field, nearly degenerate pairs of spin states can be mixed through the isotropic and anisotropic components of the hyperfine interaction. This is indicated schematically in Figure 2.23. The muon polarization oscillates between the two mixing states, and this leads to a loss of time-integrated asymmetry.

There are three types of resonances, which are characterized by the selection rule $|\Delta M| = 0, 1$ and 2, where M is the sum of the m_z quantum numbers of the muon, electron and proton spins. The resonances are referred to as Δ_0, Δ_1 and Δ_2 resonances, respectively. The Δ_2 resonance is extremely weak and is rarely observed. There can be only one Δ_1 resonance for each type of radical, but there can be as many Δ_0 resonances as there are nuclei with $I > 0$ in the radical, although in practice the number of observable Δ_0 resonances is less than the maximum as many of these nuclei can have small hfcs, which results in very small resonance amplitudes.

The Δ_0 resonance is due to mixing between spin states that have the same electron spin but opposite muon and proton spins. Δ_0 resonances can be observed for muoniated radicals in the solid, liquid or gas phases, although they are often weak and difficult to observe in solids. The Δ_0 resonance field depends on both the muon

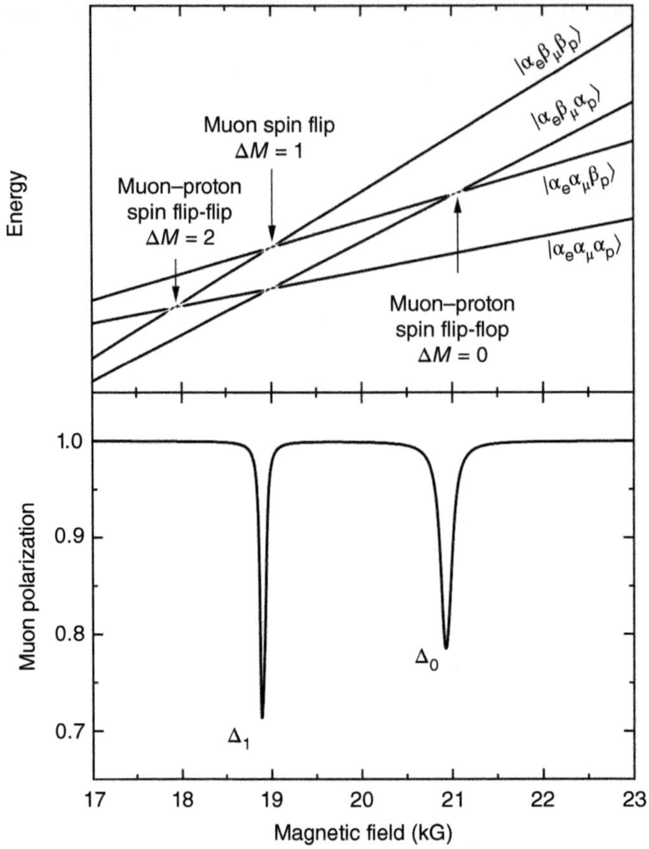

Figure 2.23 Energy-level diagram for a three-spin system of an electron e, positive muon μ and proton p in the high-field limit. The energy spacings have been exaggerated to clearly show the level crossings. Muon avoided level-crossing resonances occur when states with opposite muon spins become near degenerate in energy and mix, resulting in oscillations of their populations.

hfc and the nuclear hfc, A_X, and is given by

$$B_{\text{res}}^{\Delta_0} = \frac{1}{2}\left[\frac{A_\mu - A_X}{\gamma_\mu - \gamma_X} - \frac{A_\mu + A_X}{\gamma_e}\right], \tag{2.32}$$

where γ_X is the nuclear gyromagnetic ratio. An advantage of the ALC-μSR technique is that one can determine both the magnitude and relative sign of the nuclear hfc. The amplitude of the Δ_0 resonance is given by

$$\text{Amplitude} \propto \frac{P_z^0 \omega_{\text{LCR}}^2}{\lambda_{\text{eff}}^2 + \omega_{\text{LCR}}^2}, \tag{2.33}$$

where ω_{LCR} is the frequency at resonance and λ_{eff} is the effective relaxation rate (muon decay rate plus any additional relaxation contribution),

$$\omega_{\text{LCR}} = \frac{\sqrt{c}\pi A_\mu A_X}{\gamma_e B_{\text{res}}^{\Delta_0}}, \tag{2.34}$$

and c is given by

$$c = [I_X(I_X+1) - M(M-1)]^{\frac{1}{2}}, \quad (2.35)$$

where $I_X \geq M \geq -(I_X+1)$. The full width at half maximum, $\Delta B_{1/2}$, is

$$\Delta B_{1/2} = \frac{\sqrt{\omega_{\mathrm{LCR}}^2 + \lambda^2}}{\pi(\gamma_\mu - \gamma_p)}. \quad (2.36)$$

ALC-μSR spectra of the muoniated cyclohexadienyl radical are shown in Figure 2.24. There are four resonances in the spectrum of C_6H_6Mu in neat benzene. These are proton Δ_0 resonances of the four types of protons in the radical. The Δ_0 resonance fields and the corresponding proton hfcs are listed in Table 2.1. Four additional resonances are observed in the ALC-μSR spectrum of the $^{13}C_6H_6Mu$ radical. These are ^{13}C Δ_0 resonances of the four types of ^{13}C nuclei in the radical. The Δ_0 resonance fields and the corresponding ^{13}C hfcs are listed in Table 2.2.

The Δ_1 resonance field is given by:

$$B_{\mathrm{res}}^{\Delta_1} = \frac{1}{2}\left[\frac{A_\mu}{\gamma_\mu} - \frac{A_\mu}{\gamma_e}\right], \quad (2.37)$$

where A_μ is the muon hfc, γ_μ is the muon gyromagnetic ratio and γ_e is the electron gyromagnetic ratio. The Δ_1 resonance arises from mixing between spin states with the same electron and nuclear spins but different muon spin directions. These spin states are only mixed in the presence of anisotropy, so the presence of a Δ_1 resonance can be considered to be diagnostic of a frozen state or of anisotropic motion. The

Table 2.1 Analysis of ALC-μSR data for the ^1H hyperfine coupling constants in C_6H_6Mu at 298 K.

Nucleus	A_μ (MHz)	$B_{\mathrm{res}}^{\Delta_0}$ (kG)	A_p (MHz)
H(6)	514.48	20.774	126.11
H(2,4)	514.48	27.177	7.47
H(1,5)	514.48	28.937	−25.14
H(3)	514.48	29.533	−36.19

Source: Data from Ref. [25]. The numbering of the nuclei is shown in Figure 2.24.

Table 2.2 Analysis of ALC-μSR data for the ^{13}C hyperfine coupling constants in $^{13}C_6H_6Mu$ at 298 K.

Nucleus	A_μ (MHz)	$B_{\mathrm{res}}^{\Delta_0}$ (kG)	A_C (MHz)
C(3)	512.28	18.262	53.81
C(1,5)	512.28	18.841	39.42
C(6)	512.28	21.797	−34.05
C(2,4)	512.28	21.843	−35.19

Source: Data from Ref. [26]. The numbering of the nuclei is shown in Figure 2.24.

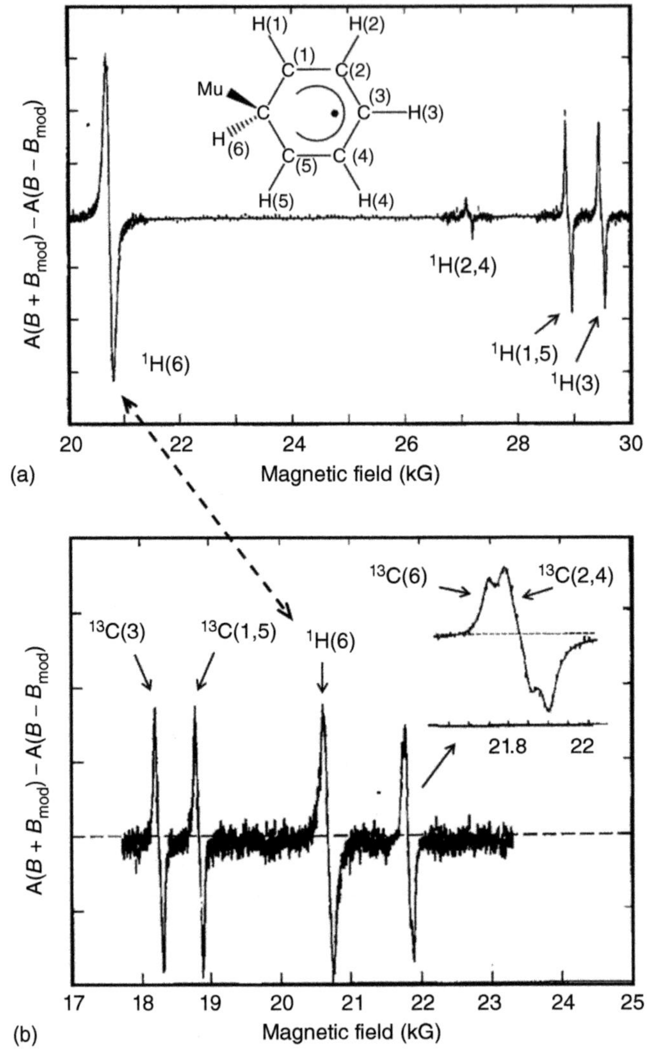

Figure 2.24 (a) ALC-μSR spectrum of the muoniated cyclohexadienyl radical (C_6H_6Mu) in neat benzene at 300 K. The inset shows the structure of the C_6H_6Mu radical and the numbering of the atoms. Source: Adapted from Percival et al. [25] with permission from Elsevier. (b) ALC-μSR spectrum of the muoniated ^{13}C-labelled cyclohexadienyl radical ($^{13}C_6H_6Mu$) in cyclohexane at 300 K. Source: Adapted from Kiefl et al. [26] with permission from Elsevier. The spectra were obtained using square-wave field modulation, which results in the differential lineshape.

resonances are averaged out by isotropic motion on a critical time scale given by the inverse of the hyperfine anisotropy (e.g. ~50 ns for a cyclohexadienyl-type radical).

The Δ_1 resonance is suitable for studying anisotropic reorientational motion of radicals in single crystals as well as in polycrystalline or amorphous states. The lineshape of the Δ_1 resonance is extremely sensitive to and characteristic of the type of reorientational motion that the muoniated radical is undergoing. This will be

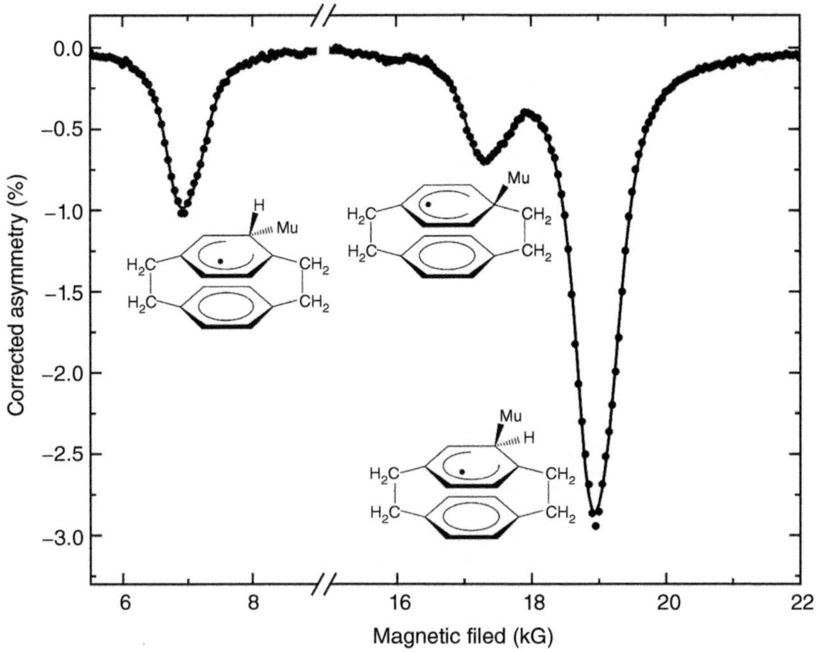

Figure 2.25 ALC-μSR spectrum of [2.2]-paracyclophane powder at 298 K [27]. The resonances are of Δ_1 type and the corresponding radical structures are shown beside each resonance.

discussed in detail in Chapter 6. The ALC-μSR spectrum of [2.2]-paracyclophane powder is shown in Figure 2.25 [27]. There are three Δ_1 resonances due to the three types of muoniated radical that were formed by addition of Mu to different sites on the aromatic ring.

2.9 RF Muon Spin Resonance of Muoniated Radicals

Among the μSR techniques, RF-μSR most resembles conventional magnetic resonance, since it relies on resonance between an exciting RF field and one of the energy transitions discussed in Section 2.6.1. This technique is more time consuming than TF-μSR and is consequently used only in situations where TF-μSR or ALC-μSR cannot be applied, such as measuring A_μ of radicals in dilute solution. There is no loss of polarization going from Mu to the muoniated radical as the muon polarization is initially aligned along the applied magnetic field.

The experimental setup is the same as for ALC-μSR except that a linearly polarized radio frequency field is applied perpendicular to the external magnetic field. The RF induces transitions between spin states with opposite muon spins when the frequency of the RF (ν_{RF}) matches the transition frequency. This results in a reduction of the time-integrated muon spin polarization at a resonance field. In the high-field limit (i.e. where the transition frequencies are given by Eq. (2.25) and Eq. (2.26)) the

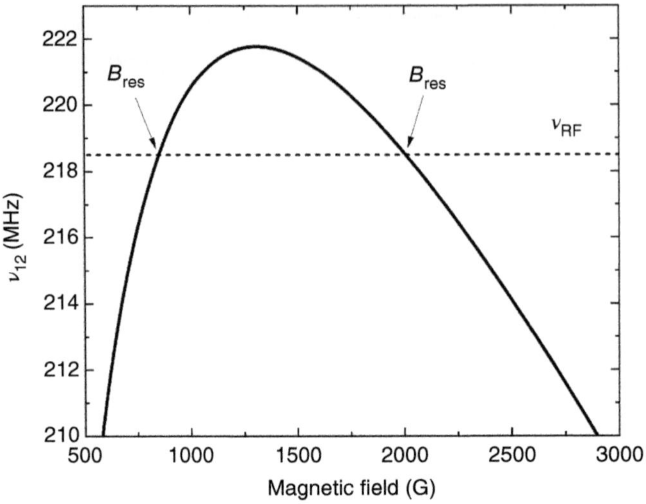

Figure 2.26 Longitudinal magnetic field dependence of the ν_{12} frequency for a radical with $A_\mu = 514.8$ MHz, the value for the C_6H_6Mu radical in benzene at 293 K. The dashed horizontal line corresponds to $\nu_{RF} = 218.5$ MHz.

resonance field is given by

$$B_{res} = \frac{\nu_{RF} \pm \frac{1}{2}A_\mu}{\gamma_\mu}. \tag{2.38}$$

The magnetic field dependence of the ν_{12} frequency for a radical with $A_\mu = 514.8$ MHz is shown in Figure 2.26. Resonances occur when ν_{12} equals ν_{RF}, which for $\nu_{RF} = 218.5$ MHz would be 849.3 G.

Complications arise when the resonances occur at a low magnetic field. In this case one must include the non-linear field dependence of the radical frequencies and hyperfine couplings to other nuclei in the radical, which causes the resonance to split. A portion of the RF-μSR spectrum of the C_6H_6Mu radical in benzene is shown in Figure 2.27 [28]. The splitting is primarily due to the methylene proton, which has the largest hfc after the muon. Additional resonances should occur at 2.1 kG but were outside the magnetic field range of the spectrometer. An example of how the RF frequency can be resonant at two fields can be seen in a study of the $C_{60}Mu$ radical [29].

2.10 Longitudinal-Field Repolarization Studies of Muoniated Radicals

Measuring LF repolarization curves is an alternative way to roughly determine muon hfcs. The technique is frequently used at pulsed facilities where one cannot perform high TF-μSR measurements and is significantly less time consuming than RF-μSR. This method has a long history and remains an important technique

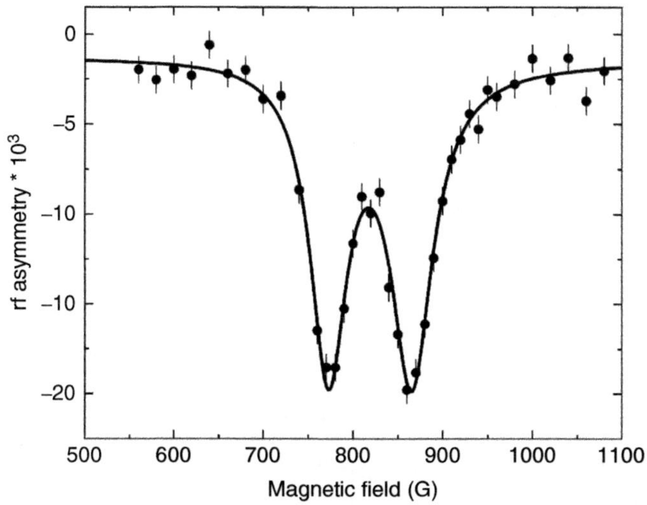

Figure 2.27 RF-μSR spectrum of the C_6H_6Mu radical in liquid benzene at 293 K with $\nu_{RF} = 218.5$ MHz [28].

for characterizing unknown systems, particularly for those with state-to-state transitions or dynamical properties, and has been useful as well in characterizing anisotropic muon environments in the solid state.

The basis of the LF repolarization technique for studying muoniated radicals is similar to that of muonium, which was described in Section 2.6.2. Increasing the strength of the magnetic field changes the relative magnitudes of the Zeeman and hyperfine terms and leads to an apparent decoupling of the muon and electron spins. The muon polarization can be fully recovered at high enough fields, above the relevant contact hyperfine field, $B_0 = A_\mu/\gamma_e$. The classic sigmoidal curve with the polarization being 1/2 at zero field is only valid for muonium or a muoniated radical with no other spin-active nuclei. In muoniated radicals where there is significant hyperfine coupling with nuclear spins the initial polarization can be below 50%. This is because of the presence of nuclear moments and the redistribution of muon polarization among a number of eigenstates at these fields.

At high enough reactant concentrations, muoniated radicals can be formed essentially instantaneously, i.e. on a sub-ns timescale (even ps in liquids) comparable to that of the muon slowing down process. In such cases one observes a single-component repolarization curve that resembles the muonium repolarization curve, except that the inflection point is at a lower field due to the smaller A_μ value for the radical. The repolarization curve can be fit to an equation of the form:

$$P = C_1 + C_2 \frac{x_B^2}{1 + x_B^2}, \tag{2.39}$$

where C_1 and C_2 are adjustable parameters and $x_B = (\gamma_e + \gamma_\mu)B/A_\mu$.

In dilute systems the formation of the muoniated radical can be slow compared with the hyperfine period (A_μ^{-1}), though still on a fast (~ns) time scale. In such

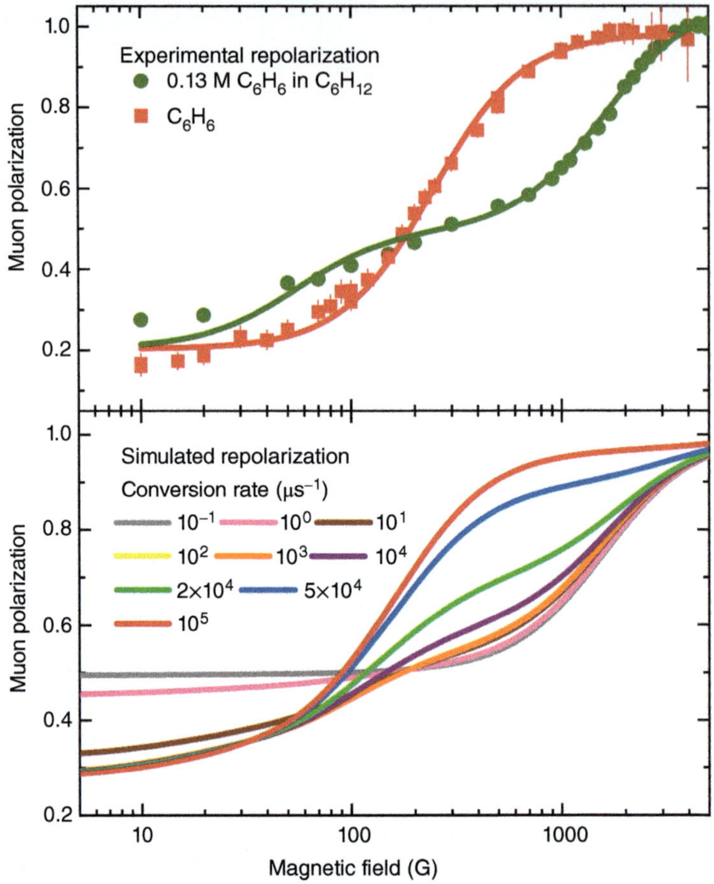

Figure 2.28 (Top) The magnetic field dependence of the initial polarization, i.e. repolarization curves, of pure benzene and a dilute solution of benzene in cyclohexane. (Bottom) Monte Carlo simulations for different formation rates for Mu addition to benzene. The muon hfcs in the radical and in Mu were fixed at their known gas-phase values, 514.3 and 4463 MHz, respectively. Proton hfcs for the ipso and para protons were included. Source: The data are taken from Fleming et al. [20].

cases a two-component repolarization curve is measured, with a shape indicative of the product of the polarizations for the prompt initial (Mu) and final state (radical) species.

LF repolarization curves are shown for pure benzene and a 0.13 M benzene solution in cyclohexane in Figure 2.28 [20]. These are compared with simulated LF repolarization curves for different conversion rates from Mu to the muoniated radical. In neat benzene the C_6H_6Mu radical is formed within ps and the repolarization curve has one component. The low-field polarization is approximately half that seen for muonium due to hyperfine coupling to other nuclear moments. The inflection point for the C_6H_6Mu repolarization curve is ~180 G due to the lower A_μ value compared with Mu. There are clearly two components in the LF repolarization curve of the dilute solution. The conversion rate from Mu to C_6H_6Mu is given by $k_{Mu}[C_6H_6]$ and is 480 μs^{-1}. The simulated repolarization curves are very similar for conversion rates between 100 and 1000 μs^{-1} and match the experimental curve in that there are two components. The low-field component is due to the C_6H_6Mu radical and the high-field component is due to the Mu precursor.

The shape of the measured repolarization curve is critically dependent on the accuracy with which the diamagnetic component can be determined over the full LF range of the spectrometer (a non-trivial exercise) and the concomitant effect of the field on both the incident muons and decay positron trajectories, affecting both the parameter α and the intrinsic asymmetry of the spectrometer. These issues can result in a considerable error in the measured A_μ values. Whenever possible, it is certainly preferable to determine the muon hfc for paramagnetic species in high transverse field, where the errors will generally be much smaller, typically 1% or less. Nevertheless, LF repolarization is frequently used due to its simplicity and in cases where there is no alternative method to determine A_μ.

References

1 Nagamine, K. (2003). *Introductory Muon Science*. Cambridge University Press.

2 Beveridge, J.L., Doornbos, J., Garner, D.M. et al. (1985). *Nucl. Instrum. Methods Phys. Res. A* 240: 316–322.

3 Bakule, P. and Morenzoni, E. (2004). *Contemp. Phys.* 45: 203–225.

4 Strasser, P., Ikedo, Y., Makimura, S. et al. (2014). *J. Phys. Conf. Ser.* 551: 012065.

5 Holzschuh, E. (1983). *Phys. Rev. B* 27: 102–111.

6 Giblin, S.R., Cottrell, S.P., King, P.J.C. et al. (2014). *Nucl. Instrum. Methods Phys. Res. A* 751: 70–78.

7 Bueno, J.F., Arseneau, D.J., Bayes, R. et al. (2011). *Phys. Rev. B* 83: 205121.

8 Blundell, S.J., De Renzi, R., Lancaster, T., and Pratt, F.L. (2021). *Muon Spectroscopy: An Introduction*, Chap. 5. Oxford University Press.

9 Pratt, F.L., Blundell, S.J., Marshall, I.M. et al. (2003). *Physica B* 326: 34–40.

10 Hayano, R.S., Uemura, Y.J., Imazato, J. et al. (1979). *Phys. Rev. B* 20: 850–859.

11 Dalmas de Réotier, P. and Yaouanc, A. (1997). *J. Phys.: Condens. Matter* 9: 9113.

12 Rigden, J.S. (2002). *Hydrogen: The Essential Element*, Chap. 12, 113–124. Harvard University Press.

13 Bartusiak, M. (1987). *The New York Times Book Review*, vol. 27.

14 Scheuermann, R., Dilger, H., Roduner, E. et al. (2000). *Physica B* 289–290: 698–701.

15 Dilger, H., Roduner, E., Scheuermann, R. et al. (2000). *Physica B* 289–290: 482–486.

16 Percival, P.W., Fischer, H., Camani, M. et al. (1976). *Chem. Phys. Lett.* 39: 333–335.

17 Percival, P.W., Brodovitch, J.-C., Ghandi, K. et al. (1999). *Phys. Chem. Chem. Phys.* 1: 4999–5004.

18 Ghandi, K., Brodovitch, J.-C., Addison-Jones, B. et al. (2000). *Physica B* 289–290: 476–481.

19 McKenzie, I., Dilger, H., Roduner, E. et al. (2008). *J. Phys. Chem. B* 112: 3070–3076.

20 Fleming, D.G., Cottrell, S.P., McKenzie, I., and Macrae, R.M. (2012). *Phys. Chem. Chem. Phys.* 14: 10953–10966.

21 Roduner, E., Percival, P.W., Han, P., and Bartels, D.M. (1995). *J. Chem. Phys.* 102: 5989–5997.

22 Roduner, E. (1988). *The Positive Muon as a Probe in Free Radical Chemistry*, Chap. 4, Lecture Notes in Chemistry, vol. 49. Springer-Verlag.

23 McKenzie, I., Brodovitch, J.-C., Ghandi, K. et al. (2007). *J. Phys. Chem. A* 111: 10625–10634.

24 Reid, I.D. and Roduner, E. (1991). *Struct. Chem.* 2: 419–431.

25 Percival, P.W., Kiefl, R.F., Kreitzman, S.R. et al. (1987). *Chem. Phys. Lett.* 133: 465–470.

26 Kiefl, R.F., Percival, P.W., Brodovitch, J.-C. et al. (1988). *Chem. Phys. Lett.* 143: 613–618.

27 McKenzie, I., Scheuermann, R., and Sedlak, K. (2012). *J. Phys. Chem. A* 116: 7765–7772.

28 McKenzie, I., Scheuermann, R., Cottrell, S.P. et al. (2013). *J. Phys. Chem. B* 117: 13614–13618.

29 Percival, P.W., Addison-Jones, B., Brodovitch, J.-C., and Sun-Mack, S. (1996). *Appl. Magn. Reson.* 11: 315–323.

3

Formation of Chemical States Incorporating Muons

3.1 μ^+ Charge Exchange and Mu Formation in the Gas Phase

Muon beams are produced at nuclear accelerators with MeV kinetic energies, whereas chemical bond energies are in the eV range, so it is important to understand how energy-loss mechanisms affect the formation of chemical states incorporating muons. The discussion starts with consideration of muonium formation in low-pressure gas, where the mean free path between collisions is long compared to typical molecular diameters. The situation for dense media is more complicated and discussion is deferred until Section 3.2.

As indicated in Figure 3.1 there are three distinct energy/time regimes describing the stopping and thermalization of μ^+ beams in a gas [1]: (i) the muon loses energy by ionization and electronic excitation interactions with the medium, as described by the Bethe–Bloch stopping-power formula [2, 3]; (ii) muonium is first formed when the muon kinetic energy falls below 100 keV or so, where it enters a series of charge-exchange cycles, repeatedly losing and gaining an electron by collisions with the moderator until the energy falls to about 10 eV; (iii) further thermalization is by elastic and inelastic collisions of the muon/muonium with molecules of the medium.

Charge exchange, indicated by the double arrow in Figure 3.1, is not an equilibrium but a cyclic sequence of reaction steps:

$$\mu^+ + \mathrm{M} \xrightarrow{\sigma_{10}} \mathrm{Mu} + \mathrm{M}^+,$$

$$\mathrm{Mu} + \mathrm{M} \xrightarrow{\sigma_{01}} \mu^+ + \mathrm{M} + \mathrm{e}^-,$$

where σ_{10} and σ_{01} are energy-dependent cross sections for electron capture and loss, respectively. The eventual distribution of muons among charge states depends on the mass and ionization potential of the collision partner M and other factors that affect the competition between charge exchange and the various energy-moderating collisional processes. The foregoing description assumes that the moderator M is chemically inert. However, in the presence of a suitable reaction partner 'epithermal' muonium may also undergo hot-atom reactions [4] prior to thermalization.

Muon Spin Spectroscopy: Methods and Applications in Chemistry and Materials Science, First Edition.
Donald G. Fleming, Iain McKenzie, and Paul W. Percival.

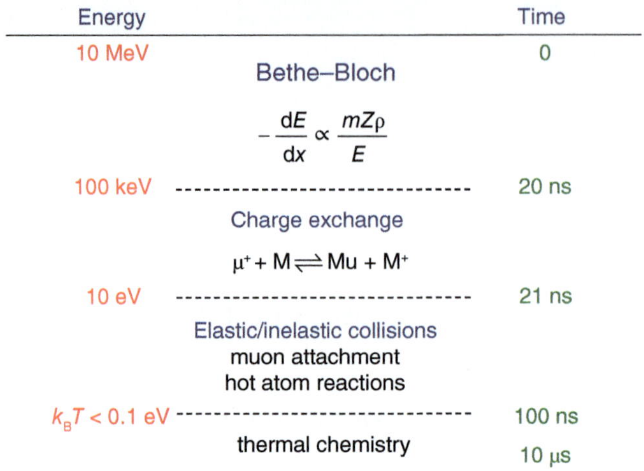

Figure 3.1 Muon thermalization regimes with times appropriate for a low-pressure gas. Z and ρ refer to the atomic number and density of the medium (M), while m and E denote the mass and energy of the muon.

Energy loss of charged particles in the Bethe–Bloch regime is well understood, and the reader is referred elsewhere for details [2, 3]. The only important point relevant to our discussion of muon thermalization is that the rate of energy loss depends on projectile speed, a feature that gives rise to the concept of 'velocity scaling' [5] in comparing stopping powers for charged particles of different masses.

Particle speed is also a crucial factor in the charge-exchange regime, since capture and loss is most efficient when the muon speed matches that of an electron in an outer atomic orbital. For example, applying the simple Bohr atomic model to argon we estimate an orbital speed of 6×10^6 m s^{-1} for an $n = 3$ electron, which corresponds to a muon kinetic energy of about 20 keV, near the start of the muon charge-exchange regime. There may be as many as 100 charge-exchange cycles before the final Mu or μ^+ emerges at the end of this regime [6].

For the periods when muonium exists there is evolution of the muon spin polarization due to hyperfine coupling of the muon spin to the electron spin. The net effect is a phase shift in the muonium precession signal, as depicted in Figure 3.2. The slowing-down time t_{SD} is the cumulative period during which μ^+ exists and precesses with a frequency 103 times smaller than that of muonium in a low magnetic field. This concept has been applied to determine muon slowing-down times in several pure gases and gas mixtures [1], as shown in Figure 3.3, and the results are in good agreement with velocity scaling of proton thermalization data. In 1 bar argon and nitrogen, for example, t_{SD} is about 14 ns for 2.8 MeV muons. Most of this time is spent in Bethe–Bloch thermalization; the cyclic charge-exchange period is only of order 1 ns. It is important to note, however, that t_{SD} is not the total slowing-down time; thermalization of muonium or the muon over the last 10 eV may extend to 100 ns or so at 1 bar pressure [1]. Of course, thermalization is much faster in dense gases, and in condensed matter the total slowing-down time is in the picosecond range.

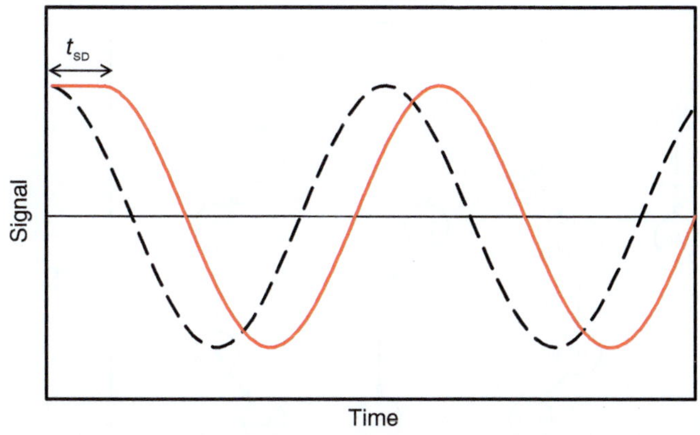

Figure 3.2 Simulated precession signal for Mu with zero slowing-down time (dashed curve), compared to the signal for Mu with slowing-down time t_{SD} (solid line), as predicted for muons thermalizing in a low-pressure gas [1].

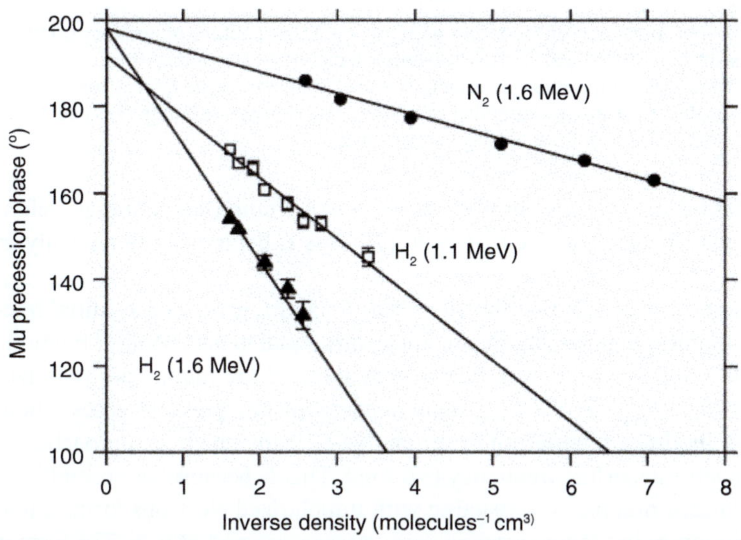

Figure 3.3 A plot of the phase precession of the Mu atom during its total slowing-down process in N_2 and H_2 at two different energies, as a function of inverse density. The slopes are proportional to t_{SD} at 1 bar pressure. The different intercept at infinite density (where $t_{SD} = 0$) for the 1.1 MeV muon kinetic energy is due to its different stopping range in the gas target. Source: Senba et al. [1] figure 3, with permission from the American Physical Society.

From the chemical point of view, the key result of muon thermalization is the partition between muonium and diamagnetic species. In principle the latter could be a bare muon, but in practice it is either a molecular ion (e.g. N_2Mu^+) or a muoniated molecule (e.g. MuH). These fractions, f_{Mu} and f_D where $f_{Mu} + f_D = 1$, are determined by the *final* charge-exchange interaction. This crucial event depends on the ionization energies of muonium and the moderator, and competition between charge-exchange and moderation processes, i.e. how fast the Mu energy falls below

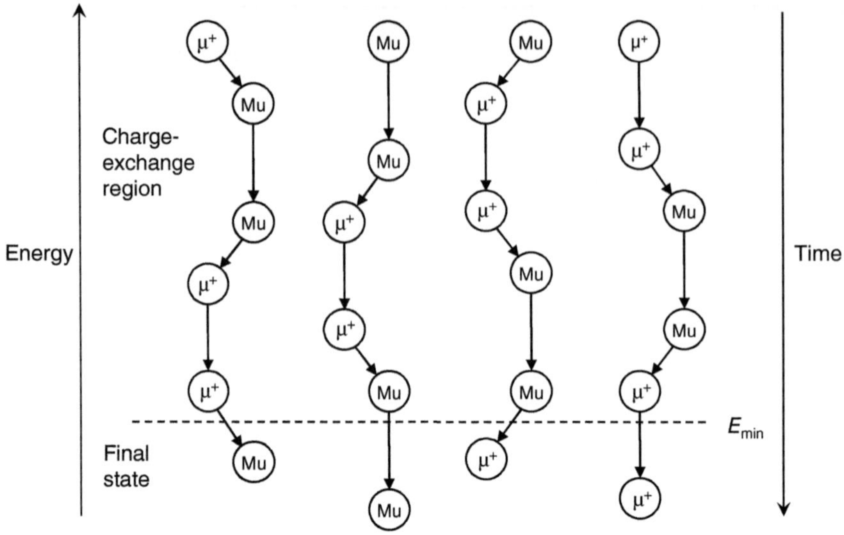

Figure 3.4 The final formation of muonium depends on either capture of an electron by a positive muon with energy just above the threshold (depicted by the dotted line), or crossing of the threshold by muonium involved in non-charge-exchange energy loss. Thus, the final charge state of the muon depends not only on the charge-exchange cross sections but also on the probabilities of other energy-loss events.

the electron-loss threshold, and how slowly a positive muon loses energy before it falls below the electron-capture threshold [7]. This is depicted schematically in Figure 3.4.

It is important to appreciate that the *observable* fractions, expressed as initial fractions P_i of the muon spin polarization, are not always equal to the chemical fractions f_i. In particular, there are circumstances where some of the muon spin polarization is lost, so that $P_{\mathrm{Mu}} + P_{\mathrm{D}} = 1 - P_{\mathrm{L}}$. This happens in low-pressure gases where the cumulative lifetime of muonium in the charge-exchange regime approaches or exceeds the inverse hyperfine frequency (~0.2 ns). This is because an ensemble of 100% spin-polarized muons, $|\alpha_\mu\rangle$, reacting with unpolarized electrons forms equal amounts of muonium in the two product spin states $|\alpha_e\alpha_\mu\rangle$ and $|\beta_e\alpha_\mu\rangle|$. The former is a stationary state (an eigenstate of the spin Hamiltonian) but the latter is not, and undergoes rapid oscillation to/from the $|\alpha_e\beta_\mu\rangle$ state. Consequently, up to half of the muon spin polarization can be lost by this effect. Other mechanisms for loss of spin polarization will be discussed in Sections 3.2 and 5.5.

A selection of data for muons thermalizing in low-pressure gases is shown in Table 3.1. It is evident that ionization energy plays a significant role and that little or no muonium survives to thermal energies if the threshold for Mu formation is high (the ionization energy of the medium exceeds that of Mu). The diamagnetic fraction in the noble gases is attributed to molecular ions, HeMu^+, NeMu^+, etc., but there is more than enough transient Mu formation during thermalization to lead to partial spin depolarization (P_{L}). At higher pressures (not shown in the table) thermalization is faster and there is less depolarization. A heavy atom such as Xe is an

Table 3.1 Muon fractions (percent)[a] in gases at ~1 bar.

Gas	I.E (eV)	f_D	f_{Mu}	P_D	P_{Mu}	P_L
He	24.6	100	0	75	0	25
Ne	21.6	95	5	38	2	60
Ar	15.8	24	76	17	55	28
Kr	14.0	0	100	0	68	32
Xe	12.1	0	100	0	70	30
H_2	15.4	44	56	25	32	43
N_2	15.6	13	87	12	78	10
H_2O	12.6	7	93	4	47	49
CH_4	12.6	12	89	9	76	15

a) Data selected from Refs. [6, 8, 9].

inefficient moderator, and at pressures below 1 bar almost all the polarization is lost [10]. However, addition of helium leads to faster thermalization and full recovery of the muonium signal [11].

On the other hand, ionization energy and moderation efficiency are not sufficient to explain the data in Table 3.1. Ar, N_2 and H_2 have similar ionization energies but quite different diamagnetic fractions. The main difference is the existence of hot-atom reactions between epithermal Mu and H_2 [12]. The lower diamagnetic fractions in other molecular gases (H_2O and CH_4 in Table 3.1) suggest that epithermal reactions are less important for these materials. Nevertheless, hot-atom reactions cannot be excluded; for example there is convincing evidence that muonium adds to ethene at low pressures while still epithermal (~0.25 eV) [13].

The consideration of hot-atom reactions is important because diamagnetic fractions in liquids and solids are generally much higher than in gases, and this was ascribed to hot-atom reactions [14, 15] in advance of wide acceptation of radiolysis effects.

3.2 Mu Formation and Track Effects in Dense Media

It is evident from the examples given in Table 3.2 that muonium formation in condensed matter is different from the gas phase. In a low-pressure gas the muon interacts with individual molecules in a series of isolated events, i.e. there is a long free path between collisions. In contrast, in condensed matter there is no truly free path between interactions. Even more importantly, the excited/ionized products of the interaction are generally trapped in close proximity to the muon. The transition from gas-like to condensed matter behaviour depends on the density of the medium, and measurements have been made over a wide range of temperatures and pressures in carbon dioxide, ethane and water, from dense gas to supercritical

Table 3.2 Representative muon fractions (percent)[a] in different phases.

	Muonium fraction P_{Mu}			Diamagnetic fraction P_D		
	Solid	Liquid	Gas	Solid	Liquid	Gas
Ar	91	48	55	9	0.02	17
N_2			78			12
H_2O	52	20	90	48	62	10
CH_3OH		23	85		62	15
CH_4	18	18	76	58	58	9
CCl_4		0	48		100	52

a) Data selected from Refs. [8, 16, 17].

fluids. The muon fractions are plotted as a function of density in Figure 3.5. There are some similarities in behaviour, given that different cross sections are expected for muon interaction with different materials. However, some new paradigm is necessary to explain the complex behaviour.

The spur model of muonium formation was originally introduced to explain the muon fractions in water [16] and was subsequently extended to other liquids [19, 21]. The basic tenet of the model is that the muon thermalizes close enough to the terminal spur of its radiation track to interact with transient species formed by radiolysis of the medium (Figure 3.6). Muonium is formed by combination of the positive muon with an excess electron, in direct analogy with the spur model of positronium formation [22]. In addition, encounters of muonium with electrons or other paramagnetic radiolysis products can account for loss of muon spin polarization, i.e. the missing fraction P_L [16, 23].

For water, the principal competing reactions are:

Muonium formation:	$\mu^+ + e^- \rightarrow Mu$
Muon attachment:	$\mu^+ + H_2O \rightarrow MuH_2O^+$
Proton transfer:	$MuH_2O^+ + H_2O \rightarrow MuOH + H_3O^+$
Cross reactions:	$Mu + e^-_{aq} \rightarrow MuH + OH^-$ or spin-depolarized Mu
	$Mu + {}^{\cdot}OH \rightarrow MuOH$ or spin-depolarized Mu

Prompt muonium formation may well involve ‘dry’ (pre-solvated) electrons, but subsequent reactions should be limited to hydrated electrons, since the timescale of solvation (rotation of water molecules) is shorter than that of molecular diffusion.

Reactions between radicals (including electrons [24]) are spin-dependent, and it is usually only those encounter pairs which are in an electronic singlet state which react to form stable products. Of the non-reactive encounters, a fraction will undergo spin exchange leading (through the hyperfine interaction) to partial muon spin depolarization [23]. The distribution of muons into polarized and unpolarized fractions is depicted in Figure 3.7.

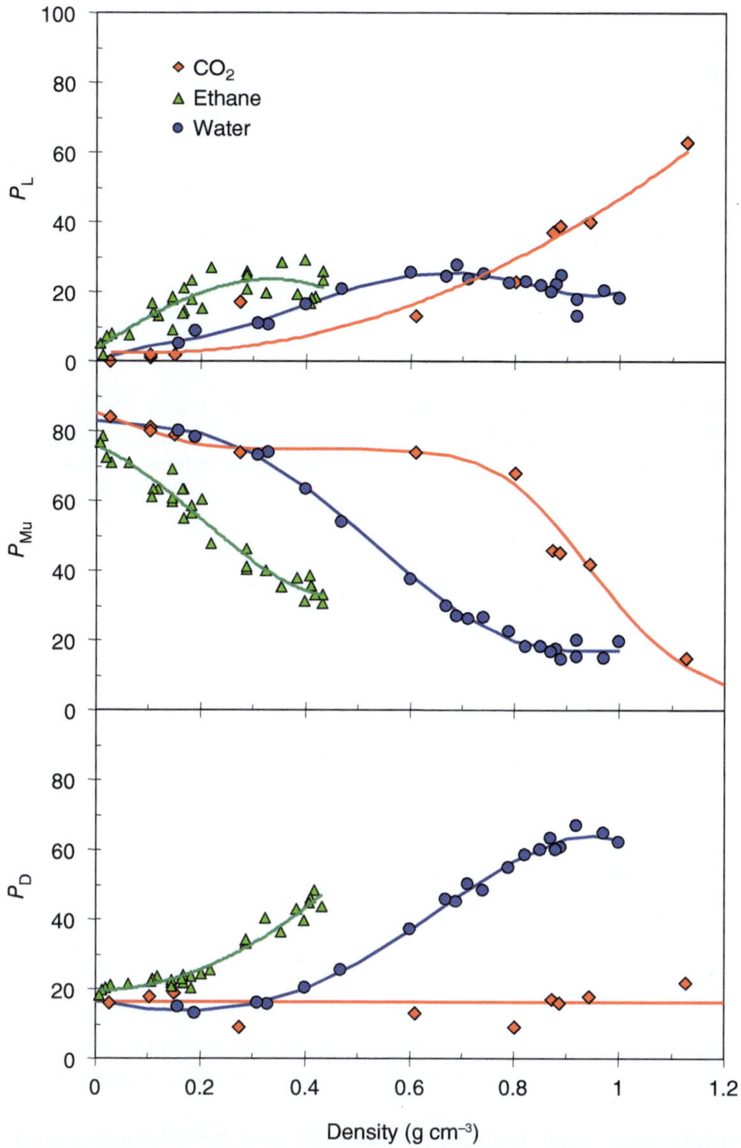

Figure 3.5 Muon polarization fractions (%) determined in fluid carbon dioxide, ethane and water. Data selected from references [18-20] and plotted without error bars for clarity. The lines through the points are to guide the eye only.

Evidence in favour of the spur model includes the inhibition of muonium in aqueous solutions of electron scavengers [16], and separation of the diamagnetic fraction into a prompt component (MuOH) and a slower forming one (MuH) by analysing the magnetic-field dependence of P_D [23].

Arguments against the spur model [25] are based on radiation chemistry data, such as the probability of intra-spur reactions and the related escape probabilities.

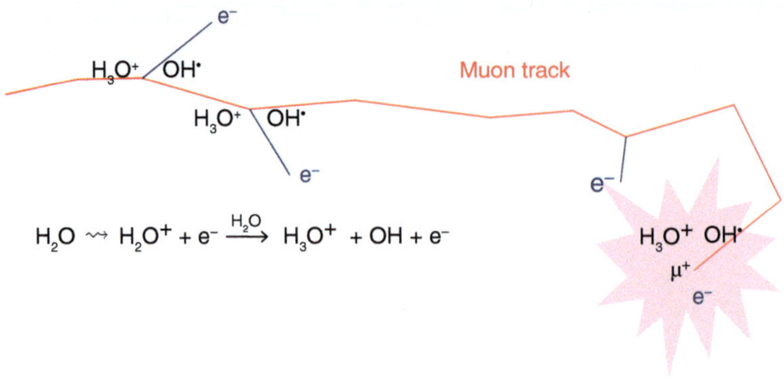

Figure 3.6 Schematic representation of the end of a muon track. H_3O^+ and $^•OH$ are formed from H_2O^+ in $\sim 10^{-13}$ s, well before any diffusion, so excess electrons and hydroxyl radicals are the primary radiolysis products in the vicinity of the muon in the terminal radiation spur.

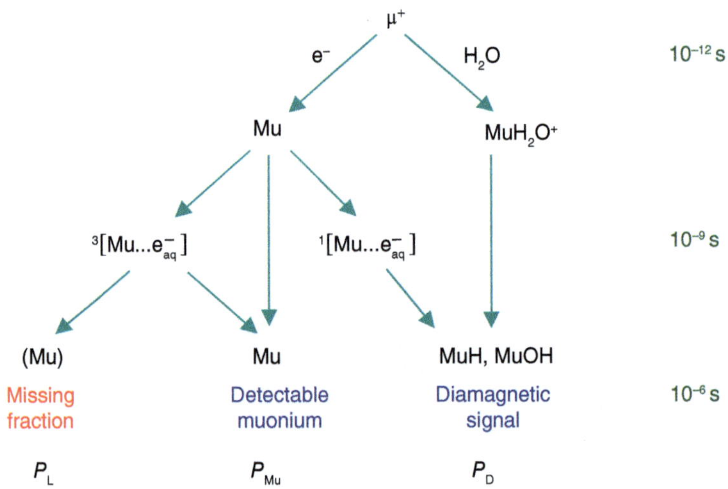

Figure 3.7 The spur model of muonium formation in liquid water. The square brackets denote encounter pairs of paramagnetic species whose reactions are spin-dependent.

However, such data refer to radiolysis events integrated over the radiation track, not at the terminus. Furthermore, and perhaps most critical, is the failure to take into account the limited time period during which muonium can be created through charge neutralization. This is limited by proton transfer from MuH_2O^+, but the equivalent reaction of H_3O^+, has no net effect, so that H atoms can be formed as long as the hydronium ion encounters an electron.

The same competition between muon attachment and muonium formation by excess electron capture occurs in other hydrogenated material, e.g. alkanes (RH):

$$\mu^+ + e^- \rightarrow Mu,$$

$$\mu^+ + RH \rightarrow RMuH^+.$$

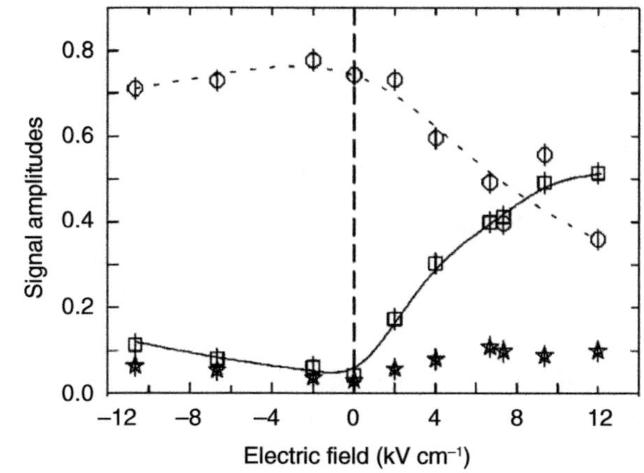

Figure 3.8 Electric field dependences of the various signal components in liquid argon at 84 K. Circles and dotted line: muonium. Squares and solid line: diamagnetic muon fraction. Stars: a fraction of muons forming Mu slowly by capture of localized electrons. Source: Eshchenko et al. [27] figure 7, with permission from the American Physical Society.

Once again it is the subsequent proton transfer

$$RMuH^+ + RH \rightarrow RMu + RH_2{}^+$$

that rapidly traps the muon in a diamagnetic molecule, since there is little probability that the single RMu molecule will encounter another protonated species. In contrast, nitrogen and noble gases lack this reaction channel, and there exists the possibility of delayed muonium formation, e.g.

$$\mu^+ + N_2 \rightarrow N_2Mu^+,$$

$$N_2Mu^+ + e^- \rightarrow Mu + N_2.$$

This has been convincingly demonstrated in μSR experiments in which the sample was subjected to an electric field. In particular it was found that muonium formation can be inhibited or enhanced according to the direction of the applied electric field [26, 27], as shown in Figure 3.8.

3.3 Chemical Processes Forming Muoniated Molecules

Sections 3.1 and 3.2 dealt with how muonium is formed. We now consider chemical reactions which result in the formation of muoniated molecules. If the product is a muoniated free radical, and there is significant interaction between the muon and the unpaired electron spin, it is generally possible to identify the radical by muon spin spectroscopy and study its properties. Such studies are described in chapter 6. In contrast, when Mu becomes incorporated in a diamagnetic molecule the muon spin precesses at the muon Larmor frequency, which is not very instructive: lifetime broadening of signals in muon spin spectroscopy precludes the resolution of the small chemical shifts and nuclear spin–spin couplings so useful in NMR spectroscopy. Thus diamagnetic molecules are not generally distinguishable, and most of the reaction types considered below result in muoniated free radicals.

H abstraction by Mu

A common reaction considered in free radical chemistry is abstraction of an atom (typically an H atom) to generate a free radical. Abstraction by muonium is not very revealing, however, because the free radical formed is not muoniated:

$$Mu + RH \rightarrow MuH + R^{\bullet}.$$

As will be explained in Chapter 4, abstraction by Mu is generally slower than by H and certainly slower than addition to unsaturated molecules. Nevertheless, abstraction cannot be discounted as a contribution to the diamagnetic fraction in some systems.

Mu Addition to Alkenes, Dienes and Allenes

Mu addition to C=C bonds is rapid and close to the diffusion-controlled limit in solution [28]. Where there are multiple reaction sites the regioselectivity of Mu addition is similar to that of H, i.e. Mu adds to the side of the double bond that generates the more stable muoniated radical, usually the less substituted side of the double bond [29]. For example, Mu adds to the terminal CH_2 of propene:

$$Mu + CH_3CH{=}CH_2 \rightarrow CH_3\dot{C}HCH_2Mu$$

to give the muoniated propyl radical [29]. The primary radical, $CH_3CHMuCH_2^{\bullet}$, is not observed. Similarly, Mu adds to the end of pent-1-ene, but there is no selectivity in reaction with pent-2-ene, because both products are secondary radicals (Figure 3.9).

In the case of dienes and allenes the more stable products are those with extended unpaired spin distribution: they are said to be 'resonance-stabilized'. Thus only two of four isomeric radicals arise from 2-methylpentadiene-1,3 (Figure 3.10) because these contain allyl structures [29].

Similarly, Mu addition to allenes occurs preferentially to the *central* carbon because this results in a substituted allyl, as shown for 1,1-dimethylallene in Figure 3.11 [30].

Mu Addition to Benzene and Aromatic Compounds

Mu adds to benzene to produce the muoniated cyclohexadienyl radical, C_6H_6Mu. It was one of the first radicals observed by muon spin spectroscopy [31], and it has since

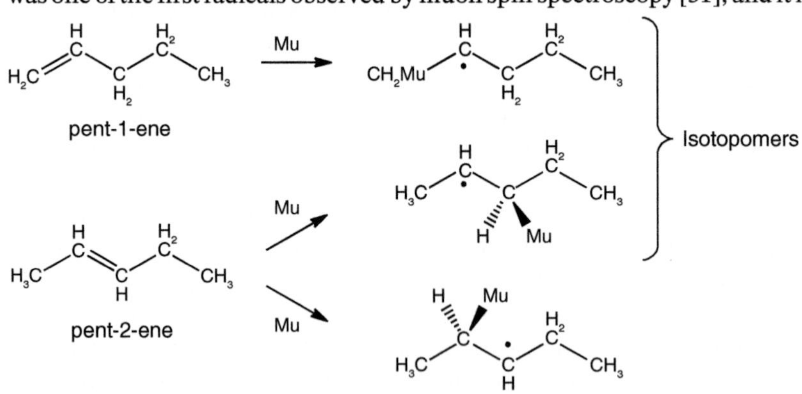

Figure 3.9 Products of the reactions of Mu with pent-1-ene and pent-2-ene [29].

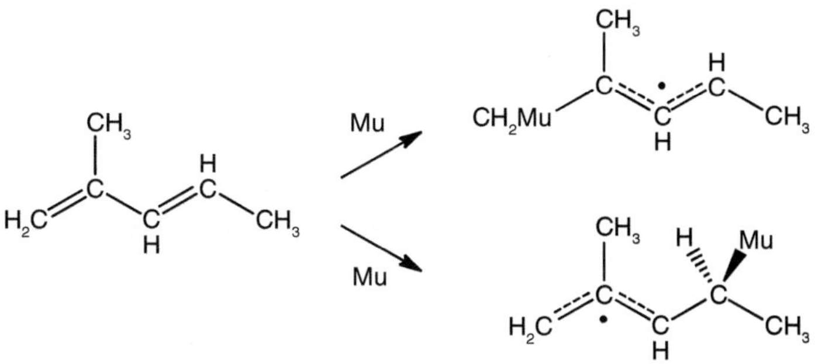

Figure 3.10 Products of the reactions of Mu with 2-methylpentadiene-1,3 [29].

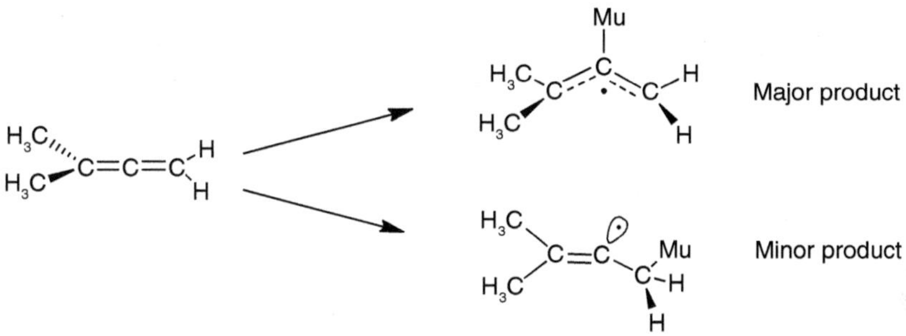

Figure 3.11 Products of the reactions of Mu with 1,1-dimethylallene [30].

been studied in a wide variety of media: solid [32], liquid [33], gas [34], supercritical fluids [35], on surfaces [36], in zeolites [37] and in a clathrate hydrate [38].

Addition of Mu to mono-substituted benzenes generally results in three isomers, as shown for toluene in Figure 3.12. Addition at the *ortho* (C2, C6) and *para* (C3) sites is favoured over *meta* addition because these radicals have unpaired spin density on the tertiary carbon (C1). Furthermore, given that there are two ortho but only one para site it is clear from the yields that the reactivity increases in the order *meta*, *ortho*, *para* [39]. The regioselectivity of Mu addition depends on the nature of the substituent, and this has been extensively studied for mono-substituted benzenes and also in some cases for multiple substituents [40].

There have been several studies of the Mu adducts of polyaromatic hydrocarbons (PAHs) such as naphthalene [41], pyrene [42], fluoranthene [43] and triphenylene [44]. In all cases Mu adds only to the secondary carbons at the edges of the fused ring systems. This is because addition at a tertiary carbon would entail an energy penalty associated with disruption of the carbon skeleton to achieve tetrahedral geometry at the addition site. Distortion is less severe for dodecahydrotriphenylene (Figure 3.13), but in this case there are only tertiary sites available [44]. A clearer example of the enhanced reactivity of ‘bent’ tertiary carbon sites is shown in Figure 3.14. All three muoniated radicals have been detected in the pure solid paracyclophane [45].

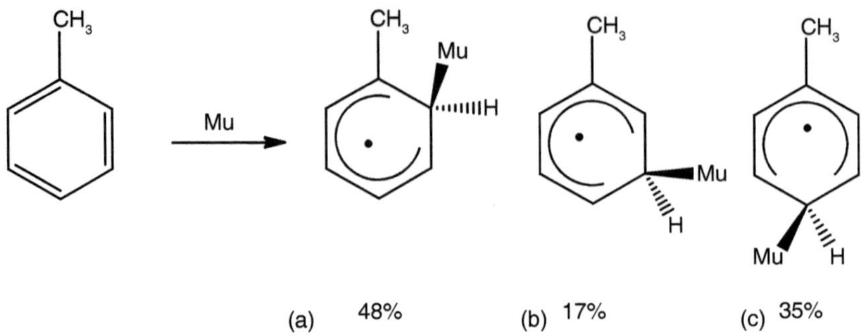

Figure 3.12 Products and relative yields of the reaction of Mu with toluene: (a) *ortho* isomer; (b) *meta* isomer; (c) *para* isomer [39].

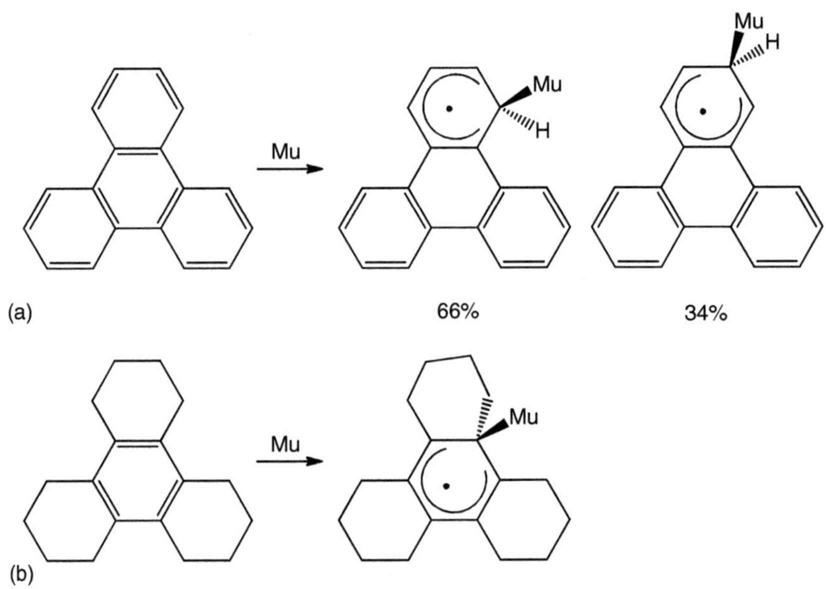

Figure 3.13 Radicals formed by Mu addition to (a) triphenylene (with relative yields) and (b) dodecahydrotriphenylene [44].

Curvature of the PAH framework also increases reactivity at tertiary carbon sites. This will be discussed with respect to fullerenes in chapter 8.

Mu Addition to Alkynes

Mu reacts with triple bonds to produce vinyl radicals. However, the muoniated vinyl radical itself, the Mu adduct of acetylene, has not been detected. It is assumed that this is because it behaves like $H\dot{C}{=}CH_2$ which is known to undergo rapid interconversion between two tautomeric forms in solution [46]. Rapid exchange between the *cis* and *trans* tautomers (Figure 3.15) modulates the unpaired spin density at the muon and consequently broadens the radical precession frequencies beyond

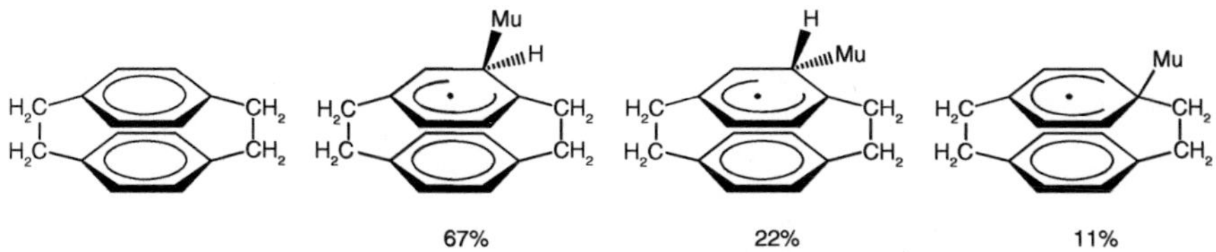

Figure 3.14 [2.2]Paracyclophane and the radicals (with relative yields) formed by Mu addition [45].

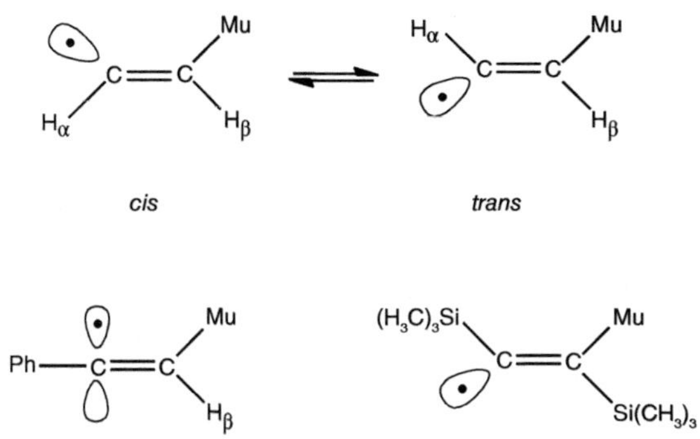

Figure 3.15 Muoniated vinyl radicals.

detection. It is possible that the radical could be observed at very low temperatures, where the conversion rate is slow enough that all three non-equivalent protons can be resolved by ESR [47].

It is possible to detect muoniated vinyl radicals if there are substituents which prevent rapid interconversion of tautomers. The first such observation was for PhĊ=CH(Mu), formed by Mu addition to the triple bond of phenylacetylene (there are also minor amounts of ring-adducts) [48]. In this case the unpaired electron occupies an orbital with primarily $2p$ character, so that it is resonance-stabilized by interaction with the aromatic π system. The resulting linear structure has no tautomers (Figure 3.15). On the other hand, the Mu adduct of 1,2-di(trimethylsilyl)acetylene has a single non-linear vinyl structure. In this case the large steric interaction between the substituents stabilizes the *trans* conformation [49].

Mu Addition to Carbonyls and Thiocarbonyls

The reaction of Mu with carbonyl groups is considerably slower than reaction with alkenes, alkynes and aromatics. This is evident from cases where two functional groups exist in the same molecule. For example, Mu addition to benzylacetone, $PhCH_2COCH_3$, occurs exclusively at the aromatic ring [50]. In the absence of other reactive groups Mu addition occurs at the O atom. However, *keto-enol* tautomerism provides another route for reaction. Thus, two different muoniated radicals have

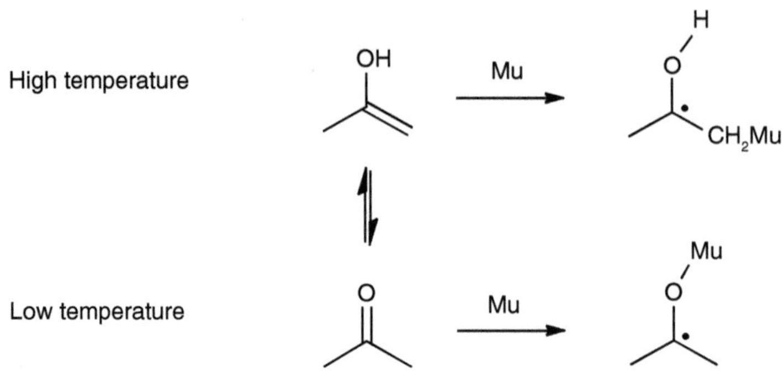

Figure 3.16 Reactions of Mu with acetone in water above and below 250 °C [51].

been characterized in aqueous solutions of acetone [51]. At room temperature the equilibrium is overwhelmingly in favour of the *keto* form, and only the ketyl radical can be detected. However above 250 °C the competition favours $CH_3\dot{C}(OH)CH_2Mu$. This does not imply that the equilibrium favours the enol, however, merely that the product of rate constant and reactant concentration favours addition to the C=C bond over C=O. Another point of note is that the two muoniated radicals (Figure 3.16) are isotopomers; they are only distinguishable because of the muon labelling.

The case for Mu addition to C=S bonds is not yet settled. On the basis of muon hyperfine constants it was suggested that for some molecules Mu adds to the carbon atom to produce a thiyl radical, $R_1R_2CMu\text{-}S^{\cdot}$ [52, 53]. However a subsequent computational study found that although the thiyl radicals are energetically favoured, the measured hfcs are consistent with Mu addition to S [54].

Mu Reactions with Diazo Compounds and Ketenes

In all the examples discussed above, Mu is incorporated in the radical product at an atom adjacent to the unpaired electron. These are known as β-muoniated radicals. Radicals with the muon attached directly to the radical centre, α-muoniated radicals, are much less common. They can be created by the addition of Mu to ketenes and diazo compounds, followed by rapid decomposition of the intermediate radical [55]. In order for muon spin polarization to be transferred to the α-muoniated radical, the decomposition must occur within a few nanoseconds; this can occur when there is a good leaving group like N_2 or CO present, such as in molecules with diazo and ketene functional groups. Mu addition occurs at the terminal carbon of the diazo or ketene functional group. The rapid decomposition occurs because the barrier to decomposition is lower than the energy of the transition state for Mu addition (see Figure 3.17).

Attempts to produce the muoniated methyl radical by the reaction of Mu with diazomethane have been unsuccessful, most likely due to the extreme instability of diazomethane at the high concentrations required for TF-μSR experiments. However, a substituted muoniated methyl radical has been made by the reaction of Mu with a more stable diazo compound, trimethylsilyldiazomethane [56]. The

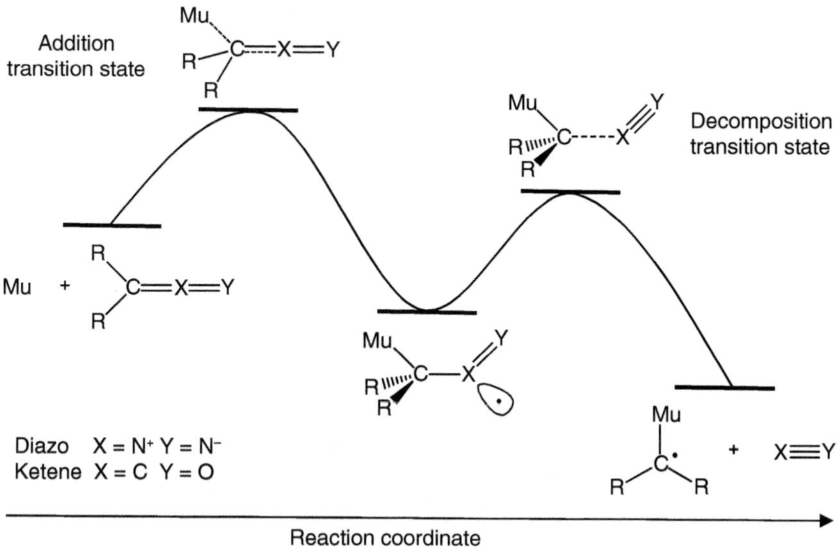

Figure 3.17 Schematic representation of the reaction pathway for the reaction of Mu with diazomethane and ketene.

muoniated methyl radical ($^{\bullet}CH_2Mu$) itself can be produced by the reaction of Mu with ketene [57].

Mu Addition to Carbenes and Related Electron-Deficient Molecules
Unlike the double- and triple-bonded molecules discussed above, carbenes have their electron deficiency concentrated at a single carbon, so that they have the general structure R_1R_2C:, i.e. a divalent carbon with only six valence electrons. They can exist in electronic singlet or triplet states, but both are usually highly reactive. However, under certain circumstances it is possible to synthesize singlet carbenes which are stable in the absence of moisture and oxygen. The first, and most common, class of stable carbenes are the N-heterocyclic carbenes (NHC). Although there may be other reactive sites in the molecule, reaction with muonium occurs exclusively at the carbene centre [58]. The result is an α-muoniated radical, as shown in Figure 3.18.

Silylenes (R_2Si:) and germylenes (R_2Ge:) are heavier analogues of carbenes. Their reactions with muonium are discussed in Section 6.7.1.

Ionic Mechanisms Forming Muoniated Radicals
All the reactions discussed so far involve muonium as the precursor to muoniated radicals. However, alternative ionic mechanisms have been proposed [59]. The most likely involves capture of a muon, followed by charge neutralization, either by charge transfer or capture of a solvated electron produced by radiolysis.

$$\mu^+ + A \rightarrow \text{Mu-A}^+,$$

$$\text{Mu-A}^+ + e^- \rightarrow \text{Mu-A}^{\bullet}.$$

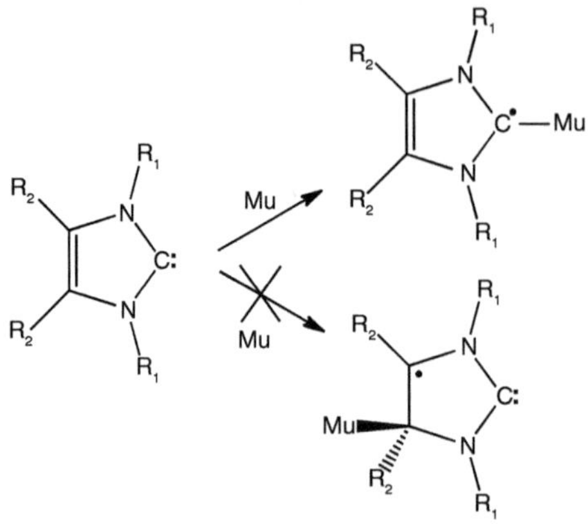

Figure 3.18 Reaction of Mu with N-heterocyclic carbenes. An example with R1 = $-CH(CH_3)_2$ and R2 = $-CH_3$ is 1,3-bis-(isopropyl)-4,5-dimethylimidazol-2-ylidene [58].

3.4 μ^- Capture and Muonic Atoms

In marked contrast to μ^+, μ^- particles at energies of tens of keV, are captured into high angular momentum atomic orbitals and then rapidly cascade down to the muonic 1s state in a fraction of a nanosecond [60, 61]. Energy is lost in the cascade process by both radiative transitions and the emission of Auger electrons [62]. The latter typically results in the formation of a positive ion, which is subsequently neutralized by electron capture from the medium. Unlike the case for μ^+ thermalization, charge exchange occurs at near-thermal energies, so it is necessary to choose dopants with an ionization energy less than that of the neutral atom, 13.6 eV. The overall time for thermalization in low-pressure gas (a few bar) is ~20 ns (Figure 3.19), very similar to that for μ^+ (Figure 3.1).

As explained in Chapter 1, the Bohr atomic radius is inversely proportional to the reduced mass of a two-particle system (Eq. (1.2)), so the effective orbiting radius for μ^- in the 1s state of a muonic atom is 207 times smaller than that of an electron, e.g. 0.013 Å for muonic helium. Since the μ^- resides in such a tight orbital close to the nucleus it effectively screens part of the nuclear charge, so that the muonic atom formed from an element with atomic number Z behaves as if it has only $Z-1$ protons – transmutation of the element! Furthermore, the muon can be captured by the nucleus, thereby reducing the mean lifetime. The capture rate is strongly dependent on nuclear charge, so although the μ^- lifetime in muonic hydrogen is almost as long as for μ^+, it falls dramatically for heavier elements, e.g. 0.33 μs for muonic calcium [62].

Another significant difference between positive and negative muons is that the cascade process results in loss of most of the μ^- spin polarization. The residual polarization for μ^- in a 1s orbital is at most 1/6, less if the nucleus has spin [61]. In particular, for the chemically interesting heavy hydrogen isotope Heμ, the maximum spin polarization is only about 5% [60]. Thus μSR experiments with negative

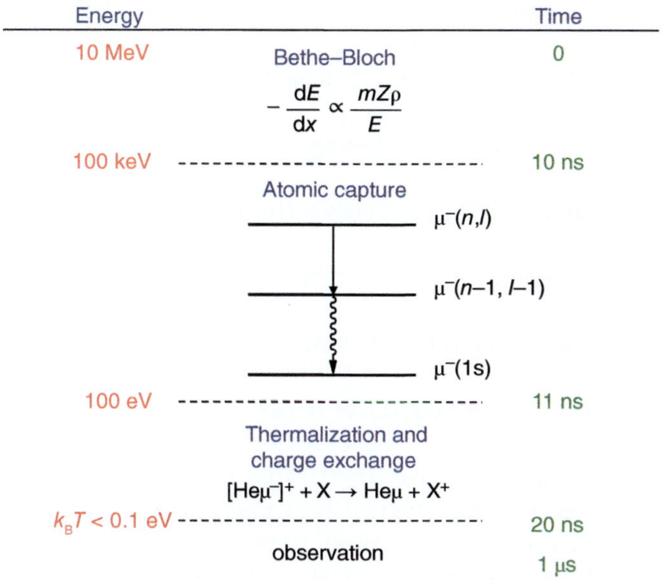

Figure 3.19 Negative muon thermalization with times and energy regimes appropriate for the example of muonic helium in a gas at ~1 bar. Z and ρ refer to the atomic number and density of the medium, while m and E denote the mass and energy of the muon. X is the charge-exchange partner, typically added to a buffer gas (the medium).

muons typically have to contend with signal amplitudes an order of magnitude smaller than their positive muon counterparts.

The combination of small spin polarization and short lifetime puts significant practical limits on negative muon spin spectroscopy, so that significant chemical applications have been limited to studies of Heμ kinetics [63, 64], which will be discussed in chapter 4. However, in principle μ^- lifetimes can be used for elemental analysis, although a more widely used technique employs the characteristic X-rays emitted during the atomic capture process [62].

References

1 Senba, M., Arseneau, D.J., Pan, J.J., and Fleming, D.G. (2006). *Phys. Rev. A* 74: 042708.

2 Horvath, D. and Vértes, A. (2011). *Handbook of Nuclear Chemistry* (ed. A. Vértes, S. Nagy, Z. Klencsár, et al.), 363–393. Boston, MA: Springer US.

3 Mozumder, A. (1999). *Fundamentals of Radiation Chemistry*. San Diego, CA: Academic Press.

4 Yoshihara, H.K. and Sekine, T. (2011). *Handbook of Nuclear Chemistry* (ed. A. Vértes, S. Nagy, Z. Klencsár, et al.), 1333–1378. Boston, MA: Springer US.

5 Bransden, B.H. and Forster, C. (1990). *J. Phys. B.* 23: 115.

6 Fleming, D.G., Mikula, R.J., and Garner, D.M. (1982). *Phys. Rev. A* 26: 2527–2544.

7 Senba, M. (1988). *J. Phys. B* 21: 3093–3111.

8 Arseneau, D.J., Garner, D.M., Senba, M., and Fleming, D.G. (1984). *J. Phys. Chem.* 88: 3688–3696.

9 Johnson, C., Cottrell, S.P., Ghandi, K., and Fleming, D.G. (2005). *J. Phys. B* 38: 119–134.

10 Senba, M., Turner, R.E., Arseneau, D.J. et al. (1986). *Hyperfine Interact.* 32: 795–799.

11 Kempton, J.R., Senba, M., Arseneau, D.J. et al. (1991). *Hyperfine Interact.* 65: 801–809.

12 Senba, M., Fleming, D.G., Arseneau, D.J., and Mayne, H.R. (2000). *J. Chem. Phys.* 112: 9390–9403.

13 Percival, P.W., Brodovitch, J.-C., Arseneau, D.J. et al. (2003). *Physica B* 326: 72–75.

14 Brewer, J.H., Gygax, F.N., and Fleming, D.G. (1973). *Phys. Rev. A* 8: 77–86.

15 Jean, Y.C., Ng, B.W., Brewer, J.H. et al. (1981). *J. Phys. Chem.* 85: 451–454.

16 Percival, P.W., Roduner, E., and Fischer, H. (1978). *Chem. Phys.* 32: 353–367.

17 Jean, Y.C., Ganti, R., and Stadlbauer, J. (1987). *J. Phys. Chem.* 91: 1270–1273.

18 Ghandi, K., Bridges, M.D., Arseneau, D.J., and Fleming, D.G. (2004). *J. Phys. Chem. A* 108: 11613–11625.

19 Kempton, J.R., Senba, M., Arseneau, D.J. et al. (1991). *J. Phys. Phys.* 94: 1046–1059.

20 Percival, P.W., Brodovitch, J.-C., Ghandi, K. et al. (1999). *Phys. Chem. Chem. Phys.* 1: 4999–5004.

21 Mogensen, O.E. and Percival, P.W. (1986). *Radiat. Phys. Chem.* 28: 85–89.

22 Morgensen, O.E. (1974). *J. Chem. Phys.* 60: 998–1004.

23 Leung, S.K., Brodovitch, J.-C., Percival, P.W. et al. (1988). *Chem. Phys.* 121: 393–403.

24 Ichino, T. and Fessenden, R.W. (2007). *J. Phys. Chem. A* 111: 2527–2541.

25 Walker, D.C. (1983). *Muon and Muonium Chemistry*. Cambridge University Press.

26 Storchak, V., Brewer, J.H., Morris, G.D. et al. (1999). *Phys. Rev. B* 59: 10559–10572.

27 Eshchenko, D.G., Storchak, V.G., Brewer, J.H. et al. (2002). *Phys. Rev. B* 66: 035105.

28 Percival, P.W. (1979). *Radiochim. Acta* 26: 1–14.

29 Roduner, E., Strub, W., Burkhard, P. et al. (1982). *Chem. Phys.* 67: 275–285.

30 Rhodes, C.J., Symons, M.C.R., Roduner, E., and Scott, C.A. (1987). *Chem. Phys. Lett.* 139: 496–498.

31 Roduner, E., Percival, P.W., Fleming, D.G. et al. (1978). *Chem. Phys. Lett.* 57: 37–40.

32 Roduner, E. (1981). *Hyperfine Interact.* 8: 561–569.

33 Kiefl, R.F., Percival, P.W., Brodovitch, J.-C. et al. (1988). *Chem. Phys. Lett.* 143: 613–618.

34 Fleming, D.G., Arseneau, D.J., Pan, J.J. et al. (1997). *Appl. Magn. Reson.* 13: 181–194.

35 Percival, P.W., Ghandi, K., Brodovitch, J.-C. et al. (2000). *Phys. Chem. Chem. Phys.* 2: 4717–4720.

36 Reid, I.D., Azuma, T., and Roduner, E. (1990). *Nature* 345: 328–330.

37 Rhodes, C.J., Reid, I.D., and Roduner, E. (1993). *J. Chem. Soc. Chem. Commun.* 512–513.

38 Mozafari, M., Chandrasena, L., McKenzie, I. et al. (2020). *Radiat. Phys. Chem.* 168: 108532.

39 Roduner, E., Brinkman, G.A., and Louwrier, P.W.F. (1982). *Chem. Phys.* 73: 117–130.

40 Roduner, E. (1988). *The Positive Muon as a Probe in Free Radical Chemistry*, Chap. 7, Lecture Notes in Chemistry, vol. 49. Springer-Verlag.

41 Reid, I.D. and Roduner, E. (1991). *Struct. Chem.* 2: 419–431.

42 Percival, P.W., Addison-Jones, B., Brodovitch, J.-C. et al. (1999). *Can. J. Chem.* 77: 326–331.

43 Brodovitch, J.-C., Addison-Jones, B., Ghandi, K. et al. (2003). *Can. J. Chem.* 81: 1–6.

44 Brodovitch, J.-C., Ghandi, K., McKenzie, I. et al. (2006). *Physica B* 374–375: 310–313.

45 McKenzie, I., Scheuermann, R., and Sedlak, K. (2012). *J. Phys. Chem. A* 116: 7765–7772.

46 Fessenden, R.W. and Schuler, R.H. (1963). *J. Chem. Phys.* 39: 2147–2195.

47 Cochran, E.L., Adrian, F.J., and Bowers, V.A. (1964). *J. Chem. Phys.* 40: 213–220.

48 Geeson, D.A., Symons, M.C.R., Roduner, E. et al. (1985). *Chem. Phys. Lett.* 116: 186–191.

49 Rhodes, C.J., Symons, M.C.R., Scott, C.A. et al. (1987). *J. Chem. Soc. Chem. Commun.* 447–448.

50 Stride, J.A., Jayasooriya, U.A., and Reid, I.D. (2002). *J. Phys. Chem. A* 106: 244–250.

51 Percival, P.W., Brodovitch, J.-C., Ghandi, K. et al. (2005). *J. Am. Chem. Soc.* 127: 13714–13719.

52 Rhodes, C.J., Symons, M.C.R., and Roduner, E. (1988). *J. Chem. Soc. Chem. Commun.* 3–4.

53 Barnabas, M.V. and Walker, D.C. (1990). *Chem. Phys. Lett.* 168: 9–13.

54 Macrae, R.M. and Carmichael, I. (2001). *J. Phys. Chem. A* 105: 3641–3651.

55 McKenzie, I., Brodovitch, J.-C., Ghandi, K. et al. (2003). *Physica B* 326: 76–80.

56 Addison-Jones, B., Percival, P.W., Brodovitch, J.-C., and Ji, F. (1997). *Hyperfine Interact.* 106: 143–149.

57 McKenzie, I., Addison-Jones, B., Brodovitch, J.-C. et al. (2002). *J. Phys. Chem. A* 106: 7083–7085.

58 McKenzie, I., Brodovitch, J.-C., Percival, P.W. et al. (2003). *J. Am. Chem. Soc.* 125: 11565–11570.

59 Hill, A., Symons, M.C.R., Cox, S.F.J. et al. (1985). *J. Chem. Soc., Faraday Trans. 1* 81: 433–448.

60 Souder, P.A., Crane, T.W., Hughes, V.W. et al. (1980). *Phys. Rev. A* 22: 33–50.

61 Nagamine, K. (2003). *Introductory Muon Science*, Chap. 3. Cambridge University Press.

62 Nagamine, K. (2003). *Introductory Muon Science*, Chap. 4. Cambridge University Press.

63 Fleming, D.G., Arseneau, D.J., Sukhorukov, O. et al. (2011). *J. Chem. Phys.* 135: 184310.

64 Arseneau, D.J., Fleming, D.G., Li, Y. et al. (2016). *J. Phys. Chem. B* 120: 1641–1648.

4

Chemical Reactivity and Dynamics in the Gas Phase

4.1 Muon Spin Spectroscopy Applied to Chemical Kinetics

A chemical reaction involves the rearrangement of atoms into new molecular forms, and the rate at which this occurs provides valuable clues to the process. Since hydrogen is the simplest atom the study of its reactions is paramount to a general understanding of chemical reactivity. The most fundamental studies in this area involve comparison of measured rate constants with the predictions of rate theories. Such measurements are most accurately carried out in the gas phase, ideally at low pressure, where the mean free path between collisions is long compared to molecular diameters, so that only bimolecular collisions need be considered. Even then, many details are difficult to pin down, so tests of rate theories often focus on kinetic isotope effects, i.e. the differing rates for reactions of isotopic species. As noted in Section 1.1, muons greatly extend the range of isotope effect studies, providing a remarkable mass range of 36 from Mu to muonic helium, $^{4}\mathrm{He}\mu$. This chapter focuses on reactions in the gas phase, while Chapter 5 concerns the kinetics of muonium and muoniated radicals in the condensed phase.

A particular advantage of muon spin spectroscopy is that there are only a few muons present in the sample at any one time. Thus all reactions of muonium (or other muon-containing species) have first-order kinetics. This is usually expressed as

$$\mathrm{Mu} + \mathrm{A} \rightarrow \text{products} \tag{4.1}$$

$$-\frac{\mathrm{d[Mu]}}{\mathrm{d}t} = k_{\mathrm{Mu}}[\mathrm{A}][\mathrm{Mu}] = \lambda[\mathrm{Mu}] \tag{4.2}$$

where k_{Mu} is the bimolecular rate constant and λ is known as the *pseudo*-first-order decay rate. Since the reactant A is in large excess its concentration is constant, and integration of Eq. (4.2) reveals exponential decay (Figure 4.1) with time constant $\tau = 1/\lambda$.

[X] usually means the concentration of X at a given time. But how does this apply to Mu if there is only one Mu atom in the system at a time? The answer lies in the Ergodic Principle, in which a time-averaged ensemble is equivalent to a

Muon Spin Spectroscopy: Methods and Applications in Chemistry and Materials Science, First Edition.
Donald G. Fleming, Iain McKenzie, and Paul W. Percival.

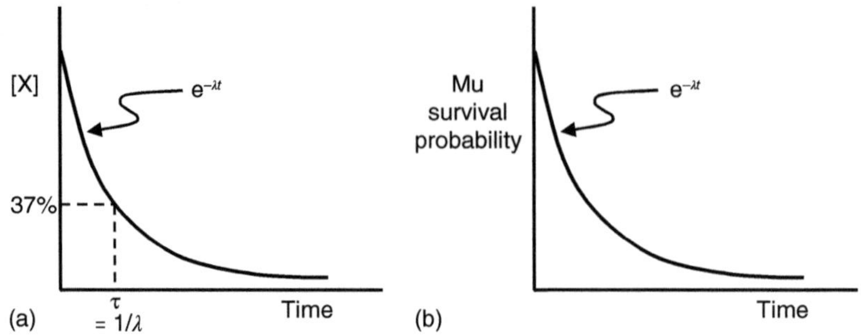

Figure 4.1 (a) First-order decay of some reactant X; and (b) the equivalent survival curve for Mu.

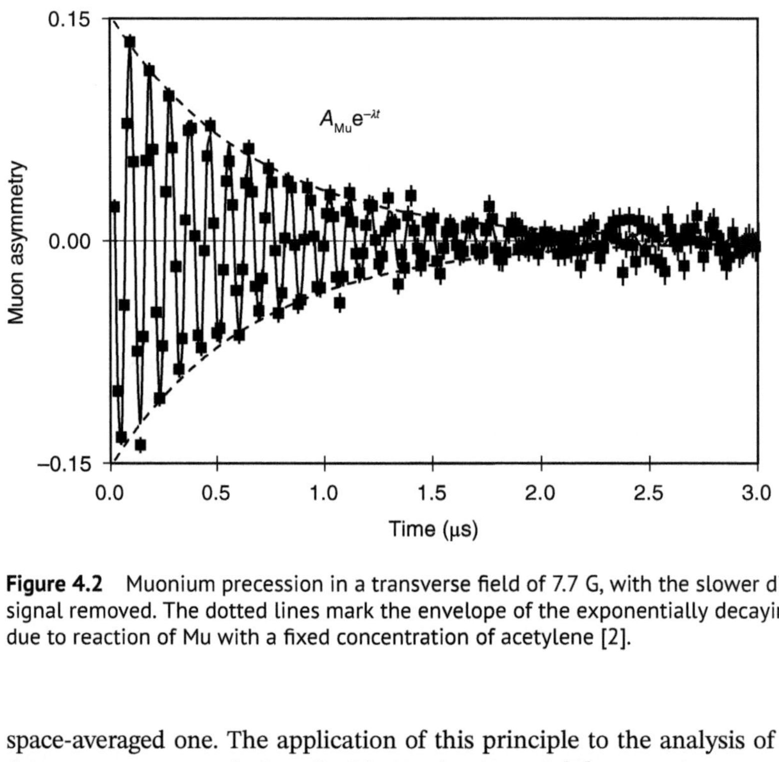

Figure 4.2 Muonium precession in a transverse field of 7.7 G, with the slower diamagnetic signal removed. The dotted lines mark the envelope of the exponentially decaying signal due to reaction of Mu with a fixed concentration of acetylene [2].

space-averaged one. The application of this principle to the analysis of muonium decay measurements is described by Duchovic et al. [1].

The most straightforward way to study muonium kinetics is to measure the decay of the precession signal in low transverse fields (corresponding to the degenerate ν_{12} and ν_{23} transitions described in Section 2.6.1). The exponential decay of the Mu signal is clearly seen in Figure 4.2. In practice, both the diamagnetic and muonium signal are fitted simultaneously and the asymmetry plot looks like Figure 2.14.

The second-order rate constant k_{Mu} can be found from a plot of the fitted decay rate as a function of reactant concentration, as expressed by Eq. (4.3):

$$\lambda = \lambda_0 + k_{Mu}[A], \tag{4.3}$$

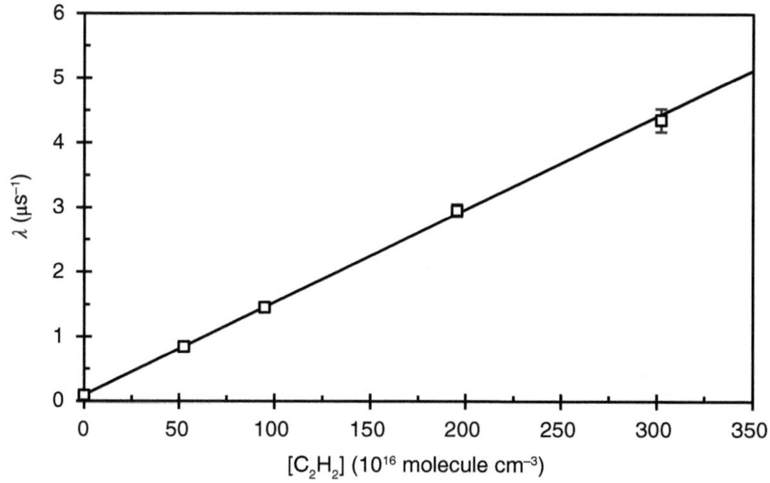

Figure 4.3 Muonium decay rate λ as a function of the reactant concentration. The slope of the fitted straight line gives the rate constant k_{Mu} for the reaction between muonium and acetylene [2].

where λ_0 is the muonium relaxation rate in the absence of reactant. An example of such an analysis is shown in Figure 4.3. In this case dilute concentrations of the reactant (acetylene) were studied in a much larger amount of nitrogen gas, to a total pressure of about 1.3 atm. The nitrogen acts as a *moderator* to ensure muon stops in the sample vessel. Since muon thermalization occurs in about 20 ns at this pressure (Section 3.1), we can be sure that such rate measurements on the microsecond timescale refer to thermal reactions.

4.2 Potential Energy Surfaces and Quantum Mass Effects

To understand how reaction rates depend on the isotopic composition of the reactants we need to discuss the potential energy surface (PES) for a reaction. In principle this displays the total electronic energy for a set of atoms as a function of all nuclear coordinates. The inherent separation of electronic and nuclear motion is a consequence of the Born–Oppenheimer Approximation. A full PES encompasses all configurations of the atoms (all nuclear positions), including reactants, products and all pathways between them. Since the electronic energy does not depend on translation or rotation of the complete system of nuclei, only the $3N-6$ internal coordinates need be considered ($3N-5$ for a linear molecule). Nevertheless, even a triatomic ($N = 3$) reaction system involves three nuclear coordinates, whereas the best we can visualize is a 3D plot of electronic energy as a function of two nuclear coordinates. Thus, the PES for the simple H-atom-exchange reaction

$$H_A + H_BH_C \rightarrow H_AH_B + H_C \tag{4.4}$$

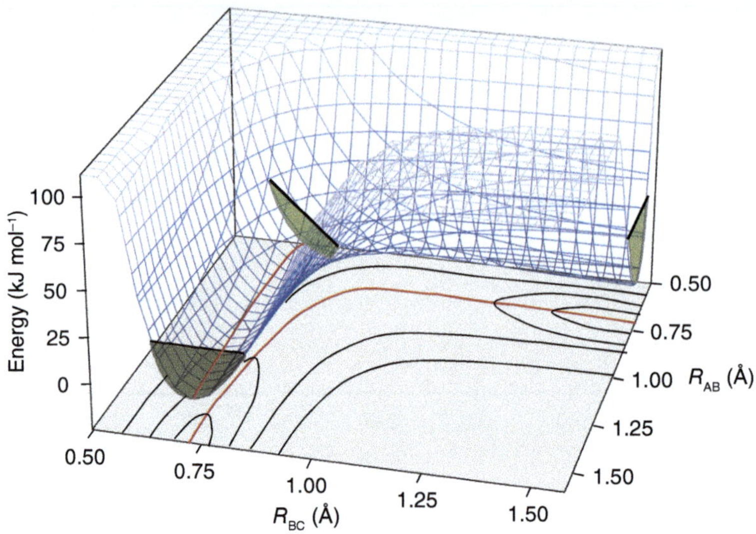

Figure 4.4 Potential energy surface for the collinear H-atom-exchange reaction. A 2D contour diagram is shown below the 3D mesh plot. The red line in each display represents the minimum-energy reaction path from reactants ($H_A + H_BH_C$) to products ($H_AH_B + H_C$). The vertical (energy) axis is truncated to enhance features at the bottom of the potential energy surface, such as the zero-point energy levels (thick black lines) at three positions along the reaction path, in the entrance channel, at the saddle point and in the exit channel.

is usually displayed for the collinear arrangement of the atoms, as shown in Figure 4.4.

The coordinates used here are the nuclear separations H_A–H_B and H_B–H_C, corresponding to the approach of H_A and the departure of H_C, respectively. In this collinear case the third coordinate H_A–H_C is simply the sum of the other two. In the general case this third variable is unconstrained and can be conveniently expressed as the angle of approach of H_A to the H_B–H_C molecule. Such bent systems will be ignored in the subsequent discussion of the H-atom-exchange reaction, since they have higher energies than the collinear case. This is not always true for other reactions. Note, also, that in the limit of large separations of the atom and molecule the PES evolves into those of the reactant H_BH_C and product H_AH_B molecules (often expressed as 1-dimensional Morse potentials).

Within the Born–Oppenheimer Approximation the PES is common to all isotopes of a given reaction, since it represents the electronic energy of the system. However, vibrations of the nuclei introduce additional zero-point energy (ZPE), which varies according to isotopic mass and the curvature of the surface at any point. This is indicated by the thick black lines (ZPE levels) and green slices in the reactant and product channels of the surface shown in Figure 4.4, as well as at the saddle-point state (the maximum along the minimum-energy path). The ZPE at the saddle point is less than for the reactants and product molecules. Note that the stretching vibrations of the H_BH_C (reactant) and of H_AH_B (product) molecules are perpendicular to the reaction path, as is the symmetric stretch of H_A–H_B–H_C at the saddle point.

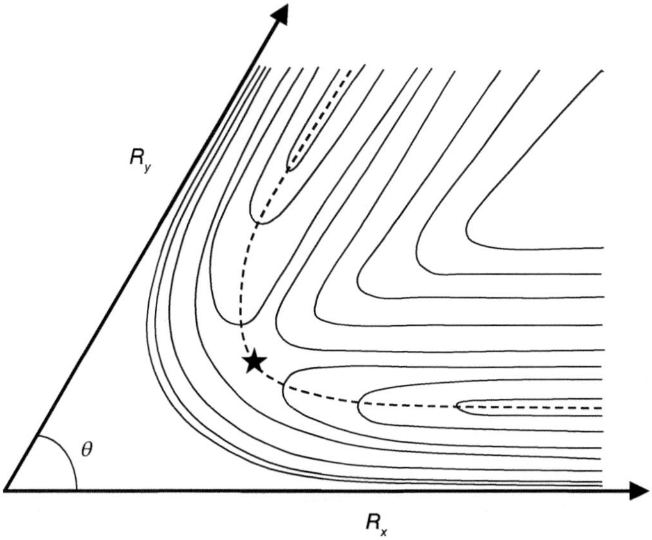

Figure 4.5 Contour plot of a potential energy surface in isoinertial coordinates. The star denotes the saddle point along the minimum-energy path (dotted line).

The latter motion is along the diagonal in the plot. The corresponding asymmetric stretch would be along the reaction coordinate, a surface with negative curvature. This shows up as an imaginary frequency in quantum calculations of the saddle point.

Ideally, we would like to conceive of a reaction as the motion of a single quasi-particle (the collection of atoms) over the potential energy surface. However this is complicated by the change of the effective mass when the surface is defined in Cartesian coordinates. Instead, it is more convenient to define a set of mass-scaled coordinates for each atom:

$$R_q = \left(\frac{m}{\mu}\right)^{1/2} q; q = x, y, z, \tag{4.5}$$

where $q = x, y, z$ refers to the Cartesian coordinates with respect to the centre of mass of the system, and $\mu = m_A m_{BC}/(m_A + m_{BC})$ is the reduced mass of the reactants. When expressed in such *isoinertial* coordinates, the PES becomes skewed, as shown schematically in Figure 4.5.

The skew angle θ is given by [3]

$$\theta = \cos^{-1}\left[\left(\frac{m_A m_C}{(m_A + m_B)(m_B + m_C)}\right)^{1/2}\right] \tag{4.6}$$

For equal masses, such as H_3, the skew angle is 60°, but introduction of a light atom results in a larger angle, 77° for the MuH_2 system.

Since vibrational frequencies, and hence ZPE contributions, depend on nuclear mass the overall energy profile of a reaction depends on the isotopic composition. This is conveniently displayed as a plot of potential energy as a function of distance

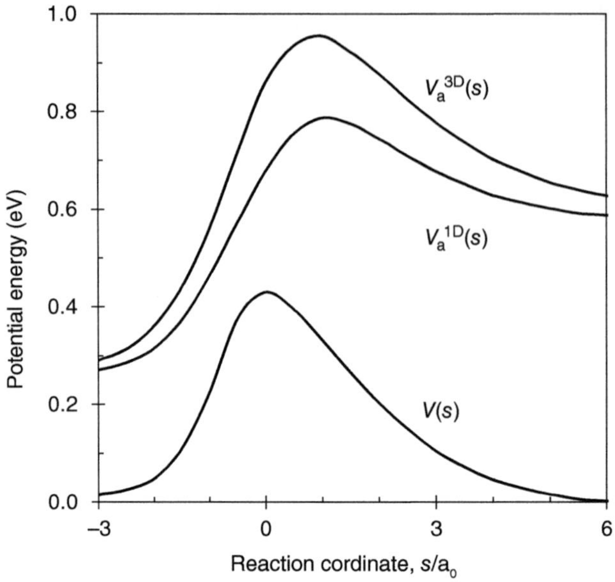

Figure 4.6 One-dimensional potential energy profiles for the reaction Mu + H_2 → MuH + H. $V(s)$ show the potential in the absence of zero-point energy. The vibrationally adiabatic potential $V_a^{1D}(s)$ accounts for zero-point energy in stretch vibrations for the collinear reaction, whereas $V_a^{3D}(s)$ also includes bending vibrations. Source: Mielke et al. [4].

along the reaction coordinate (Figure 4.6). Since the Mu-H product has ~0.32 eV more ZPE than H_2 the reaction Mu + H_2 → MuH + H is markedly endoergic, as indicated by the vibrationally adiabatic potentials $V_a^{1D}(s)$ and $V_a^{3D}(s)$. The former includes the symmetric stretch of Mu-H-H, while the latter (3D) has an additional contribution from the two bending vibrations. Zero-point vibrations affect not only the height of the reaction barrier but also its position and width. Thus, as shown in Figure 4.6, the Mu + H_2 reaction has a later barrier than H + H_2.

Knowledge of the height and shape of the potential barrier to reaction is essential for accurate calculations of tunnelling, i.e. transmission *through* the barrier instead of over the top. Like zero-point vibrations, this is a quantum effect that favours light atoms. We might therefore expect enhanced tunnelling in reactions of Mu compared to H. However, this is not always the case. In addition to the particle mass, quantum penetration depends on the height and the width of the barrier. It is evident from Figure 4.6 that the vibrationally adiabatic potential curves are considerably broader than the classical reaction profile for the Mu + H_2 system. Thus, the two quantum contributions to kinetic isotope effects, tunnelling and zero-point vibrations, are linked, and should not be treated as independent factors in a complete rate theory. A further complication is that tunnelling is not restricted to the one-dimensional minimum-energy path. Accurate calculations must also consider corner-cutting pathways, particularly where the skew angle of the isoinertial surface is small. Straying off the path (dotted line in Figure 4.5) results in a higher potential than at the saddle point, but this can be compensated by a narrower barrier.

It should be clear from the above that rigorous calculation of rate constants requires knowledge of the complete PES. The best-characterized surface is, not surprisingly, the simplest: H_3. But even there, small non-Born–Oppenheimer effects must be included, notably the Born–Oppenheimer diagonal correction [5]. Since this accounts for the motion of nuclei the resulting PES is mass dependent. It is therefore imperative that any successful theory of reaction rates must address isotopic mass effects, and it should not be surprising that muonium rate data have long been used as a benchmark to test theoretical developments [6–12].

4.3 Theoretical Background to Rate Calculations for Bimolecular Reactions

There are two broad approaches for the calculation of chemical reaction rates: *Reaction Dynamics* and *Transition-State Theory*. The former has its origin in *Collision Theory*, which in its simplest form treats reactants as structureless hard spheres. The collision rate depends on the relative velocity of the reaction partners,

$$\bar{v} = \left(\frac{8k_B T}{\pi\mu}\right)^{1/2}, \tag{4.7}$$

and the cross section for the collision, which is simply $\sigma = \pi d^2$, where d is the sum of the radii of the two reacting atoms or molecules. However, not all collisions lead to reaction, either because their relative orientation is wrong, or because the kinetic energy of the reactants along their line of centres does not meet some minimum-energy threshold for reaction, E^*. Thus, the standard textbook expression for the bimolecular rate constant is [13, 14]

$$k_{\text{coll}} = \pi d^2 \left(\frac{8k_B T}{\pi\mu}\right)^{1/2} p \exp\left(-\frac{E^*}{k_B T}\right), \tag{4.8}$$

where p is an empirical factor introduced to account for any orientation dependence of the reaction rate (hence often called the *steric factor*), and because the hard-sphere approximation typically overestimates the cross section. It should be noted that this equation employs the molecular units widely used in gas-phase kinetics. Thus the cross section has units cm^2, the relative velocity is in cm s^{-1} and the Boltzmann constant k_B is used instead of the molar gas constant R. The result is a rate constant expressed in units of cm^3 molecule^{-1} s^{-1}.[1]

From this simple expression for the rate constant it can be seen that the threshold energy E^* is not quite the same as the experimental activation energy:

$$E_a = k_B T^2 \frac{d(\ln k_{\text{exp}})}{dT} \quad \text{(by defintion)} \tag{4.9}$$

$$\text{but} \quad k_B T^2 \frac{d(\ln k_{\text{coll}})}{dT} = \frac{1}{2} k_B T + E^*. \tag{4.10}$$

1 Although these non-S.I. units are commonly used in gas-phase kinetics and therefore here, Chapter 5 will report bimolecular rate constants in S.I. units (litre mol^{-1} s^{-1}, or simply M^{-1} s^{-1}) in line with standard practice for condensed-phase kinetics.

In addition, the velocity term introduces temperature dependence to the pre-exponential factor. Because of this, experimental data are often analysed with a modified Arrhenius expression:

$$k(T) = BT^{1/2} \exp(-E_a/RT) \tag{4.11}$$

instead of the usual $A \exp(-E_a/RT)$ for which A is temperature independent. Other theories also predict temperature-dependent pre-exponential factors, so an even more general form $B'T^n \exp(-E_a/RT)$ is sometimes used, where the exponent n is an adjustable fit variable.

More realistic treatments of collision theory recognize that reactions occur on potential energy surfaces, such as described in Section 4.2. The dynamical effects are contained in the reaction cross section, which is inevitably energy-dependent. Thus Eq. (4.8) needs to be replaced with the more general expression

$$k_{\text{coll}} = \frac{1}{k_B T}\left(\frac{8}{\pi \mu k_B T}\right)^{1/2} \int_0^\infty \sigma(E) E \exp\left(-\frac{E}{k_B T}\right) \mathrm{d}E. \tag{4.12}$$

In the reactive hard-sphere model

$$\sigma(E) = \pi d^2 \left(1 - \frac{E^*}{E}\right). \tag{4.13}$$

Classical equations of motion can be used to find reaction probabilities and cross sections. Such trajectory calculations can be visualized as shooting marbles up a hill (the activation barrier), with the probability of reaction given by the fraction of shots that get over the peak. *Quasiclassical* trajectories (QCT) include vibrational contributions to the energy surface, while still treating the dynamics classically. If trajectories based on quantum scattering theory are used instead of classical mechanics, the approach is known as a quantum mechanical trajectory (QMT) calculation. It should be noted that both QCT and QMT calculations require knowledge of the full potential energy surface (PES).

The details affecting reaction rates are often more transparent in transition-state theory (TST), where it is only the barrier region of the PES that is important. In classical TST the dividing line between reactants and products is placed at the saddle point, the maximum on the minimum-energy path, and the rate of reaction is given by the one-way flux of transition states crossing this line (more generally a surface perpendicular to the reaction coordinate). TST involves the quantum-statistical properties (partition functions) of the reactants and the transition state. The standard textbook expression for the rate constant is [13, 14]

$$k_{\text{TST}}(T) = \kappa(T) \frac{k_B T}{h} \frac{[Q^{\ddagger}_{\text{TS}}]}{[Q_A Q_B]} \exp\left(-\frac{E^{\ddagger}_0}{k_B T}\right), \tag{4.14}$$

where Q_A and Q_B are total partition functions for the reactants and $Q^{\ddagger}_{\text{TS}}$ applies to the transition state but omitting the degree of freedom corresponding to the reaction coordinate. $E^{\ddagger}_0$ is the difference in energy between the lowest energy levels (including zero-point energy) of the transition state and the reactants. The transmission coefficient, $\kappa(T)$, is an ad hoc correction to account for quantum tunnelling and any other adjustments to what is otherwise a classical formalism.

The tunnelling probability can be estimated by solving the Schrödinger equation for a one-dimensional barrier which approximates the potential along the reaction coordinate. The resulting permeability $G(E)$ must be integrated over a Boltzmann distribution to give the tunnelling correction:

$$\kappa(T) = \exp\left(\frac{E_0^{\ddagger}}{k_B T}\right)\int_0^{\infty}\frac{1}{k_B T}G(E)\exp\left(-\frac{E_0^{\ddagger}}{k_B T}\right)\mathrm{d}E. \tag{4.15}$$

Although there are analytical solutions for some simple barrier shapes, $G(E)$ is usually found by numerical methods. Both analytical and numerical methods confirm that tunnelling is enhanced for light particles and narrow barriers, as mentioned in Section 4.2.

More advanced variants of TST take account of multiple barrier crossings by optimizing the position of the transition state dividing surface (hence *variational* transition-state theory, VTST); multiple reaction paths and multi-dimensional quantum tunnelling [3].

4.4 Early Experimental Studies: Mu + Halogens

The first direct measurements of muonium kinetics in the gas phase took place at the Lawrence Berkeley Laboratory, using the newly developed 'Arizona' beam [15], which is now known as the surface muon beam (Section 2.1). The relatively low energy of such a beam permits muon stops in low-density gas samples, ~1 bar. Even then the muon range is about 30 cm, so large sample vessels are necessary.

The Mu + Br_2 reaction was chosen for this first study [16], since expectations based on the equivalent H atom reaction suggested an easily measured Mu decay rate at room temperature. Subsequent work at TRIUMF extended measurements to other halogens: Cl_2 [17] and F_2 [18]. Since these were some of the first experiments carried out at TRIUMF the data were sparse and characterized by relatively low statistics. However, these reactions, thought to be halogen (X) abstractions,

$$\mathrm{Mu} + \mathrm{X}_2 \rightarrow \mathrm{MuX} + \mathrm{X}^{\bullet}, \tag{4.16}$$

provided valuable data to compare with early theoretical calculations [19, 20]. The same three reactions were revisited in 1989, with much better beam conditions and over a wider range of temperatures [21]. The bimolecular rate constants, $k_{\mathrm{Mu}}(T)$, from that study are shown in the Arrhenius plots in Figure 4.7. The marked curvature seen at the lower temperatures in both the Mu + Cl_2 and F_2 reactions is noteworthy and is indicative of quantum tunnelling. In contrast, the apparent *negative* activation energy seen for the Mu + Br_2 reaction suggests that the kinetic data are associated with a multi-step reaction, e.g. one involving an intermediate complex (see Section 4.8).

The most extensive data set is for the Mu + F_2 reaction, and this has been used to test the accuracy of the PES for H + F_2. The quantum reactive scattering calculations of Tanaka and Takayanagi [22] agree with experiment data for both Mu and H (Figure 4.8), and emphasize the extreme sensitivity of Mu to quantum tunnelling.

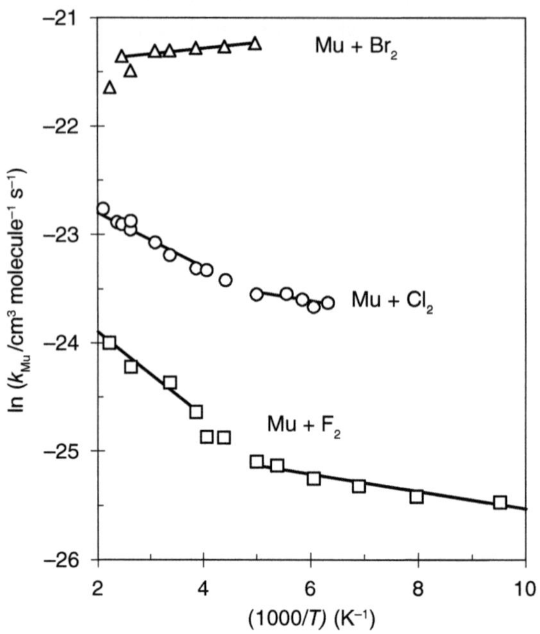

Figure 4.7 Arrhenius plots of the rate constants k_{Mu} for the reactions of Mu with Br_2, Cl_2 and F_2. The lines indicate separate fits to lower- and higher-temperature data [21].

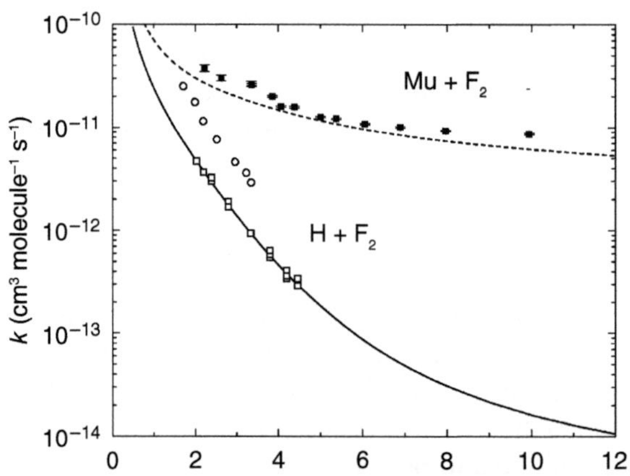

Figure 4.8 Comparison of quantum scattering calculations with rate constant data for the Mu + F_2 reaction [21] and the H + F_2 reaction (open circles, Albright et al. [23]; open squares, Homann et al. [24]). The calculated rate constants are in excellent agreement (solid line) with the latter data for H + F_2 and successfully reproduce (dotted line) the large curvature of the Arrhenius plot for Mu + F_2. Source: Adapted from a figure by Tanaka and Takayanagi [22], with permission from Elsevier.

4.5 H Atom Abstraction Reactions

The growing success of rate constant calculations prompted Fleming et al. to extend their experimental investigations to the fundamental $H + H_2$ system, first [25] via the isotopic variants

$$Mu + H_2 \rightarrow MuH + H^\bullet, \tag{4.17}$$

$$Mu + D_2 \rightarrow MuD + D^\bullet, \tag{4.18}$$

but also, more recently, for the much heavier muonic helium atom [26]:

$$He\mu + H_2 \rightarrow He\mu H + H^\bullet. \tag{4.19}$$

The Mu reactions with H_2 and D_2 are far too slow to be measured near room temperature, requiring temperatures of ~500 K or more, and relatively high reactant partial pressures of order 15 bar, to achieve decay rates approaching 1 μs^{-1}. The competing demands of large volume, high temperature and pressure, yet an entrance window thin enough to pass surface muons, posed significant technical barriers [25]. Nevertheless, it is apparent from Figure 4.9 that these early experimental results already provided excellent data to test rate theories.

The technical difficulties of studying reaction (4.19), $He\mu + H_2$, were even more extreme [26]. As explained in Section 3.4, μ^- capture results in loss of most of the muon spin polarization. Thus, the precession signals in Heμ are only about 5% of those observed for Mu, requiring very long acquisition times for adequate statistics. Another notable difference is that a backward muon beam had to be used to produce μ^-, which in turn demanded a high-pressure gas target (~500 bar He) to stop these high-momentum (~70 MeV/c) muons. Some examples of the weak μ^- precession signals can be seen in Figure 4.10.

As evident from Figure 4.11, Heμ reacts faster than Mu at a given temperature. This might seem surprising, given its 36 times greater mass, but it can be readily explained in terms of the isotope effects on the PES. There is only slight curvature in the Arrhenius plots, so tunnelling can be discounted as a major factor. For Mu, zero-point vibration is higher in the MuH product than the H_2 reactant, resulting in an endoergic reaction. In contrast, for Heμ the product has *less* zero-point energy than H_2, and the reaction is exoergic. (This is also the case for the $D + H_2$ reaction.) Reaction energies and activation energies for isotopically related reactions are compared in Table 4.1. Because of these differences, $Mu + H_2$ has a late barrier (Figure 4.6), while $He\mu + H_2$ has an early barrier.

Abstraction of H atoms from polyatomic molecules is expected to be similar to abstraction from H_2, i.e. an inverse isotope effect (Mu reaction slower than H), consistent with the enhanced ZPE in the product and the associated late activation barrier. The experimental results for Mu + methane and ethane [27] are in accord with this expectation, but higher than expected rates were found for Mu + propane [28]. Tunnelling has been suggested as an explanation [28], and although this was disputed by a theoretical study [29], more recent calculations show much better agreement with the experimental data and emphasize the importance of including

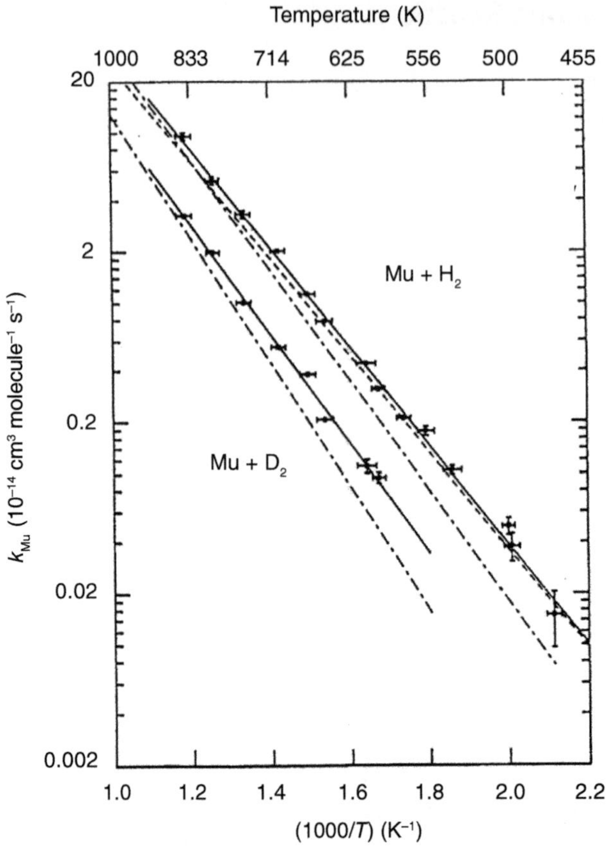

Figure 4.9 Arrhenius plots of the rate constants k_{Mu} for the reactions of Mu with H_2 and D_2. The experimental data [25] are denoted by points and solid lines. Theoretical predictions are shown by the dashed line (3D quantum-coupled states [9]) and dot-dash lines (variational transition-state theory [8]). Source: Adapted from a figure by Reid et al. [25] with the permission of AIP Publishing.

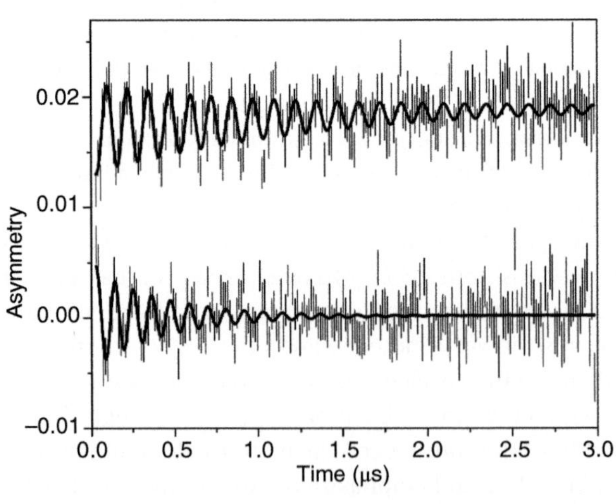

Figure 4.10 Negative muon precession signals for Heμ in a 6.8 G field at 500 bar He pressure, with 4.8 bar H_2 at 405 K (top) and 2.9 bar H_2 at 500 K (bottom) [26].

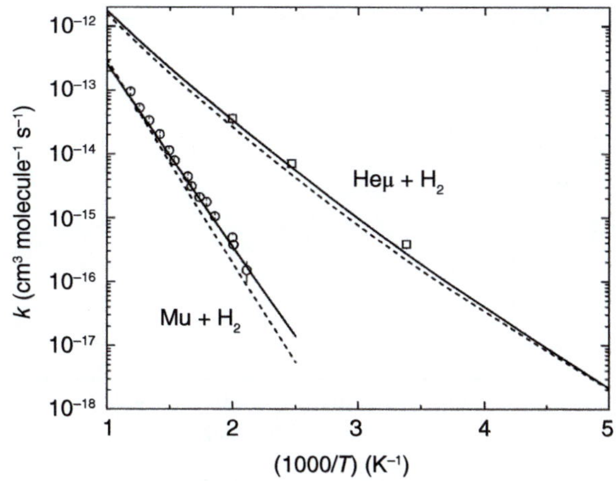

Figure 4.11 Arrhenius plots of the rate constants for the Heμ + H_2 and Mu + H_2 reactions compared with accurate QM rate calculations (solid lines) and the predictions of variational transition-state theory (dotted lines) [26].

Table 4.1 Reaction energies[a)] and activation energies[b)] for isotopically related reactions on the H_3 PES.

Reaction	ΔH (kJ mol^{-1})	E_{act} (kJ mol^{-1})
$Mu + H_2 \rightarrow MuH + H$	31.8	55.6 ± 0.8
$H + H_2 \rightarrow H_2 + H$	0.0	35.6 ± 2.1
$D + H_2 \rightarrow HD + H$	−5.0	33.5 ± 1.7
$He\mu + H_2 \rightarrow He\mu H + H$	−5.9	26.8 ± 1.3
$Mu + D_2 \rightarrow MuD + D$	36.8	61.5 ± 1.7
$H + D_2 \rightarrow HD + H$	2.5	41.4 ± 1.3

a) Zero-point energies calculated for harmonic vibrations.
b) Experimental data.

vibrational anharmonicity in the transverse modes along the whole reaction path, thereby decreasing both the height and width of the vibrationally adiabatic potential barrier [30].

4.6 State-selected Reactivity: Mu + H_2(v = 1)

All of the reaction dynamics discussed above are for thermal reactions, and the corresponding calculations involve integrals over Boltzmann distributions of reactant states. A more fundamental insight is provided by the study of state-to-state dynamics [31]. The contribution of muon spin spectroscopy to this area is still in its infancy and is currently limited to the study of Mu + H_2(v = 1), i.e. the reaction of muonium with hydrogen in its first vibrationally excited state [32]. Not surprisingly, the rate of this reaction is much faster than for ground state H_2; indeed, this is a prerequisite of the method, because the desired reactant, H_2(v = 1), was prepared in relatively low concentration in an excess of H_2(v = 0). Excitation was achieved

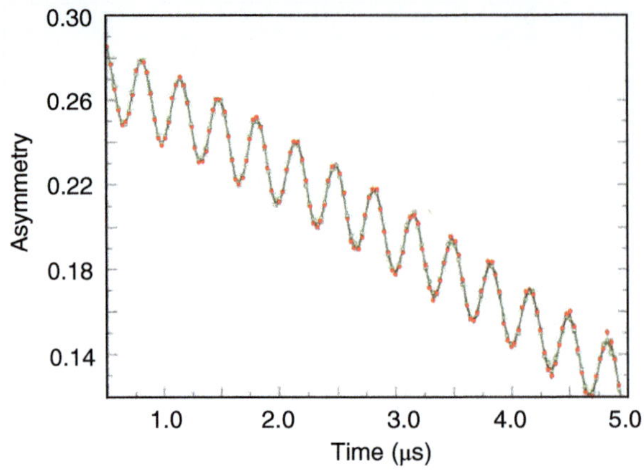

Figure 4.12 Muonium precession in 50 bar H_2 at ~300 K, with laser on (green points and line) and laser off (red points). The difference is small but significant, as shown in Figure 4.13. Source: Adapted with permission of IOP Publishing from a figure by Bakule et al. [32]; permission conveyed through Copyright Clearance Center, Inc.

by stimulated Raman pumping, induced by 532 nm light from the 2nd harmonic of a Nd:YAG laser. Although this is an established method for photoexcitation of H_2, coupling this technology with a muon spin precession experiment posed considerable technical challenges.

To optimize the temporal overlap of muon arrival and photoexcitation, the experiment was carried out with a pulsed beam at the RIKEN-RAL facility at ISIS. The laser was operated at 25 Hz, synchronized to the ISIS accelerator operating at 50 Hz. In this way muon data could be accumulated alternately into laser-on and laser-off data sets. Furthermore, because the ISIS beam has a double-pulse structure (each ~80 ns long, separated by ~330 ns) a transverse field of 2.15 G was chosen, so that the period of the 3 MHz muonium precession just matched the pulse separation. The resulting Mu signal can be seen in Figure 4.12.

Spatial overlap was another concern, in order to achieve sufficient reactant to give a measurable decay of the Mu precession signal. This involved various compromises to best match the laser profile with the muon-stopping distribution in the 50 bar sample gas. Each laser pulse produced about 4×10^{16} excited H_2 molecules, as determined from optical measurements, but it is the *concentration* that matters for kinetics, so estimation of the effective volume of the reaction is crucial for determination of the decay rate. Taking the laser-pumped volume as $1.25\,cm^3$ the resulting concentration of $H_2(v = 1)$ was estimated to be 3×10^{16} molecules cm^{-3}. This is not a uniform concentration, however, and the full data analysis [32] took into account the calculated muon-stopping distribution and the measured laser profile.

Yet another complication of the experiment is the collisional deactivation of $H_2(v = 1)$ back to the ground state. This reduces the concentration of the reactant (Mu reaction with the ground state is negligible at the temperature of the experiment), so the Mu decay rate, λ_M, is not constant; it slows with time. A simplified version of the kinetic equations follows. Of course, the actual concentration of Mu is vanishingly small at any one time, but within the Ergodic Principle (Section 4.1) repeated pulse experiments can be treated as if an ensemble concentration $[Mu]_0$ exists at $t = 0$.

$$\mathrm{Mu} + \mathrm{H_2}(v = 1) \xrightarrow{k_1} \mathrm{MuH}(v = 0) + \mathrm{H}^\bullet, \tag{4.20}$$

$$H_2(v = 1) + H_2(v = 0) \xrightarrow{k_C} H_2(v = 0) + H_2(v = 0). \tag{4.21}$$

The molecular product in Eq. (4.20) has been explicitly specified as $v = 0$ to emphasize the fact that the $v = 1$ state is far too high (over 1 eV above the ground state) to be populated. For simplicity of writing this designation will be dropped in the equations below, and $H_2(v = 1)$ will be written as $H_2{}^*$.

The deactivation process expressed in Eq. (4.21) is pseudo-first-order, since $[H_2{}^*] \ll [H_2]$.

Thus,

$$-\frac{d\left[H_2{}^*\right]}{dt} = k_C\left[H_2{}^*\right][H_2] = \lambda_C\left[H_2{}^*\right],$$
$$\left[H_2{}^*\right]_t = \left[H_2{}^*\right]_0 e^{-\lambda_C t}. \tag{4.22}$$

Similarly, the Mu decay can be expressed as

$$-\frac{d[\mathrm{Mu}]}{dt} = k_1\left[H_2{}^*\right]_t[\mathrm{Mu}] = k_1\left[H_2{}^*\right]_0 e^{-\lambda_C t}[\mathrm{Mu}] = \lambda_1 e^{-\lambda_C t}[\mathrm{Mu}], \tag{4.23}$$

where $\lambda_1 = k_1[H_2{}^*]_0$. Integrating Eq. (4.23) results in

$$[\mathrm{Mu}]_t = [\mathrm{Mu}]_0 \exp\left\{-\lambda_1\left(1 - e^{-\lambda_C t}\right)/\lambda_C\right\}. \tag{4.24}$$

If deactivation is negligible ($\lambda_C \ll \lambda_1$), the Mu signal decays exponentially, as $\exp(-\lambda_1 t)$, but otherwise there is a reduced relaxation rate as time increases. Mu reaction is not obvious in the laser-on Mu precession signal of Figure 4.12, but the difference signal (laser-on minus laser-off) shows the time dependence in the envelope of the oscillations:

$$S(t) = S_{\mathrm{Mu}}(0)\left[\exp\left\{-\lambda_1\left(1 - e^{-\lambda_C t}\right)/\lambda_C\right\} - 1\right], \tag{4.25}$$

where $S_{\mathrm{Mu}}(0)$ represents the Mu precession signal in the absence of reaction, i.e. light-off. The results are displayed in Figure 4.13.

Detailed analysis of multiple data sets resulted in $k_1 = 9.9^{+1.7}_{-1.4} \times 10^{-13}\,\mathrm{cm}^3$ molecule^{-1}s^{-1} for the reaction of Mu with $H_2(v = 1)$ [32], in remarkable agreement with a rigorous quantum calculation [33]. Combining this result with the data of Figure 4.11 reveals an unprecedented level of agreement between exact theory and experiment, extending over a factor of 36 in atomic mass and four orders of magnitude in rate constants.

4.7 Addition Reactions

All the H/Mu reactions discussed so far in this chapter involve abstraction of an atom from the other reactant. We now consider addition. Combination of two species to form a single product might seem simple, but this is not the case in the gas phase. A complication arises from the possibility of back reaction:

$$\mathrm{H} + \mathrm{A} \underset{k_d}{\overset{k_a}{\rightleftharpoons}} (\mathrm{HA})^* \xrightarrow[\mathrm{M}]{k_s} \mathrm{HA}. \tag{4.26}$$

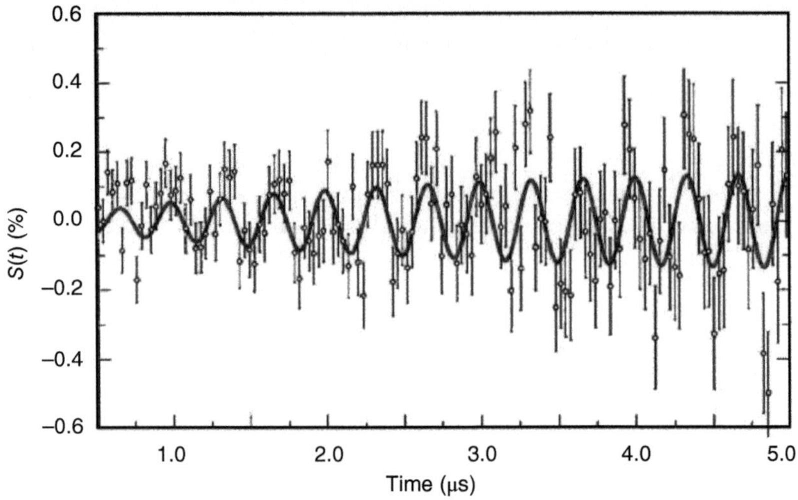

Figure 4.13 The difference signal (laser-on minus laser-off) of muonium precession in 50 bar H_2 at ~300 K. Note that $S(0) = 0$, but the time axis in the figure starts at 0.5 μs. Source: Adapted with permission of IOP Publishing from a figure by Bakule et al. [32]; permission conveyed through Copyright Clearance Center, Inc.

Because the initial combination of the H atom and reactant A results in a new bond, the process is exoergic, leading to the formation of an excited molecule, HA*. This can either fall apart (unimolecular rate constant k_d) or be stabilized by energy transfer to some moderator M (bimolecular rate constant k_s). The rate of the stabilization reaction depends on the concentration of M, which could be A itself or the unreactive moderator gas. Thus, the competition between dissociation and stabilization of (HA)* depends on [M], and the overall rate of reaction (H → HA) depends on pressure. This simple model is described by the Lindemann expression,

$$k_{app} = \frac{k_a k_s[M]}{k_d + k_s[M]}, \tag{4.27}$$

although more advanced treatments take into account additional complications, such as the dependence of k_s on the excitation energy in the Lindemann–Hinshelwood theory. (Further details can be found in standard texts on chemical kinetics [13, 14], under the heading *Unimolecular Reactions*.) However, Eq. (4.27) is sufficient to demonstrate that the apparent rate constant, k_{app}, depends linearly on pressure if $k_s[M] \ll k_d$, but has a constant value when $k_s[M] \gg k_d$, giving the high-pressure limiting value k_a, the bimolecular rate constant needed for comparison with the results of transition state and collision theories calculated for potential energy surfaces, as described in Section 4.3. The dissociation constant, k_d, can be calculated by RRKM theory [13], which includes, among other parameters, the number of vibrational degrees of freedom of the excited molecule. As a consequence, it is predicted that excited polyatomic molecules with many degrees of freedom will live longer than small molecules and thus have lower dissociation rates.

In principle Lindemann kinetics also applies to Mu addition reactions. However, muon spin spectroscopy monitors the spin polarization of muoniated species, not their concentration. Therefore, if polarization is completely lost in the initial addition step it does not matter that the back reaction recreates muonium, since it would not be observable [1]. As explained in Section 4.1, the standard method to study muonium kinetics is to measure the decay of the Mu precession signal in low transverse fields. This implicitly acknowledges that the muon precession frequencies of the product are significantly different from those of Mu. Quantitative treatment of polarization transfer will be deferred until Sections 5.4 and 6.6.3. For the specific case of muonium addition under consideration here, spin coherence is lost under typical conditions for which the [MuA]* lifetime is ~1 ns or longer [1]. Thus $k_{app} = k_a$ for all practical pressures. The conclusion is that measurements of the decay of Mu precession give directly the high-pressure limiting rate constant. This is a huge advantage compared to experimental studies of H-atom kinetics, where this limit is usually difficult to establish.

Experimental investigations of Mu reaction with ethylene [34] and acetylene [2] have indeed confirmed that the measured rate constants are pressure-independent. Since the rate constants are much higher than those measured for H abstraction reactions it can be safely assumed that they represent the addition reactions Mu + CH_2=CH_2 and Mu + HC≡CH. This is also in accord with theoretical expectations, since H abstraction reactions are endoergic, due to the high ZPE of the MuH product, whereas the addition reactions are exoergic.

Rate constant data for these two reactions are displayed as Arrhenius plots in Figure 4.14. The lines through the data points correspond to fits using the modified Arrhenius expression given in Eq. (4.11). The activation energies so determined are E_a = 2.6 kJ mol^{-1} for Mu + HC≡CH (red) and E_a = 2.1 kJ mol^{-1} for Mu + CH_2=CH_2 (blue), both much smaller than for the equivalent H atom reactions. This is taken as an indication of significant quantum tunnelling in these reactions [2].

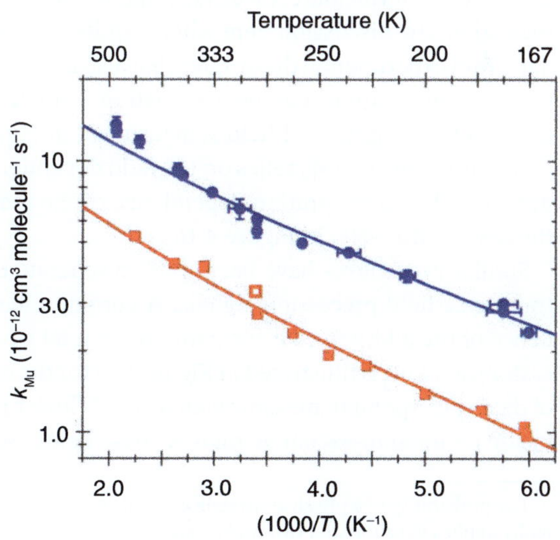

Figure 4.14 Temperature dependence of rate constants for the reactions Mu + CH_2=CH_2 [34] (blue) and Mu + HC≡CH [2] (red). The open red square denotes a measurement made with over 20 times higher moderator pressure, demonstrating the lack of pressure dependence. Source: Adapted from a figure by Arseneau et al. [2] with permission from the American Chemical Society.

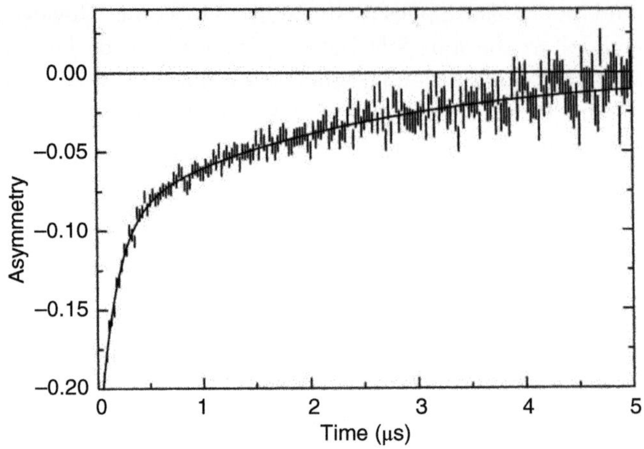

Figure 4.15 Longitudinal field relaxation curve for Mu + O_2 in the presence of a nitrogen gas moderator. The faster decay is due to Mu addition and the slower one is ascribed to spin relaxation in the MuO_2 radical. Source: Adapted from a figure by Himmer et al. [35] with permission from the American Chemical Society.

In contrast to the above reactions of Mu with alkenes, the measured rate constants for addition to O_2 [35], NO [36] and CO [37] are strongly pressure-dependent. In terms of the Lindemann model, $k_{app} = (k_a k_s/k_d)$ [M], and the proportionality factor $(k_a k_s/k_d)$ has the units of a termolecular process.[2] This marked difference in behaviour for the small molecules is a result of the low density of states in the triatomic intermediates (MuA)*, which leads to inefficient stabilization (a small value for k_s).

Another difference is the interplay of chemical reaction and spin relaxation. Both O_2 and NO are paramagnetic, and their spin-exchange interactions with Mu lead to muon spin relaxation. For CO, which is diamagnetic, it is electron spin relaxation in the excited intermediate, (MuCO)*, that matters. Since Chapter 7 is devoted to spin relaxation, the discussion here will be limited to extraction of the chemical reaction data from experimental decay rates that also include spin relaxation. In optimal cases the two contributions can be extracted in a single fit, as in the double exponential decay of the longitudinal-field signal shown in Figure 4.15.

Quantitative analysis relies on the field dependence of the spin relaxation component and the concentration dependence of the reaction rate. This is illustrated for the case of Mu + NO in Figure 4.16.

Similar procedures have been used to separate the contributions to the decay of transverse-field precession signals. A common feature of the results is the dependence of the addition rate constant on the total gas pressure (mostly the inert moderator gas). This is illustrated in Figure 4.17 for the case of Mu + CO [37]. The two sets of data correspond to measurements at 5 G (lower rates) and 55 G. Theory predicts a factor 1.5 for spin-exchange rates at these fields (Section 7.4) and the intercepts are

2 The probability of a physical three-body collision is very low, so the designation 'termolecular' really applies to sequential two-body collisions.

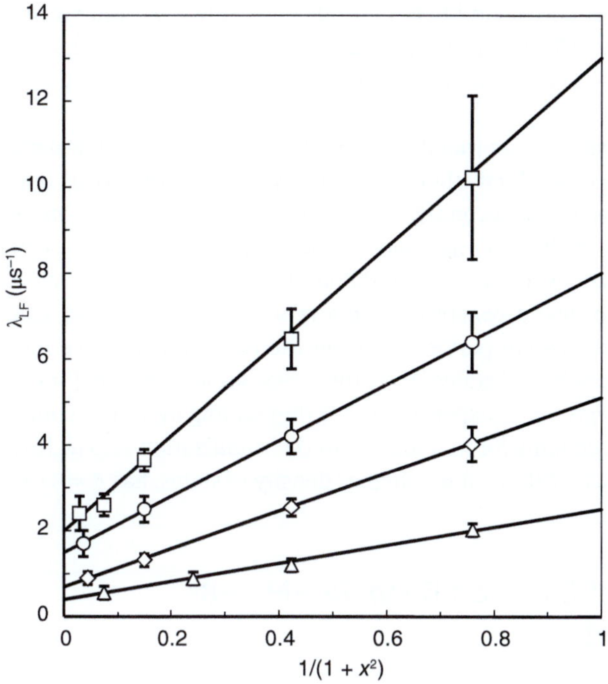

Figure 4.16 Longitudinal-field relaxation rates for Mu + NO in the presence of nitrogen, to a total gas pressure of 151 bar. Relaxation due to spin exchange is magnetic-field dependent (*x* is a dimensionless field parameter) and the different data sets correspond to different concentrations of NO. The chemical contribution is given by the intercepts. Source: Data taken from Pan et al. [36].

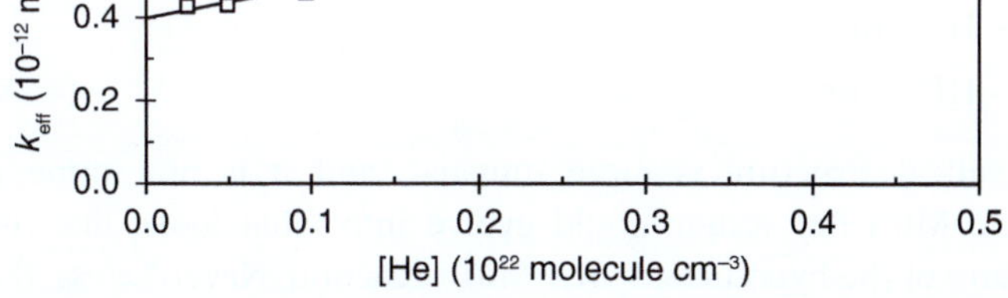

Figure 4.17 Effective rate constants for Mu + CO determined from the decay of muonium precession at 5 G (squares) and 55 G (diamonds) [37]. The slope of the lines gives the termolecular rate constant and the intercepts correspond to the spin-flip rate constant.

consistent with this, if a small allowance is made for additional effects, such as field inhomogeneity. The slopes correspond to the termolecular rate constant. Since the reaction product, the radical MuCO, lacks spin-active nuclei (ignoring the small natural abundance of ^{13}C) the only mechanism for muon spin relaxation is via electron spin relaxation.

The Mu + CO system exhibits remarkable isotope effects related to the low density of states of the MuCO radical. First, there are only two $J = 0$ vibrational bound states, compared to 15 for HCO. Secondly, MuCO has a huge rotational constant, 181 cm^{-1} for rotation about the C–O axis (compared to 24 cm^{-1} for HCO), which results in higher rotational levels $J > 0$ being unbound. Statistical theories, such as RRKM, are inadequate to treat a system with so few bound states. Instead, calculations of termolecular rates were performed using the isolated resonance model (IRM), which includes quasibound states above the dissociation threshold [38]. The predictions of the model agree with experiment [37] after scaling by a weak collision factor, $\beta = 0.001$, which accounts for inefficiency in the stabilization step (k_s[M] in Eq. (4.26). In contrast, HCO, with its much higher density of states, has $\beta = 0.08$.

4.8 A New Type of Chemical Bond: Br—Mu—Br Vibrational Bonding

As shown in Figure 4.7 the apparent rate constant for the reaction of Mu with Br_2 falls with temperature, in contrast to the reactions of H and D which show conventional Arrhenius behaviour [39]. The negative activation energy found for Mu [21] may indicate the existence of an intermediate complex, for which there are two main contenders: a van der Waals complex, Mu$\cdots$$Br_2$ (where the dots indicate a very weak bond), and the symmetrical triatomic radical Br—Mu—Br. Van der Waals complexes are well known in other systems, and indeed calculations for H + Br_2 indicate a shallow minimum in the entrance channel of the HBrBr potential energy surface [40]. The reaction scheme is described by Eq. (4.28) and the entrance region of the potential energy surface is represented schematically in Figure 4.18. Stabilization of the van der Waals molecule requires loss of energy by collision with a moderator (M). Tunnelling through the weak barrier would then lead to the expected reaction products.

$$\begin{aligned} H + Br_2 + M &\rightarrow H\cdots Br_2 + M \\ H\cdots Br_2 &\rightarrow HBr + Br \end{aligned} \tag{4.28}$$

The symmetrical radical structure is more unusual, and it is not immediately obvious how the Mu + Br_2 system could evolve into what looks like the transition-state structure of the hypothetical Br + MuBr reaction. Nevertheless, the concept of *vibrational bonding* in heavy–light–heavy triatomic molecules has been discussed theoretically for many decades, first for I–H–I [41] but soon thereafter for the Br–X–Br system (X = Mu, H, D) [42]. More recent, high-level quantum calculations support the notion of van der Waals bonding for HBr_2 and for all its

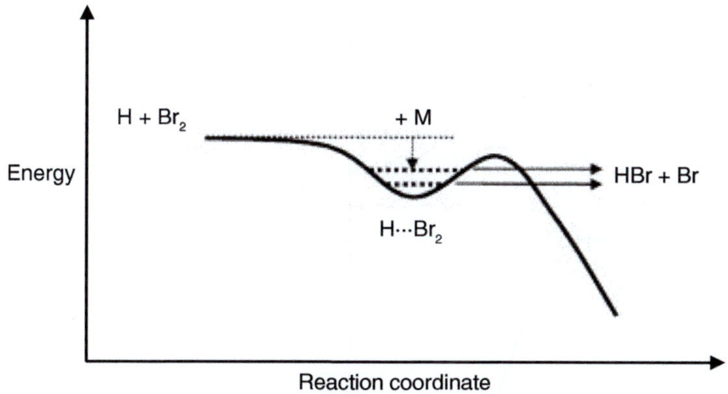

Figure 4.18 Schematic representation of part of the potential energy surface for the reaction of H with Br_2, showing quasi-bound states in the local minimum which corresponds to a van der Waals complex. Decay of the complex to final products occurs via tunnelling. Source: Adapted from a figure by Takayanagi [40] with permission from Elsevier.

heavier isotopomers, but vibrational bonding (the Br–X–Br intermediate) only in the Mu + Br_2 system [43]. Similar results have been found for the H(Mu) + I_2 system [44, 45]. The reason for this *unprecedented* switch in type of chemical bonding can be explained with the help of Figure 4.19. Although the Born–Oppenheimer potential energy surface has a maximum at Br–Mu–Br, the vibrationally adiabatic surface looks quite different. Specifically, the *reduction* in zero-point vibrational energy at the saddle point (compared to MuBr) is larger in magnitude than the potential energy barrier, so that a minimum ensues (the dashed line in Figure 4.19). Thus, calculations show that the probability density for Mu is symmetrically placed between the Br atoms [43]. Heavier isotopes have smaller vibrational contributions and there is no such potential energy minimum.

Regarding a mechanism for formation of Br–Mu–Br, it may be pertinent that calculations reveal two van der Waals structures for H···Br_2, one with a bent geometry at a slightly lower energy than the linear one. If Mu passes through a similar bent structure this might facilitate its insertion between the Br atoms.

Both the van der Waals complex and the proposed Br–Mu–Br molecule are radicals, and this might allow for experimental confirmation of their existence. They are too short-lived for unambiguous detection by transverse-field μSR (Section 2.7), but the longitudinal-field repolarization technique is a viable alternative. As explained for muonium in Section 2.6.2, the shape of a repolarization curve is characteristic of the muon hyperfine constant. In the absence of additional effects (e.g. spin relaxation) the repolarization curve for muonium has an inflexion point at 1585 G. Inspection of Eq. (2.24) shows that radicals with smaller hfc values than muonium should have their inflexion point at lower fields. A preliminary report in 2006 suggested that the radical intermediate in the $MuBr_2$ system has a muon hfc of 1770 MHz [46]. However later work, which included an extended range of magnetic field and a thorough study of systematic effects, concluded that the hfc is

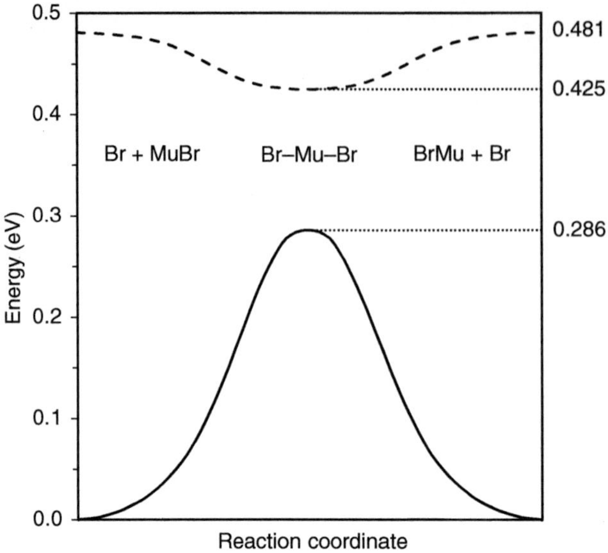

Figure 4.19 Calculated potential energy profile for the formation of Br–Mu–Br (solid curve) and the effect of including zero-point energy (dashed curve) [43].

closer to 3300 MHz [47]. Such a high value, close to muonium itself, suggests that the radical species is more likely to be the linear van der Waals complex rather than the sought-after example of vibrational bonding [48].

References

1 Duchovic, R.J., Wagner, A.F., Turner, R.E. et al. (1991). *J. Chem. Phys.* 94: 2794–2806.

2 Arseneau, D.J., Garner, D.M., Reid, I.D., and Fleming, D.G. (2015). *J. Phys. Chem. A* 119: 7247–7256.

3 Fernández-Ramos, A., Miller, J.A., Klippenstein, S.J., and Truhlar, D.G. (2006). *Chem. Rev.* 106: 4518–4584.

4 Mielke, S.L., Garrett, B.C., Fleming, D.G., and Truhlar, D.G. (2015). *Mol. Phys.* 113: 160–175.

5 Mielke, S.L., Schwenke, D.W., Schatz, G.C. et al. (2009). *J. Phys. Chem. A* 113: 4479–4488.

6 Bondi, D.K., Clary, D.C., Connor, J.N.L. et al. (1982). *J. Chem. Phys.* 76: 4986–4995.

7 Blais, N.C., Truhlar, D.G., and Garrett, B.C. (1983). *J. Chem. Phys.* 78: 2363–2367.

8 Garrett, B.C. and Truhlar, D.G. (1984). *J. Chem. Phys.* 81: 309–317.

9 Schatz, G.C. (1985). *J. Chem. Phys.* 83: 3441–3447.

10 Tsuda, K., Moribayashi, K., and Nakamura, H. (1994). *Chem. Phys. Lett.* 231: 439–443.

11 Jambrina, P.G., Garcia, E., Herrero, V.J. et al. (2011). *J. Chem. Phys.* 135: 034310.

12 Suleimanov, Y.V., de Tudela, R.P., Jambrina, P.G. et al. (2013). *Phys. Chem. Chem. Phys.* 15: 3655–3665.

13 Steinfeld, J.I., Francisco, J.S., and Hase, W.L. (1999). *Chemical Kinetics and Dynamics*. Prentice Hall.

14 Wright, M.R. (2005). *An Introduction to Chemical Kinetics*. Wiley.

15 Pifer, A.E., Bowen, T., and Kendall, K.R. (1976). *Nucl. Inst. Methods* 135: 39–46.

16 Fleming, D.G., Brewer, J.H., Garner, D.M. et al. (1976). *J. Chem. Phys.* 64: 1281–1287.

17 Fleming, D.G., Garner, D.M., Brewer, J.H. et al. (1977). *Chem. Phys. Lett.* 48: 393–398.

18 Garner, D.M., Fleming, D.G., and Brewer, J.H. (1978). *Chem. Phys. Lett.* 55: 163–167.

19 Connor, J.N.L., Jakubetz, W., and Manz, J. (1978). *Chem. Phys.* 28: 219–230.

20 Connor, J.N.L., Jakubetz, W., Manz, J., and Whitehead, J.C. (1980). *J. Chem. Phys.* 72: 6209–6226.

21 Gonzalez, A.C., Reid, I.D., Garner, D.M. et al. (1989). *J. Chem. Phys.* 91: 6164–6176.

22 Tanaka, T. and Takayanagi, T. (2010). *Chem. Phys. Lett.* 496: 248–253.

23 Albright, R.G., Dodonov, A.F., Lavrovskaya, G.K. et al. (1969). *J. Chem. Phys.* 50: 3632–3633.

24 Homann, K.H., Schweinfurth, H., and Warnatz, J. (1977). *Ber. Bunsenges. Phys. Chem.* 81: 724–728.

25 Reid, I.D., Garner, D.M., Lee, L.Y. et al. (1987). *J. Chem. Phys.* 86: 5578–5583.

26 Fleming, D.G., Arseneau, D.J., Sukhorukov, O. et al. (2011). *J. Chem. Phys.* 135: 184310.

27 Snooks, R., Arseneau, D.J., Pan, J.J. et al. (1995). *J. Chem. Phys.* 102: 4860.

28 D.G. Fleming, D.J. Arseneau, S. . Cottrell and J.N.T. Peck, *Phys. Chem. Chem. Phys.*, 22, 6326-6334 (2020).

29 Laude, G., Calderini, D., Welsch, R., and Richardson, J.O. (2020). *Phys. Chem. Chem. Phys.* 22: 16843–16854.

30 Gao, L.G., Fleming, D.G., Truhlar, D.G., and Xu, X. (2021). *J. Phys. Chem. Lett.* 12: 4154–4159.

31 Yang, X. (2007). *Annu. Rev. Phys. Chem.* 58: 433–459.

32 Bakule, P., Sukhorukov, O., Ishida, K. et al. (2015). *J. Phys. B: At. Mol. Phys.* 48: 045204.

33 Bakule, P., Fleming, D.G., Sukhorukov, O. et al. (2012). *J. Phys. Chem. Lett.* 3: 2755–2760.

34 Garner, D.M., Fleming, D.G., Arseneau, D.J. et al. (1990). *J. Chem. Phys.* 93: 1732–1740.

35 Himmer, U., Dilger, H., Roduner, E. et al. (1999). *J. Phys. Chem. A* 103: 2076–2087.

36 Pan, J.J., Arseneau, D.J., Senba, M. et al. (2000). *Phys. Chem. Chem. Phys.* 2: 621–629.

37 Pan, J.J., Arseneau, D.J., Senba, M. et al. (2006). *J. Chem. Phys.* 125: 014307.

38 Bowman, J.M. (1998). *J. Phys. Chem. A* 102: 3006–3017.

39 Wada, Y., Takayanagi, T., Umemoto, H. et al. (1991). *J. Chem. Phys.* 94: 4896–4899.

40 Takayanagi, T. (2007). *Chem. Phys.* 334: 109–116.

41 Manz, J., Meyer, R., Pollak, E., and Römelt, J. (1982). *Chem. Phys. Lett.* 93: 184–187.

42 Clary, D.C. and Connor, J.N.L. (1984). *J. Phys. Chem.* 88: 2758–2764.

43 Fleming, D.G., Manz, J., Sato, K., and Takayanagi, T. (2014). *Angew. Chem. Int. Ed.* 53: 13706–13709.

44 Yoshida, T., Sato, K., and Takayanagi, T. (2015). *Chem. Phys.* 457: 51–56.

45 Manz, J., Sato, K., Takayanagi, T., and Yoshida, T. (2015). *J. Chem. Phys.* 142: 164308.

46 Ghandi, K., Cottrell, S.P., Fleming, D., and Johnson, C. (2006). *Physica B* 374–375: 303–306.

47 Fleming, D.G., Cottrell, S.P., McKenzie, I., and Macrae, R.M. (2012). *Phys. Chem. Chem. Phys.* 14: 10953–10966.

48 Rhodes, C.J. and Macrae, R.M. (2015). *Sci. Prog.* 98: 12–33.

5

Muonium Chemistry and Chemical Kinetics in Condensed Phases

5.1 Setting the Stage: Chemical Reactivity in Liquids vs. Gases

Chapter 4 concerns the kinetics of muonium in gases, where only bimolecular collisions need be considered. In liquids and solids densities are much higher, and reactants interact with the surrounding molecules (usually the solvent) on a timescale typically shorter than that of the reaction. The principles of reaction dynamics – energetics, isotope effects, tunnelling – remain the same in condensed matter, but comparisons between theory and experiment cannot be expected to approach the same level of accuracy as in the gas phase. In particular, in this chapter we need to consider the additional effects of solvent (or lattice) structure and diffusion of the reactants. The frequency of reactant–solvent collisions in liquids means that most reactions occur in thermodynamic equilibrium with their surroundings. State-selected reactions (Section 4.5) are generally not possible, nor do we need to consider pressure-dependent rate constants (Section 4.6). Of course, gases and liquids constitute limiting cases, and at intermediate densities the situation is more complicated. This can be studied in supercritical fluids (Section 5.3).

For gas-phase reactions the mean free path between collisions is large compared to molecular dimensions. Not so in liquids, where the reactants have to squeeze their way through a crowd of solvent molecules before they can react. It is convenient to separate the two processes, diffusion and reaction:

$$\mathrm{A} + \mathrm{B} \underset{k_{-\mathrm{D}}}{\overset{k_\mathrm{D}}{\rightleftharpoons}} \{\mathrm{AB}\} \xrightarrow{k_\mathrm{R}} \text{products}. \tag{5.1}$$

The designation {AB} denotes an encounter pair, which may either react (with first-order rate constant k_R) or separate without reaction. Note that the formation of {AB} is a bimolecular process (second-order rate constant k_D), whereas the reverse process is characterized by a first-order rate constant $k_{-\mathrm{D}}$.

Simple steady-state kinetics analysis shows that the overall rate constant for production of products is given by

$$k_\mathrm{eff} = \frac{k_\mathrm{D} k_\mathrm{R}}{k_{-\mathrm{D}} + k_\mathrm{R}}. \tag{5.2}$$

Muon Spin Spectroscopy: Methods and Applications in Chemistry and Materials Science, First Edition.
Donald G. Fleming, Iain McKenzie, and Paul W. Percival.

In the limit of slow diffusion ($k_{-D} \ll k_R$), $k_{eff} \to k_D$, and the reaction is said to be diffusion-limited. In the other extreme ($k_{-D} \gg k_R$), $k_{eff} \to (k_D k_R / k_{-D})$, which is conveniently replaced by k_{act}, denoting the effective second-order rate constant for an 'activated' reaction, i.e. one whose kinetics are determined by reaction dynamics, as described in Chapter 4.

The full range of behaviour is summed up in the Noyes equation [1]:

$$\frac{1}{k_{eff}} = \frac{1}{k_D} + \frac{1}{k_{act}}. \tag{5.3}$$

The use of inverse rate constants in Eq. (5.3) shows how the effective rate constant depends on either diffusion or activation, but in the limiting cases it is the *slower* process that dominates (the so-called rate-determining step).

In the extreme of fast diffusion the equilibrium on the left of reaction (5.1) maintains a steady-state concentration of encounter pairs, so that $k_{act} = (k_D/k_{-D})k_R = Kk_R$. Thus we see that the rate constant for reaction, k_R, is modified by the dynamics of the encounter pair formation and separation. This emphasizes that the lifetime of the encounter pair is crucial to the competition between reaction and unreactive separation. In some solvents there is a well-defined cage around the reactants, and there may be multiple collisions before eventual reaction or separation.

In the diffusion-limited case the measured rate constant depends only on the encounter process. The rate constant, k_D, can be estimated with the Smoluchowski equation:

$$k_D = 4 \times 10^3 \pi\sigma N_{Av} r_{AB} D_{AB}, \tag{5.4}$$

where σ is a spin statistical factor (equal to 1 unless both reactants have spin), N_{Av} is the Avogadro number, r_{AB} is the sum of reaction radii and $D_{AB} = D_A + D_B$ is the coefficient of mutual diffusion. The individual diffusion constants depend on the hydrodynamic radii as expressed by the Stokes–Einstein equation:

$$D_i = \frac{k_B}{n\pi r_i}\left(\frac{T}{\eta}\right), \tag{5.5}$$

where η is the *solvent* viscosity and r_i is the *solute* radius. The coefficient n has the value 6 for pure hydrodynamic diffusion (a spherical solute moving in a continuous medium) but is closer to 4 when solute and solvent molecules are of similar size.

If $r_A + r_B$ are approximately equal to each other and to the corresponding hydrodynamic radii, then the size dependence of the r_{AB} and D_{AB} terms cancels in Eq. (5.4). The non-intuitive conclusion is that k_D depends only on solvent viscosity and temperature, not on the identity of the reactants. Despite the crude approximations involved, there is experimental support for this prediction – the fastest H atom reactions in solution have rate constants in the range 10^{10}–10^{11} M^{-1} s^{-1}, with an upper limit which depends only on solvent properties.

5.2 Muonium Diffusion in Water

Most data on muonium kinetics in liquids have been obtained by measurements of the decay of the Mu precession signal in low transverse fields, as described in

Section 4.1. The first such experiments employed water as the solvent [2, 3], and this has continued to be the most common medium for condensed-phase kinetics [4, 5]. What is so special about water? Apart from its ubiquity and the relevance to biological processes, there is the pragmatic fact that muonium is long lived in pure water [6]. The only significant impurity is dissolved oxygen, which is relatively easy to remove by sparging (bubbling with nitrogen or other inert gas) or repeated freeze–pump–thaw cycles. In contrast, the Mu lifetime in organic solvents is typically limited by small amounts of reactive impurities, e.g. alkenes in alkanes [7].

One of the first rate constants determined for Mu in water was $k_{\mathrm{Mu}} = 2.4 \times 10^{10}$ $\mathrm{M^{-1}\,s^{-1}}$ for reaction with permanganate [2], assumed to be the redox reaction

$$\mathrm{Mu} + \mathrm{MnO_4^-} \rightarrow \mathrm{MuH_2O^+} + \mathrm{MnO_4^{2-}}. \tag{5.6}$$

This equals the rate constant for the equivalent H atom reaction [8], which is thought to be diffusion-limited. It was originally suggested that the lack of isotope effect is due to the similarity of the H and Mu atomic radii [3], but a more accurate statement should refer to their hydrodynamic radii, which are much larger.

The effective size of Mu in water is related to its hydrophobic nature – it exists in a nano-sized bubble, as do inert gases, molecular hydrogen and H atoms. An early classical molecular dynamics (MD) simulation of H in water revealed transient clathrate-like structures, i.e. the H atom exists in a roughly spherical cage of hydrogen-bonded water molecules [9]. When these MD calculations were extended to include quantum effects it was found that Mu resides in a slightly larger cage consisting of 23 molecules, instead of 18 for H [10]. The difference arises from the mass dependence of the thermal de Broglie wavelength:

$$\lambda_T = h/(2\pi m k_{\mathrm{B}} T)^{1/2}, \tag{5.7}$$

which amounts to 3.0 Å for Mu, compared to only 1.0 Å for the H atom. λ_{T} represents the uncertainty in the localization of a particle of average thermal momentum. Various modern MD calculations [11, 12] agree that the H atom exists in a cavity of about 3 Å in water, as indicated by the solute–water radial distribution functions. Thus, there is room for H, which has a van der Waals radius of 1.0 Å, to 'rattle' around in a solvent cage, but Mu needs more space.

It has been suggested that H is small enough to reside in the natural voids in the solvent, which effectively diffuse as a result of random fluctuations of the hydrogen-bonded water molecules [11]. In this picture the H atom is simply a passenger in a moving bubble. An alternative view is that H remains in a solvent cage until an opening appears in the wall, whereupon it can jump into the neighbouring cavity. This model has much in common with the hopping mechanism for H and Mu diffusion in ice [13, 14], the difference being that ice has a well-defined lattice but liquid water has dynamic clathrate-like cavities which are only occasionally adjacent to each other. Activated hopping depends on the frequency of collisions with the potential barrier and therefore, just like the collision frequency in gas-phase kinetics, depends on $m^{-1/2}$. Quantum effects modify the hop rate in two ways: the activation barrier is affected by zero-point energy, and the swelling effect of the thermal wavelength can interfere with passage through the window from one cavity to the next.

Relative diffusion coefficients have been determined in water for Mu, H and D by measuring their diffusion-limited spin-exchange rate constants with Ni^{2+} [15]. They are in the ratio 1.43 : 1.00 : 0.89, well below the inverse square root mass dependence predicted by the pure hopping mechanism. The inference is that diffusion of these light particles depends on a combination of the mass-independent cavity diffusion and the mass-dependent hopping. This view is supported by ring polymer MD simulations [12].

The concept of a cage or bubble about Mu (and H) in water has also been used to rationalize the temperature dependence of the hyperfine constant of the atom [16, 17]. Detailed discussion of the model is deferred to Section 8.1, which deals with muonium in confined spaces. However, it is relevant to note that, with appropriate modifications, the model has also been applied to the Mu hfc in aqueous solutions of methanol. The concentration dependence of the hfc is consistent with the hydrophobic nature of Mu, which leads to preferential solvation by the methyl group of methanol [18].

The possibility of Mu reacting in a liquid microenvironment has also been considered as an explanation for the remarkable effects of micelles on muonium kinetics. The reaction of Mu with 2-propanol (assumed H abstraction) is very slow in water, somewhat faster in *n*-hexane, but over 10^4 times faster in water containing micelles [19]. The effective rate constant is at the diffusion limit for encounters of Mu with the large volume micelle, so one consideration was that Mu and propanol were mutually confined in the organic environment of the micelle until reaction. However, experiments with ionic solutes, which favour the aqueous phase, showed only small effects consistent with a Mu residence time of the order of 1 ns [20]. This view was later supported by a clever experiment involving competition between Mu reaction with the nitrate ion in the bulk aqueous phase and Mu reaction with benzene or styrene localized in the micelles [21].

5.3 Pressure and Density Dependence

It should be clear from Sections 5.1 and 5.2 that liquid-phase kinetics differs from the behaviour in gases, primarily because of the caging of reactants that leads to multiple collisions per encounter. We now consider what happens under intermediate conditions. It will be helpful to refer to changes on a generic phase diagram (Figure 5.1).

The density change from normal liquid to supercritical fluid has major implications for the clustering of molecules, and the concept of a solvent cage eventually breaks down. At high densities a pair of diffusing reactants may undergo multiple collisions when they encounter, as opposed to isolated collisions in a gas. This is indicated in Figure 5.2a. The lifetime of the solvent cage falls at lower densities, with a consequent drop in the number of collisions per encounter (Figure 5.2b). Not all collisions are reactive, so the change in collision frequency is equivalent to a

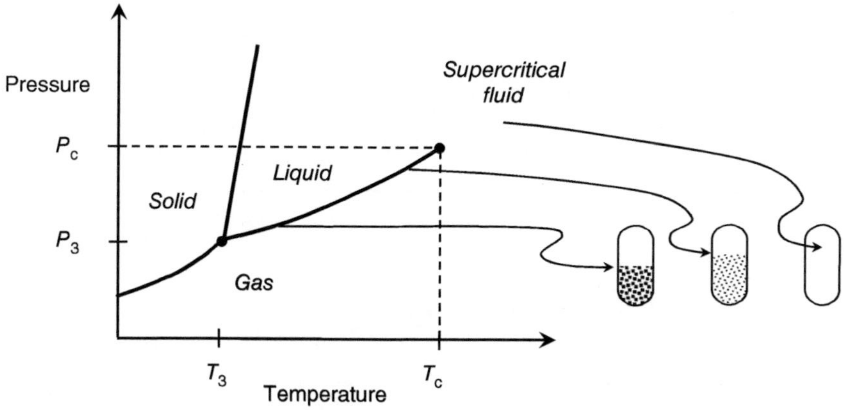

Figure 5.1 A typical single-component phase diagram showing phase stability as a function of temperature and pressure. At the triple point (T_3, P_3) three phases are in equilibrium. The vapour pressure of the liquid increases with temperature along the boiling curve and the density of the liquid falls, as indicated by the images of sealed vessels on the right. At the supercritical point (T_c, P_c) the density of the liquid becomes the same as that of the supernatant vapour. At higher temperatures there is no distinction between the two phases, and the material is said to be a supercritical fluid.

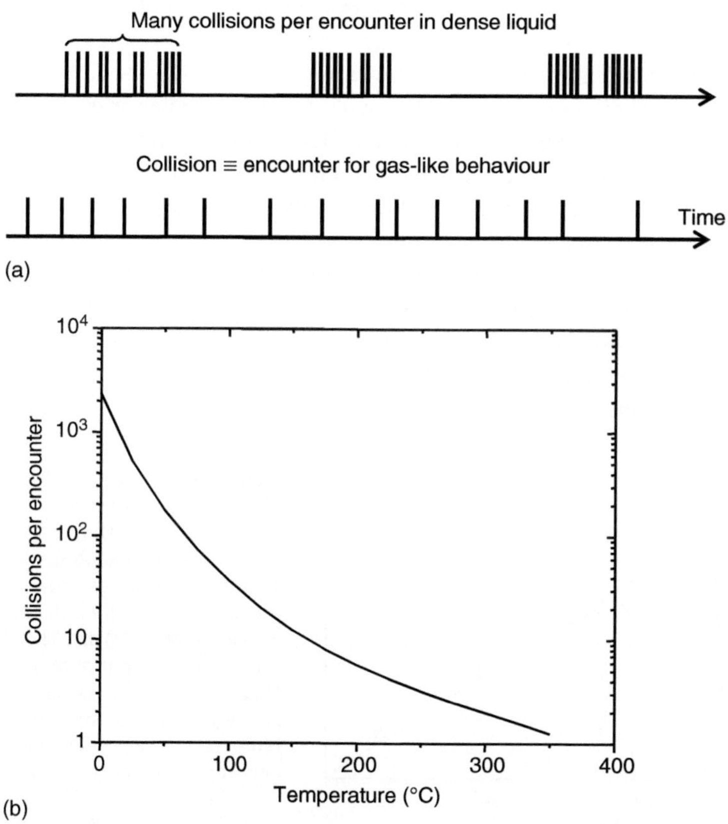

Figure 5.2 (a) A pair of reactants may collide many times in a solvent cage, unlike a gas, where there are single-collision encounters. (b) Estimated number of collisions per encounter calculated for a Mu reaction in water [22].

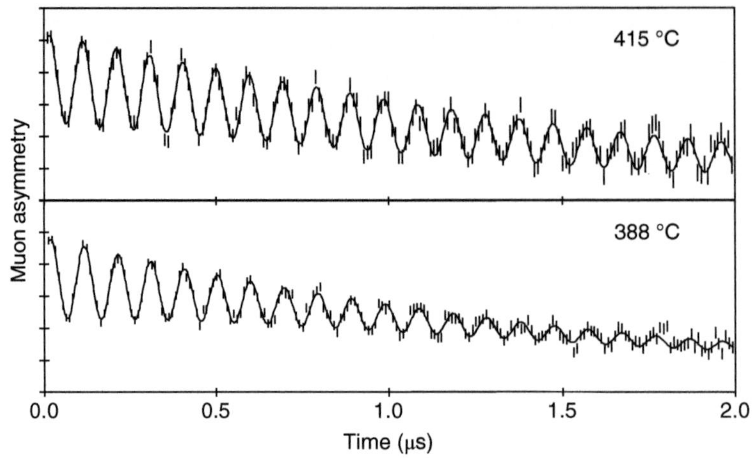

Figure 5.3 Muonium decay for an aqueous solution of 150 mM Ni^{2+} at 388 and 415 °C under ~250 bar pressure [22]. Contrary to expectations, the decay is slower at the higher temperature.

change in reaction efficiency. This applies as well to spin-exchange reactions, where the spin-flip probability depends on the mutual spin states and the applied magnetic field [23].

The effect is evident in Figure 5.3, which shows a lower muonium decay rate at 415 °C than at 388 °C. Similar behaviour has been observed for a wide variety of Mu reactions, both diffusion-limited [22] and activated [24]. At low temperatures rate constants increase with temperature, the usual Arrhenius behaviour, but close to the critical point they turn over. An example is shown in Figure 5.4 for the Mu + OH^- reaction.

Curvature in Arrhenius plots has long been recognized for fast reactions of transients (H, OH, e^-_{aq}) from the radiolysis of water, but in most cases it was explained by the transition from diffusion-limited to activated behaviour [26]. Good fits were obtained to Eq. (5.2) except for some hydrated electron reactions, which exhibited a maximum in their Arrhenius plots, which would imply negative activation energies.

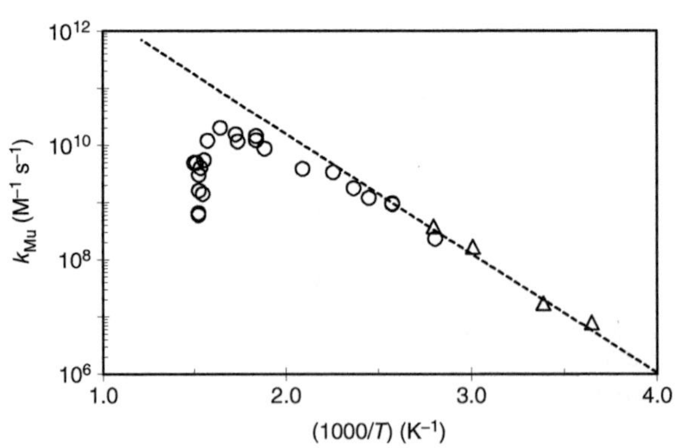

Figure 5.4 Arrhenius plot of rate constants for the reaction of Mu with the hydroxide ion in aqueous solution. The circles show the data of Ghandi et al. [24] measured at ~250 bar. The triangles represent data from Ng et al. [25] determined at ambient pressure. Arrhenius temperature dependence is denoted by the dotted line.

One possible explanation is that the relevant reactions involve multiple steps, possibly even a dielectron intermediate [27]. However, such hypotheses do not apply to muonium reactions, where decay of the Mu precession signal can only apply to the initial reaction step. Furthermore, extension of pulse radiolysis studies to higher temperatures also revealed examples where the rate constants for reactions of H atoms and OH radicals fall at high temperatures [28–30].

The non-Arrhenius kinetics observed for radiolysis reactions in near- and super-critical water can be explained by the multi-collision model [22, 31] developed to explain Mu kinetics. The key aspect of the model is modification of k_{act} in Eq. (5.3) to include an efficiency factor f_R:

$$k_{react} = f_R A \exp(-E_a/RT), \quad f_R = \frac{p_R q}{1 + p_R q}, \tag{5.8}$$

which accounts for both the orientation factor (p_R) and the number of collisions per encounter (q). The efficiency falls from 100% in dense liquids to some smaller value determined by the orientation factor in gases.

This model not only describes the maximum in rate constant in sub-critical liquids, it also describes the recovery and subsequent increase in rate constant as the density falls and the kinetic behaviour develops into that of an activated reaction in the gas phase. Prior to development of this model, rate constants used to model aqueous chemistry in water-cooled nuclear reactors were simply extrapolated from data determined at lower temperatures (mostly below 200 °C). Luckily, this resulted in overestimation of radiolysis, since the opposite error could have had serious consequences for safety in the nuclear industry [31].

The complete temperature dependence, i.e. a maximum followed by a minimum before transition to gas-like behaviour, can be seen in Figure 5.5, which shows

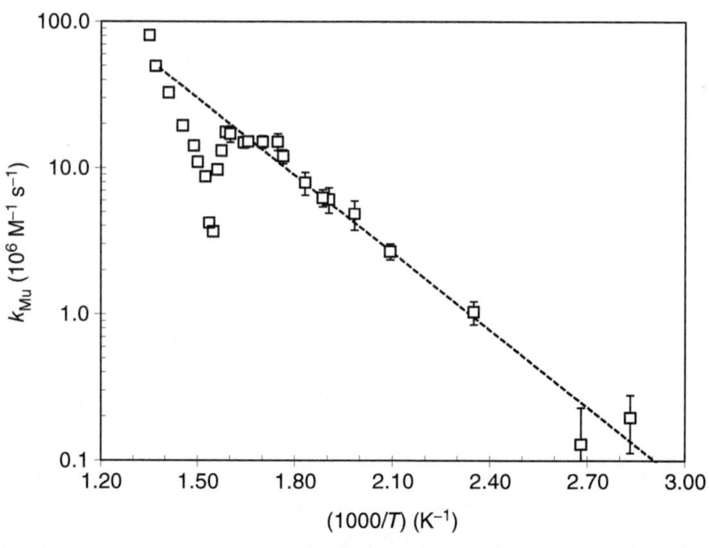

Figure 5.5 Arrhenius plot of rate constants for the reaction of Mu with methanol in aqueous solution [32]. Arrhenius temperature dependence is denoted by the dotted line.

experimental data for the reaction of muonium with methanol. The data for other reactions hint at this behaviour but do not extend to such a high temperature (470 °C).

A bimolecular reaction that does not follow the two-step process described by Eq. (5.1) is that of the H atom with water itself:

$$H + H_2O \rightarrow H_2 + \cdot OH. \tag{5.9}$$

Although this reaction is very slow at room temperature, there has been much discussion about its role in the radiolysis of water at high temperature [33]. It is common practice to add H_2 gas to pressurized-water reactor coolant systems to reduce the steady-state concentrations of O_2 and H_2O_2 in order to reduce corrosion. However, the standard models of water chemistry are unable to predict the correct amount of hydrogen. Alcorn et al. [34] determined rate constants for the reaction of Mu + H_2O up to 450 °C and confirmed that this reaction is a significant source of H_2 at high temperature. Their finding has since been validated by pulse radiolysis studies in the range 250–350 °C [35].

Since there is no diffusion step needed for reaction with the solvent, the model described by Eqs. (5.3) and (5.8) is inappropriate. Nevertheless, a small dip in rate constants close to the critical point was noted by Alcorn et al [34]. More specifically, the dip was found at the temperature of maximum compressibility of the solvent. At this point there would be significant density inhomogeneity, namely the clustering of solvent molecules leaving large voids which would accommodate a hydrophobic solute such as H or Mu. Similar behaviour has also been reported for Mu reactions in near-critical carbon dioxide [36].

5.4 Muon Spin Dephasing During Reaction

The standard method for studying muonium kinetics is to measure the decay of the Mu precession signal in low transverse fields. This relies on dephasing of the ensemble muon precession signal as individual spins suddenly precess at a different frequency in the product. The key factor is the relative magnitudes of the Mu decay rate, λ, and the change in precession frequency, $\Delta\omega$. If $\lambda \ll \Delta\omega$ there is incoherent muon precession in the product. This is the normal condition for direct measurement of Mu decay kinetics (Section 4.1). In the opposite extreme all muons pass to the product before there is significant spin precession, and no Mu is detected. The intermediate situation, when $\lambda \sim \Delta\omega$, provides an alternative means to study muonium kinetics. The essential features are shown in Figure 5.6. Because the muon spins precess for different durations (t_M) before reaction, they end up with different phases, i.e. there is incoherent precession in the product. The magnitude of the muon spin polarization is therefore reduced and the average phase is shifted.

The theory of muon spin dephasing during reaction has been treated by several authors [37–39] and will not be repeated here. Discussion of the formation of muoniated radicals will be delayed until Chapter 6; here, we consider only the reaction of

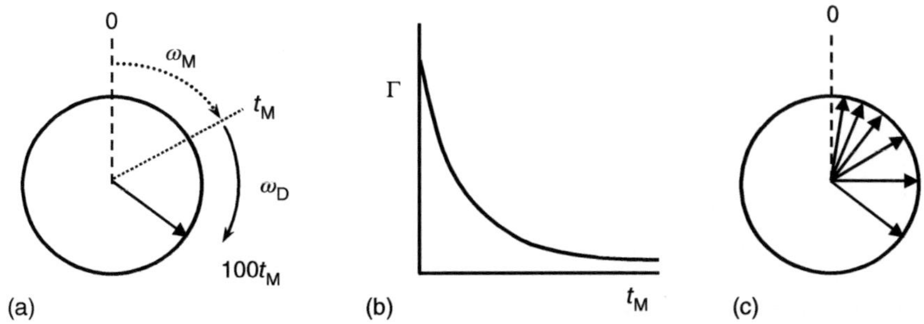

Figure 5.6 (a) A muon spin precesses at the muonium frequency until the reaction time t_M, after which precession occurs at a slower frequency. (b) The reaction probability Γ decays exponentially with t_M. (c) The ensemble of spins in the product precess at the same frequency but with different phases.

muonium to diamagnetic products. The original papers refer to the 'residual' polarization, i.e. the muon polarization precessing at the diamagnetic frequency after muonium has reacted. Both the magnitude ($P_\perp$) and the phase (ϕ) of the precessing signal are affected:

$$P_x(\mathrm{D}) = \frac{1}{2}c^2\left[\frac{1}{1+\Delta_{12}^2}+\frac{1}{1+\Delta_{43}^2}\right]+\frac{1}{2}s^2\left[\frac{1}{1+\Delta_{23}^2}+\frac{1}{1+\Delta_{14}^2}\right],$$

$$P_y(\mathrm{D}) = \frac{1}{2}c^2\left[\frac{\Delta_{12}}{1+\Delta_{12}^2}+\frac{\Delta_{43}}{1+\Delta_{43}^2}\right]+\frac{1}{2}s^2\left[\frac{\Delta_{23}}{1+\Delta_{23}^2}+\frac{\Delta_{14}}{1+\Delta_{14}^2}\right],$$

$$P_\perp = \left[{P_x}^2+{P_y}^2\right]^{1/2}, \quad \phi = \tan^{-1}(P_y/P_x), \tag{5.10}$$

where c and s are the muonium mixing coefficients (Table A.1 in the Appendix) and $\Delta_{nm} = 2\pi(\nu_{nm}+\nu_\mu)/\lambda$. Note that the numerator of Δ_{nm} corresponds to the *difference* in muonium and diamagnetic frequencies, $\Delta\omega$. The positive sign is a consequence of using unsigned values for the electron and muon Larmor frequencies. Figure 5.7 shows how $P_\perp$ varies with reaction rate and magnetic field strength. The plateau evident in the low field curves in Figure 5.7a results from the different magnitudes of λ needed to compete with the low (ν_{12} and ν_{23}) and high (ν_{14} and ν_{43}) muonium precession frequencies.

The principle of this method for studying Mu kinetics was already established by early Soviet researchers, and a zero-field plot equivalent to Figure 5.7a was published by Nosov and Yakovleva in 1963 [40]. The early applications of this method involved determining the Mu rate constant k_{Mu} by measuring P_D for a series of solution samples of varying concentration, i.e. different values of $\lambda = k_{\mathrm{Mu}}[\mathrm{X}]$. Analysis of the experimental results requires knowledge of the prompt fraction h_D (often described as the hot fraction), the diamagnetic muon polarization in the absence of reaction. Introduction of h_D simply results in a constant threshold under the curves shown in Figure 5.7. However, since the prompt fraction does not suffer the phase shift of the reacting muonium fraction, the combined signal has a much reduced value of ϕ, as shown for water ($h_\mathrm{D} = 0.62$) in Figure 5.8.

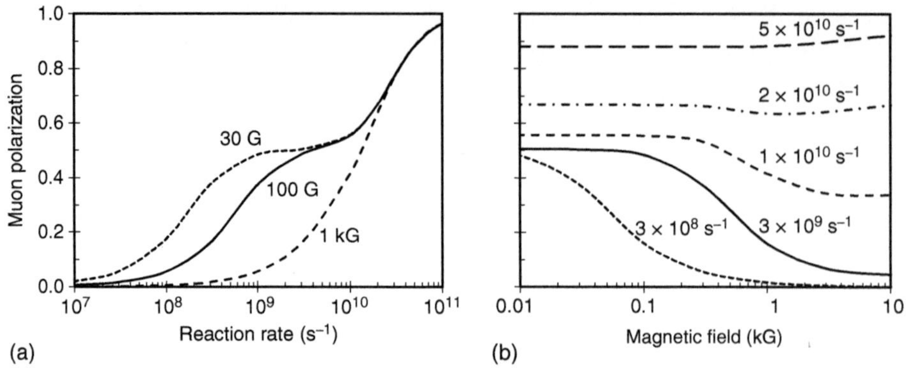

Figure 5.7 Muon spin polarization transferred from muonium to a diamagnetic product with first-order reaction rate λ. The polarization is plotted as a function of (a) reaction rate for selected transverse magnetic fields; (b) magnetic field for various reaction rates

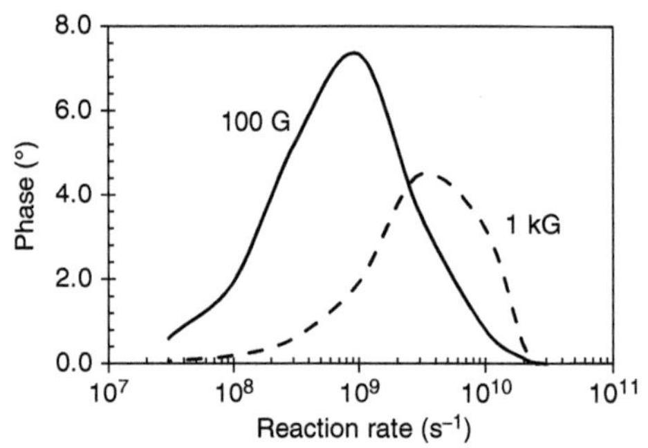

Figure 5.8 The phase shift of the diamagnetic precession signal for muonium reaction in water as a function of first-order reaction rate λ, taking into account a prompt muon fraction $h_D = 0.62$.

Brewer et al. [38, 41] used the plateau in P_D, together with the corresponding phase plot, to differentiate between direct reactions of muonium to diamagnetic product (the so-called 'proper muonium mechanism') and those that involved free radical intermediates. Since this was before the direct detection of Mu and muoniated radicals in liquids, it was a very significant development in the field of muonium chemistry.

An alternative approach is to fit the field dependence (Figure 5.7b). This has the advantage of requiring fewer sample changes, but one must be careful to avoid systematic errors in the measurement of P_D at different fields. In particular, to avoid errors due to field dependence of instrumental asymmetry, signal amplitudes should be compared with those of known standards. Early applications of the field-dependence method [42, 43] gave values for k_{Mu} close to, but somewhat higher than, those determined from Mu decay measurements at low field. This was not surprising, because decay measurements typically only require dilute solutions, while residual polarization requires much higher reactant concentrations, often in the molar range. In the latter case, when a muon stops in the sample there is a significant probability that the ensuing Mu atom is close to a reactant molecule, even forming a *contact pair*. The diffusion-limited rate constant described in Section 5.1 only applies to the limiting behaviour of relaxed reactant concentration. If one reactant (Mu) is suddenly created in the presence of a homogeneous distribution of its reaction partner then the effective bimolecular rate constant is time-dependent [43]:

$$k_M = k_\infty \left(1 + \frac{k_\infty r_{AB}}{k_D \pi (D_{AB} t)^{1/2}}\right), \tag{5.11}$$

where k_∞ is the equilibrium rate constant as measured at long times, and k_D is given by the standard Smoluchowski expression, Eq. (5.4). Although this explanation was invoked in the analysis of residual polarization for the reaction between Mu and thiosulfate in aqueous solution [43], subsequent reanalysis showed that the fitted values of λ are very close to the rate constants determined from Mu decay at low concentrations, provided one allows for a small additional contribution from the reaction of Mu with e^-_{aq} [44].

5.5 Additional Effects on Mu and H Kinetics in the Liquid Phase

There are a large number and variety of muonium reactions whose rate constants have been determined in liquids; a compilation lists over 160 [45]. The corresponding k_{Mu}/k_H isotope effects span many orders of magnitude, from very low (e.g. H abstraction reactions) to very high (e.g. addition reactions). Tunnelling favours muonium, of course, and this is no doubt the reason for the extraordinary isotope effect (at least 75 000) found for the Mu transfer reaction between the muoniated cyclohexadienyl radical and 2,3-dimethylbutadiene [46]. Strictly speaking, however, this is a free radical reaction, so discussion belongs in Chapter 6. Nevertheless, extreme isotope effects have also been reported for Mu itself as reactant, for example in the reactions with oxalic acid (very large) [47] and the formate ion (very small) [48].

Unlike gas-phase reactions, the ratio of observed rate constants for Mu and H in solution does not always lead directly to isotope effects, since other factors can affect the comparison. In particular, reactions of H atoms in aqueous solution have mostly been studied at low pH, because under these conditions (typically 0.1 M $HClO_4$) hydrated electrons produced by radiolysis are converted to H atoms. In contrast, aqueous reactions of muonium have mostly been studied in neutral solution. Comparison under like conditions is essential in situations where pH affects the nature of the reactant.

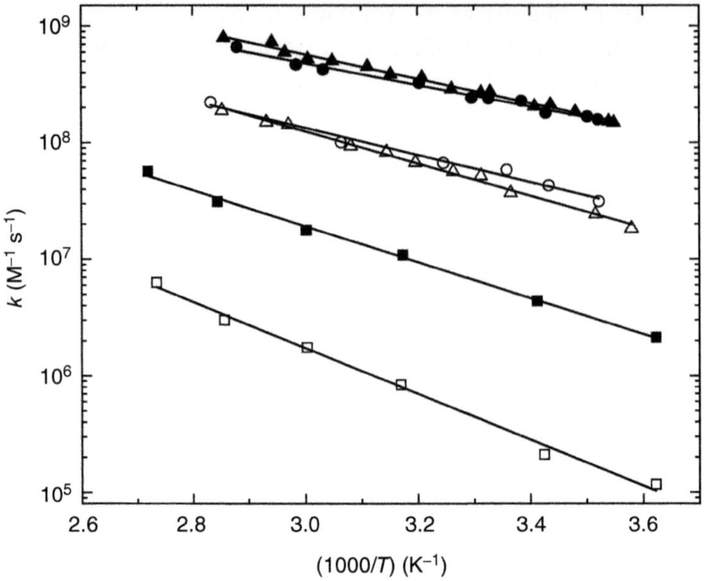

Figure 5.9 Rate constants for the reactions of Mu (squares), H atoms (triangles) and D atoms (circles) with $HCOO^-$ (closed symbols) and $DCOO^-$ (open symbols). Source: Adapted from a figure by Lossack et al. [48], with permission of the Royal Society of Chemistry.

One such example is the reaction of H + formic acid, which is slower by two orders of magnitude than the reaction of H with the formate anion ($HCOO^-$). In both cases the reaction is H abstraction from the carbon atom, but bond resonance in the anion weakens the C—H bond [49]. Careful measurements at pH 7 have provided comparable data on the reactions of Mu, H and D atoms with formate [48]. The rate constants are displayed in Figure 5.9. It is clear that abstraction of H is faster than abstraction of D, and that abstraction by Mu is much slower than H or D atoms, consistent with the isotope effects discussed in Sections 4.2 and 4.4. It is not so simple, however, as calculation of the reaction path with solvent molecules present suggests that the mechanism is closer to an electron transfer followed by proton transfer. This results in a large change of the dipole moment as the reaction proceeds. It is even suggested that the dynamic solvation of the transition state contributes to the kinetic isotope effect, as reorientation of water molecules may be less complete for the light Mu atom.

Another example of pH dependence occurs for reactions with oxalic acid, where the situation is complicated by the existence of three acid–base forms: HOCOCOOH, $HOCOCOO^-$ and $^-OCOCOO^-$. The effective rate constant is a weighted average of rate constants for three reactions (see Table 5.1), as expressed by Eq. (5.12) [47]:

$$k_{obs} = c_1k_1 + c_2k_2 + c_3k_3, \tag{5.12}$$

where c_1, c_2, c_3 are the fractional compositions of the three species, and k_1, k_2, k_3 are the rate constants for the corresponding reactions. The fractional compositions vary with pH, but the values can be calculated from the known acid dissociation

Table 5.1 Rate constants for Mu reaction with oxalic acid and its anions in water.

	Reaction	k ($M^{-1}\,s^{-1}$) at 25 °C
1.	Mu + HOCOCOOH	5.6×10^8
2.	Mu + $HOCOCOO^-$	7.1×10^7
3.	Mu + $^-OCOCOO^-$	5.4×10^6

constants. By fitting k_{obs} determined at four different pH values, Stadlbauer et al. were able to deduce the individual rate constants shown in Table 5.1 [47].

There is little doubt that the reactions involve addition of Mu to the C=O oxygen of one of the carboxylate groups. The decrease in rate constant for the basic species is consistent with loss of π-electron delocalization when Mu adds to C=O. Low activation energies suggest that the reactions are heavily influenced by tunnelling.

5.6 Can Mu React by a Different Mechanism to H?

It is widely claimed that the atomic properties of Mu are so similar to H that they must undergo the same chemical reactions. Let us examine the situation where parallel reactions are possible, taking as example the possible reactions of X (=Mu, H) with a ketone (R = H, alkyl):

$$X + RCH_2C = O \rightarrow XH + R\dot{C}HC = O, \tag{5.13}$$

$$X + RCH_2C = O \rightarrow RCH_2\dot{C} - OX. \tag{5.14}$$

The rate constant for H varies with the nature of R, but not so for Mu [50]. The implication is that while Mu adds to the carbonyl (the expected free radical is produced), H reacts predominantly by H abstraction. This means that the measured isotope effect, k_{Mu}/k_H does not tell us anything about the isotope effects of the individual reactions.

$$\frac{k_{Mu}}{k_H} = \frac{k_{Mu}^{abst.} + k_{Mu}^{addn.}}{k_H^{abst.} + k_H^{addn.}} \approx \frac{k_{Mu}^{addn.}}{k_H^{abst.}}. \tag{5.15}$$

The danger of comparing 'apples' with 'oranges' can be illustrated with the example of pyrazine. The 'anomalous' isotope effect noted for the reactions of Mu and H was initially ascribed to different behaviour of the two isotopes, namely Mu addition to a carbon atom but H addition to N [51] (Figure 5.10). However, calculations show that H addition to C is both thermodynamically and kinetically preferred [52]. A key experimental detail is that the H reaction was studied under acidic conditions, where pyrazine would be protonated at an N atom. Electron capture by the cation would then produce the N-adduct radical.

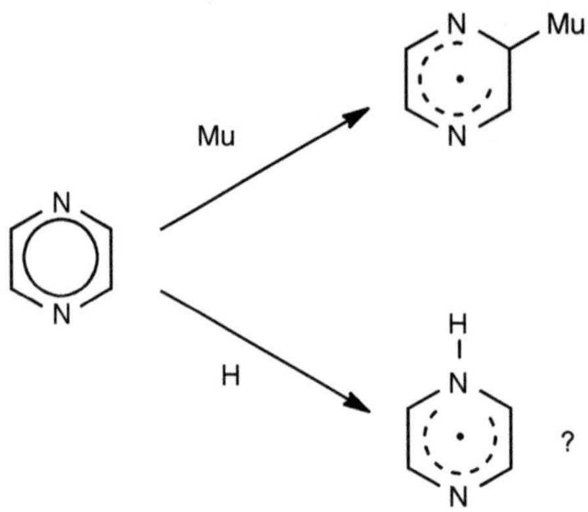

Figure 5.10 Do H and Mu react differently with pyrazine? Or is the N-adduct formed by electron addition to the protonated pyrazine?

5.7 A Case Study of a Complex Reaction System: $Mu + H_2O_2$

The twin complications of pH dependence and parallel reactions are illustrated with the case of the reactions of hydrogen atoms with hydrogen peroxide, important in fields as diverse as combustion (gas phase), radiolysis (aqueous) and astrochemistry. Two distinct reactions have been studied in the gas phase:

$$H + H_2O_2 \rightarrow H_2 + {}^{\cdot}OOH \tag{5.16}$$

and

$$H + H_2O_2 \rightarrow H_2O + {}^{\cdot}OH, \tag{5.17}$$

but models of water radiolysis usually only consider reaction (5.17). The equivalent reactions for Mu are

$$Mu + H_2O_2 \rightarrow MuH + {}^{\cdot}OOH \tag{5.18}$$

and

$$Mu + H_2O_2 \rightarrow MuOH + {}^{\cdot}OH. \tag{5.19}$$

Note that the attack of Mu on an oxygen atom (reaction (5.19)) results in the diamagnetic product MuOH, not the free radical ·OMu erroneously suggested by Brewer et al. [41].

Calculations show that the OH channel (5.17) and (5.19) has a lower reaction barrier and larger exothermicity than H abstraction (5.16) and (5.18) [53]. On the other hand, tunnelling at low temperatures favours H abstraction, and there is a crossover close to room temperature. It is not clear how well these findings carry over to aqueous reactions, since there may be differential solvent effects.

Mu decay measurements have been made for aqueous solutions of both H_2O_2 and D_2O_2 over a range of temperatures and pH (or pD) values [54]. Some selected data are shown in Figure 5.11.

The Mu decay rates clearly depend on pH; the data displayed in Figure 5.11a were chosen to show conditions where the peroxide is in acidic (H_2O_2) or basic (HOO^-)

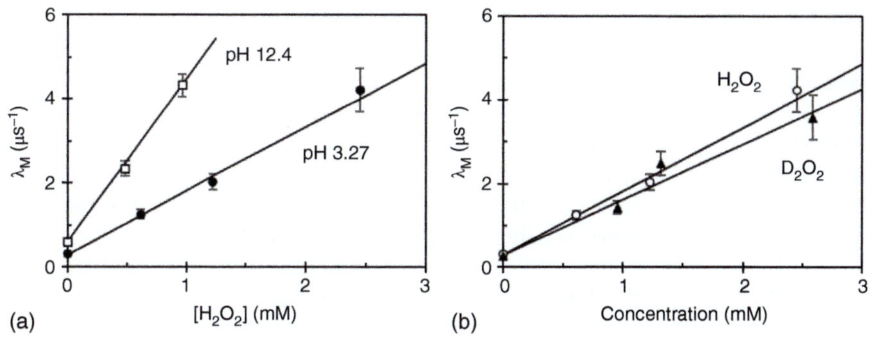

Figure 5.11 Muonium decay rates at 25 °C for aqueous solutions of hydrogen peroxide: (a) H_2O_2 under acidic (pH 3.27) and alkaline (pH 12.4) conditions; (b) H_2O_2 compared to D_2O_2 under neutral or acidic conditions. Source: Data taken from Percival et al. [54].

form. The difference between H_2O_2 and D_2O_2 is not so clear (Figure 5.11b), and since the acid dissociation constants of the two species differ [55], it is necessary to consider all contributions to Mu decay [54, 55]. Thus,

$$\begin{aligned}\lambda_{\text{tot}} &= k_1[\text{H}_2\text{O}_2] + k_2[\text{HOO}^-] + k_3[\text{OH}^-] \\ &= \frac{k_1[\text{H}^+] + k_2K_\text{a}}{K_\text{a} + [\text{H}^+]}[\text{H}_2\text{O}_2]_{\text{total}} + k_3\frac{K_\text{w}}{[\text{H}^+]}\end{aligned} \tag{5.20}$$

where K_a is the acid dissociation constant,

$$K_\text{a} = \frac{[\text{H}^+][\text{HOO}^-]}{[\text{H}_2\text{O}_2]}, \tag{5.21}$$

$K_w = [H^+][OH^-]$ is the ionic product of water, and $[H_2O_2]_{tot} = [H_2O_2] + [HOO^-]$. k_3 is the rate constant for reaction of Mu with OH^-. Although smaller than the rate constants for reactions with peroxide, its effect is evident from the different intercepts in Figure 5.11a. The intercept for the acidic data set is the usual background decay rate due to instrumental effects, λ_0, so that the observable Mu decay rate is $\lambda_{Mu} = \lambda_{tot} + \lambda_0$.

A global fit of Mu decay rates in solutions of H_2O_2 and D_2O_2 at six temperatures and five values of pH (pD) gave the rate constants and activation energies shown in Table 5.2. Various pulse radiolysis studies have found a much lower rate constant for H + H_2O_2: $3.4 \times 10^7\ M^{-1}\ s^{-1}$ at pH1 and 25 °C [56]. Some of these studies utilized product build-up kinetics focused on ·OH production, so there is little doubt that reaction (5.17) dominates under acidic conditions. However, in alkaline solution the reaction is much faster ($1.4 \times 10^9\ M^{-1}\ s^{-1}$) and might well be H abstraction [57]:

$$\text{H} + \text{HO}_2^- \rightarrow \text{H}_2 + \text{O}_2^{\cdot-}. \tag{5.22}$$

The similarity of the rate constants for Mu reaction with acidic and basic forms of H_2O_2 and D_2O_2 suggests that the abstraction reaction is dominant. Furthermore, Mu has lower activation energies than H [57], consistent with increased tunnelling in the Mu reaction.

Table 5.2 Global analysis of muonium decay measurements.

Reaction	k (M^{-1} s^{-1}) at 25 °°C	E_a (kJ mol^{-1})	k_H/k_D
$Mu + H_2O_2$	$1.52\ (4) \times 10^9$	5.8 (1.3)	1.15 (7)
$Mu + D_2O_2$	$1.32\ (4) \times 10^9$	11.1 (2.0)	
$Mu + HOO^-$	$4.30\ (12) \times 10^9$	10.4 (1.2)	0.96 (6)
$Mu + DOO^-$	$4.46\ (16) \times 10^9$	10.6 (1.8)	
$Mu + OH^-$	$1.30\ (4) \times 10^7$	43.7 (3.1)	0.64 (5)
$Mu + OD^-$	$2.03\ (11) \times 10^7$	36.1 (6.0)	

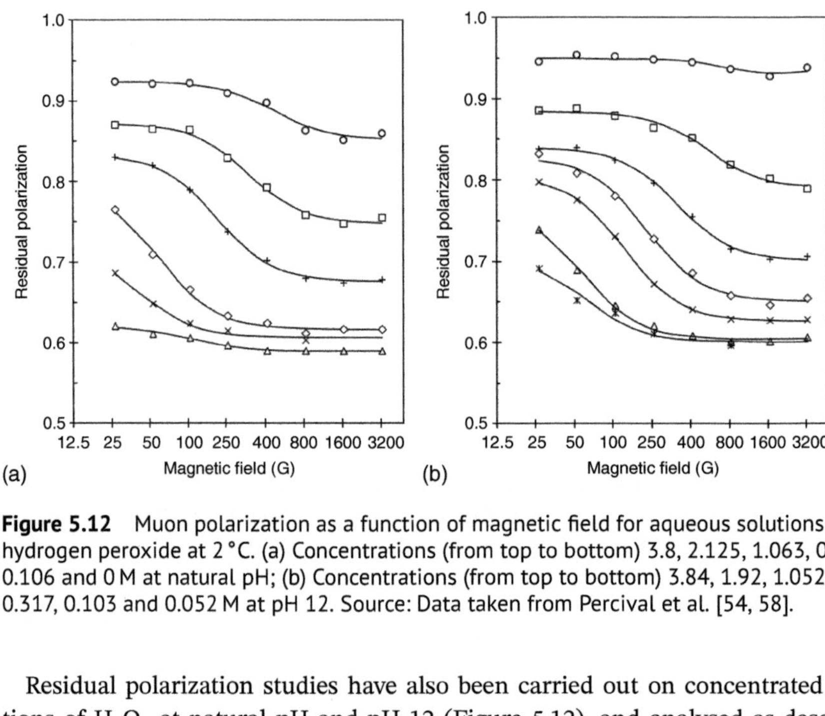

Figure 5.12 Muon polarization as a function of magnetic field for aqueous solutions of hydrogen peroxide at 2 °C. (a) Concentrations (from top to bottom) 3.8, 2.125, 1.063, 0.319, 0.106 and 0 M at natural pH; (b) Concentrations (from top to bottom) 3.84, 1.92, 1.052, 0.317, 0.103 and 0.052 M at pH 12. Source: Data taken from Percival et al. [54, 58].

Residual polarization studies have also been carried out on concentrated solutions of H_2O_2 at natural pH and pH 12 (Figure 5.12), and analysed as described in Section 5.4. The lines through the points represent 3-parameter fits to each concentration: the prompt diamagnetic fraction h_D, the muonium fraction h_M and the muonium decay rate, λ_M. h_D increases with peroxide concentration and h_M falls, consistent with Mu inhibition caused by scavenging of electrons (Section 3.2). The fitted values of λ_M are very close to the rate constants determined from Mu decay at low concentrations (Table 5.2), with a small additional contribution from the reaction of Mu with e^-_{aq} [44].

The residual polarization analysis confirms that muonium reacts with hydrogen peroxide to form diamagnetic products, but it does not help us distinguish between MuH and MuOH. In view of the room-temperature crossover in $H + H_2O_2$ reaction channels calculated by Lu et al. [53], and the additional effect of the solvent, particularly for reaction with HOO^-, it may well be that the dominant reaction channel changes with pH.

References

1 Noyes, R.M. (1961). *Prog. React. Kinet. Mech.* 1: 129–160.
2 Percival, P.W., Roduner, E., Fischer, H. et al. (1977). *Chem. Phys. Lett.* 47: 11–14.
3 Percival, P.W., Roduner, E., and Fischer, H. (1979). *Positronium and Muonium Chemistry*, vol. 175 (ed. H.J. Ache), 335–355. American Chemical Society Advances in Chemistry.
4 Barnabas, M.V. and Walker, D.C. (1991). *Can. J. Chem.* 69: 1252–1258.
5 Walker, D.C. (1998). *J. Chem. Soc., Far. Trans.* 94: 1–9.
6 Nagamine, K., Nishiyama, K., Imazato, J. et al. (1982). *Chem. Phys. Lett.* 87: 186–191.
7 Ito, Y., Miyake, Y., Tabata, Y. et al. (1982). *Chem. Phys. Lett.* 93: 361–365.
8 Buxton, G.V., Greenstock, C.L., Helman, W.P., and Ross, A.B. (1988). *J. Phys. Chem. Ref. Data* 17: 513–886.
9 Tse, J.S. and Klein, M.L. (1983). *J. Phys. Chem.* 87: 5055–5057.
10 Raedt, B.D., Sprik, M., and Klein, M.L. (1984). *J. Chem. Phys.* 80: 5719–5724.
11 Kirchner, B., Stubbs, J., and Marx, D. (2002). *Phys. Rev. Lett.* 89: 215901.
12 Markland, T.E., Habershon, S., and Manolopoulos, D.E. (2008). *J. Chem. Phys.* 128: 194506.
13 Leung, S.K., Brodovitch, J.-C., Newman, K.E., and Percival, P.W. (1987). *Chem. Phys.* 114: 399–409.
14 Bartels, D.M., Han, P., and Percival, P.W. (1992). *Chem. Phys.* 164: 421–437.
15 Walker, A., Mezyk, S.P., Roduner, E., and Bartels, D.M. (2016). *J. Phys. Chem. B* 120: 1771–1779.
16 Roduner, E., Percival, P.W., Han, P., and Bartels, D.M. (1995). *J. Chem. Phys.* 102: 5989–5997.
17 Ghandi, K., Brodovitch, J.-C., Addison-Jones, B. et al. (2000). *Physica B* 289: 476–481.
18 McKenzie, I., Dilger, H., Roduner, E. et al. (2008). *J. Phys. Chem. B* 112: 3070–3076.
19 Venkateswaran, K., Barnabas, M.V., Wu, Z.N. et al. (1988). *Chem. Phys. Lett.* 143: 313–316.
20 Venkateswaran, K., Barnabas, M.V., Ng, B.W., and Walker, D.C. (1988). *Can. J. Chem.* 66: 1979–1983.
21 Stadlbauer, J.M., Venkateswaran, K., Porter, G.B., and Walker, D.C. (1994). *Hyperfine Interactions* 87: 941–946.
22 Ghandi, K., Addison-Jones, B., Brodovitch, J.-C. et al. (2002). *Phys. Chem. Chem. Phys.* 4: 586–595.
23 Senba, M. (1995). *Phys. Rev. A* 52: 4599–4615.
24 Ghandi, K., Addison-Jones, B., Brodovitch, J.-C. et al. (2003). *Physica B* 326: 55–60.
25 Ng, B.W., Stadlbauer, J.M., and Walker, D.C. (1984). *J. Phys. Chem.* 88: 857–860.
26 Elliot, A.J., McCracken, D.R., Buxton, G.V., and Wood, N.D. (1990). *J. Chem. Soc., Far. Trans.* 86: 1539–1547.
27 Christensen, H. and Sehested, K. (1986). *J. Phys. Chem.* 90: 186–190.
28 Marin, T.W., Jonah, C.D., and Bartels, D.M. (2003). *Chem. Phys. Lett.* 373: 228–228.

29 Janik, I., Bartels, D.M., Marin, T., and Jonah, C.D. (2007). *J. Phys. Chem. A* 111: 79–88.
30 Bonin, J., Janik, I., Janik, D., and Bartels, D.M. (2007). *J. Phys. Chem. A* 111: 1869–1878.
31 Ghandi, K. and Percival, P.W. (2003). *J. Phys. Chem. A* 107: 3005–3008.
32 Percival, P.W., Brodovitch, J.-C., Ghandi, K. et al. (2007). *Rad. Phys. Chem.* 76: 1231–1235.
33 Swiatla-Wojcik, D. and Buxton, G.V. (2010). *Rad. Phys. Chem.* 79: 52–56.
34 Alcorn, C.D., Brodovitch, J.-C., Percival, P.W. et al. (2014). *Chem. Phys.* 435: 29–39.
35 Muroya, Y., Yamashita, S., Lertnaisat, P. et al. (2017). *Phys. Chem. Chem. Phys.* 19: 30834–30841.
36 Ghandi, K., McFadden, R.M.L., Cormier, P.J. et al. (2012). *Phys. Chem. Chem. Phys.* 14: 8502–8505.
37 Ivanter, I.G. and Smilga, V.P. (1969). *Soviet Physics JETP* 28: 796–801.
38 Brewer, J.H., Gygax, F.N., and Fleming, D.G. (1973). *Phys. Rev. A* 8: 77–86.
39 Percival, P.W. and Fischer, H. (1976). *Chem. Phys.* 16: 89–99.
40 Nosov, V.G. and Yakovleva, I.V. (1963). *Soviet Physics JETP* 16: 1236–1246.
41 Brewer, J.H., Crowe, K.M., Gygax, F.N. et al. (1974). *Phys. Rev. A* 9: 495–507.
42 Percival, P.W., Roduner, E., and Fischer, H. (1978). *Chem. Phys.* 32: 353–367.
43 Percival, P.W., Brodovitch, J.-C., and Newman, K.E. (1984). *Faraday Discuss. Chem. Soc.* 78: 315–326.
44 Leung, S.K., Brodovitch, J.-C., Percival, P.W. et al. (1988). *Chem. Phys.* 121: 393–403.
45 Karolczak, S. and Walker, D.C.Muonium Data Base. https://muonsources.org/resources/muon-spectroscopy-tables/
46 Roduner, E. and Münger, K. (1984). *Hyperfine Interactions* 19: 793–796.
47 Stadlbauer, J.M., Barnabas, M.V., and Walker, D.C. (1997). *J. Phys. Chem. A* 101: 2442–2444.
48 Lossack, M., Roduner, E., and Bartels, D.M. (2001). *Phys. Chem. Chem. Phys.* 3: 2031–2037.
49 Lossack, M., Bartels, D.M., and Roduner, E. (2001). *Res. Chem. Intermed.* 27: 475–483.
50 Stadlbauer, J.M., Venkateswaran, K., and Walker, D.C. (1997). *Can. J. Chem.* 75: 74–76.
51 Wu, Z., Barnabas, M.V., Stadlbauer, J.M. et al. (1991). *J. Am. Chem. Soc.* 113: 9096–9099.
52 Gelabert, R., Moreno, M., and Lluch, J.M. (1994). *J. Phys. Chem.* 98: 7858–7861.
53 Lu, X., Wang, X., Fu, B., and Zhang, D. (2019). *J. Phys. Chem. A* 123: 3969–3976.
54 P.W. Percival, J.-C. Brodovitch and K.E. Newman, *NBS Special Publication 716. NBS, Washington*, 547–559 (1986).
55 Brodovitch, J.-C., Percival, P.W., and Poteryko, D. (1988). *Can. J. Chem.* 66: 2410–2411.
56 Madden, K.P. and Mezyk, S.P. (2011). *J. Phys. Chem. Ref. Data* 40: 023103.
57 Mezyk, S.P. and Bartels, D.M. (1995). *J. Chem. Soc., Far. Trans.* 91: 3127–3132.
58 Percival, P.W. and Brodovitch, J.-C. (2023). *ChemPhysChem* 24: e202300288.

6

Muoniated Free Radicals

6.1 Isotropic Hyperfine Coupling

The spectroscopic techniques used to characterize muoniated radicals were described in detail in Chapter 2. This chapter is concerned with showing the connection between the experimentally determined parameters and the structure, dynamics and reactions of free radicals. The discussion will be limited to muoniated molecules with only one unpaired electron and has much in common with the interpretation of ESR experiments on free radicals [1, 2]. To compare hyperfine constants (hfcs) of different isotopes which have different gyromagnetic ratios (γ), it is common practice to refer to the reduced coupling constant, the proton-equivalent hfc:

$$A'_{X} = A_{X}(\gamma_{p}/\gamma_{X}). \tag{6.1}$$

In a radical the unpaired electron occupies a molecular orbital that can be delocalized over several atoms. The orbital occupied by the unpaired electron is called a singly occupied molecular orbital (SOMO). The electron density at any point in the molecule is the sum of the density of spin-up, or α, electrons and the density of spin-down, or β, electrons.

$$\rho(x,y,z) = \rho^{\alpha}(x,y,z) + \rho^{\beta}(x,y,z) \tag{6.2}$$

By contrast, the unpaired electron spin density, ρ_S, is the difference between the density of α electrons and the density of β electrons.

$$\rho_S(x,y,z) = \rho^{\alpha}(x,y,z) - \rho^{\beta}(x,y,z) \tag{6.3}$$

In a diamagnetic molecule the total numbers of α electrons and β electrons are the same, so the integrated spin density is zero. In a simple radical there is a single unpaired electron (conventionally labelled α), so integrating ρ_S over all space results in a value of 1.0. However, although there are regions of space in a radical where the local spin density is positive, there can also be regions with $\rho_S < 0$. This depends on the structure of the radical and the distribution of the electrons.

The unpaired electron interacts with any nucleus with a spin of 1/2 or greater. One mechanism is the Fermi contact interaction, which requires unpaired electron spin

Muon Spin Spectroscopy: Methods and Applications in Chemistry and Materials Science, First Edition.
Donald G. Fleming, Iain McKenzie, and Paul W. Percival.

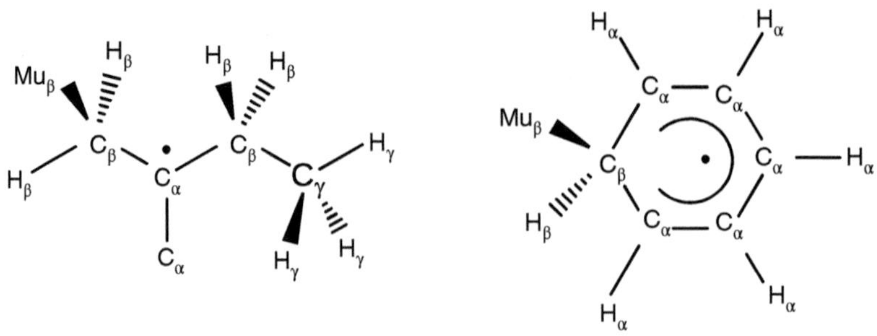

Figure 6.1 (a) The muoniated *sec*-butyl radical is formed by Mu addition to 1-butene. The unpaired electron spin density is localized on one carbon atom, so there is one α carbon. (b) The muoniated cyclohexadienyl radical is formed by Mu addition to benzene. The unpaired electron spin density is delocalized around the ring, so there are multiple α carbons.

density at the nucleus, i.e. $\rho_S(0)$.[1] The strength of the interaction with a nucleus X is the isotropic hyperfine coupling constant, a_X, given (in energy units) by:

$$a_X = \frac{2\mu_0}{3} g_e \beta g_X \beta_X \rho_S^X(0), \tag{6.4}$$

where $\mu_0 = 4\pi \times 10^{-7}$ N A^{-2} is the permeability of free space, g_e is the g-factor of the free electron, $\beta = |e|\hbar/(2m_e)$ is the Bohr magneton, g_X is the nuclear g-factor and $\beta_X = |e|\hbar/(2m_X)$ is the nuclear magneton.

Determining the structure of a radical relies on measuring as many coupling constants as possible in order to map out the distribution of the unpaired electron. Since the isotropic hfc is related to the unpaired electron spin density at a given nucleus, it depends on the geometry of the molecule. Understanding the relationship between hfcs and the geometry of organic radicals is essential to understand the effect of molecular motion on measured hfcs. A convention, borrowed from ESR spectroscopy, is to label atoms with Greek letters according to their distance from the principal unpaired electron spin density. An α atom is one with a 'significant' amount of unpaired electron spin density; an α-muon or proton is directly bonded to the α atom. A β atom is bonded to the α atom and a β-muon or proton is directly bonded to the β atom. Subsequent atoms are labelled γ, δ, ε, etc. Examples of radicals with this labelling scheme are shown in Figure 6.1. Hyperfine coupling is generally only significant at α and β nuclei, so only these classes of nuclei will be considered here.

6.1.1 Isotropic Hyperfine Coupling Constants of α Nuclei

For carbon-centred radicals the hfc of a nucleus X (= Mu, H, D) in the α position depends on spin density in the π orbital on the α-carbon and whether the radical centre is planar or not. If it is planar then X would reside in the nodal plane of the

---

1 It is important to distinguish the spin density at the nucleus, otherwise known as the Fermi contact integral, from the atomic spin population. For example, the H atom has a spin population of 1.0, but the spin density $\rho_S(0) = 1/(\pi a_0^3)$, where a_0 is the Bohr radius.

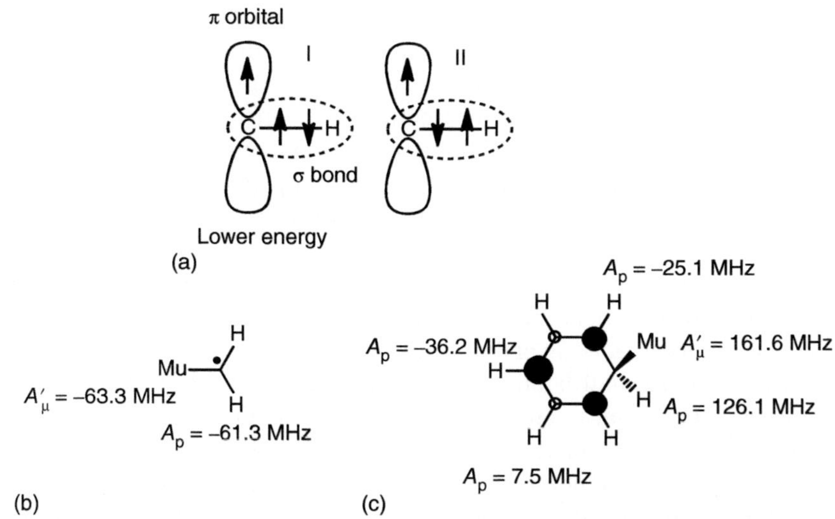

Figure 6.2 (a) Two alternative spin arrangements (I and II) of a C–H fragment with an unpaired electron in a π orbital. (b) Structure of the muoniated methyl radical and measured hyperfine coupling constants [3, 4]. (c) Structure of the muoniated cyclohexadienyl radical and measured hyperfine coupling constants [5, 6]. The circles represent fractional spin populations in the π orbital on each carbon of the six-membered ring. The circle area is proportional to ρ^π and the filled and empty circles correspond to positive and negative spin density, respectively.

SOMO, and as a first approximation there would be no spin density at the α-nucleus, leading to $A_X = 0$. A rationalization of why this is not so is that the unpaired electron in the carbon p_z orbital polarizes the spins of the electron pair making up the C—X σ bond. The origin of this effect is the exchange integral, which in atoms results in parallel spin configurations having lower energy than antiparallel spins. Consider a fragment of a molecule which has the unpaired electron in a p orbital and a C—X σ bond is formed from the overlap of a carbon sp^2 hybrid orbital with the s orbital of X (Figure 6.2a). In arrangement I the spins of the electrons more closely associated with the carbon atom are parallel, while in arrangement II the spins are antiparallel. The exchange interaction reduces the energy of I compared with arrangement II. This slight preference for I results in *negative* spin density at X and a *negative* A_X. The reduced hfc is given by

$$A'_X = Q\rho^\pi_\alpha, \tag{6.5}$$

where Q is a proportionality constant that is approximately −65 MHz and ρ^π_α is the spin population in the π orbital on the adjacent carbon atom [7].

The muoniated methyl radical, CH_2Mu (Figure 6.2b), is one of the simplest muoniated radicals to have been studied [3, 4]. The radical is planar and the unpaired electron is in a $2p_z$ orbital. The measured muon hfc is −201.37(4) MHz at 152 K, giving a reduced muon hfc of −63.26(2) MHz, which can be compared to the proton hfc for the same radical at this temperature: −61.26(4) MHz. The slight difference between A'_μ and A_p is due to the lighter mass of Mu and will be discussed in Section 6.3.1.

The muoniated cyclohexadienyl radical (Figure 6.2c) has a SOMO that is a π orbital distributed over the five sp^2-hybridized carbons of the six-membered ring. The same mechanism of spin polarization as that introduced for the α-protons in CH_2Mu is effective for C_6H_6Mu. The distribution of the unpaired electron around the ring of the radical can be estimated from the proton hfcs using Eq. (6.5). Setting Q to −72 MHz gives the corresponding values of π spin population: 0.512, −0.105 and 0.350. Taking account of degeneracy, this give the expected total spin population of 1.00.

Increasing the temperature will result in the population of excited vibrational states. A prominent vibrational mode at the α-carbon is the out-of-plane bend, which moves the substituents out of the nodal plane and into direct overlap with the SOMO. This results in a contribution of positive spin density. Thus, A_X will become less negative with increasing temperature and could eventually become positive.

The direct overlap of the α nucleus with the SOMO is important for understanding the hyperfine parameters of non-planar radicals. The hfc of the α-substituent will depend on the spin density at the radical centre as well as the out-of-plane angle, and there will be a delicate balance between spin polarization and hyperconjugation. The hfcs can change substantially with temperature due to the thermal population of vibrational states. The Mu adducts of stable singlet carbenes have very non-planar radical centres, and these have been found to have positive A_μ values [8].

Equation (6.5) is also valid for a few other nuclei, in particular ^{14}N, which has spin 1. For example, the N-muono-pyridinyl radical is one of three radicals formed by Mu addition to the nitrogen atom of pyridine [9]. This radical has a planar structure and the small A'_μ value of 1.8 MHz means that less than 3% of the unpaired spin population is located on the nitrogen.

In general it is found that hfcs depend not only on the local spin population (ρ_α) but also on those of neighbouring atoms (ρ_β). Thus, ^{13}C hfcs determined for C_6H_6Mu [10] and $C_{60}Mu$ [11] have been analysed using Eq. (6.6):

$$A_C = Q_\alpha \rho_\alpha + Q_\beta \sum_\beta \rho_\beta, \tag{6.6}$$

a simplified form of the Karplus–Fraenkel equation [12].

6.1.2 Isotropic Hyperfine Coupling Constants of β Nuclei

The vast majority of muoniated radicals ever studied have the muon in the β position with respect to the unpaired electron. The reduced hfc of a nucleus X in the β position obeys the empirical McConnell relation:

$$A'_X = (L + M\langle\cos^2\theta\rangle)\rho_S^\pi, \tag{6.7}$$

where L and M are constants, θ is the angle between the C—C bond and the axis of the singly occupied molecular orbital (SOMO) and ρ_S^π is the spin population in the π orbital on the α-carbon [13]. The angled brackets indicate this is the average over the internal rotation about the C—C bond. L accounts for the effect of spin polarization

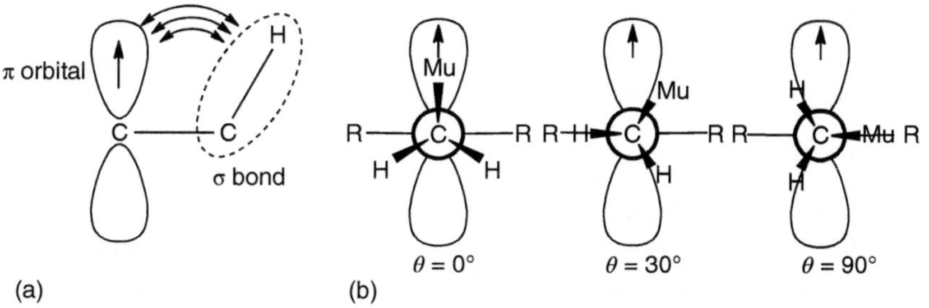

Figure 6.3 (a) Schematic of a C–C–H fragment with an unpaired electron in a π orbital. The arrows indicate hyperconjugation between the π orbital and the σ bond. (b) Newman projections of a group for different angles (θ) between the C–Mu bond and the axis of the SOMO, which in this case is depicted as a π orbital. Rotation away from $\theta = 0$ decreases the extent of hyperconjugation.

described in the previous section, and M arises from hyperconjugation (or direct overlap) between the p_z-orbital containing the unpaired electron and the C—X bond (Figure 6.3a). Values of L and M for Mu in the muoniated *tert*-butyl radical were found to be ~−22 and +86 MHz, respectively [14]. The values of L and M will be slightly different for different isotopes due to the isotope effect on the bond lengths (discussed in Section 6.2.1).

The hfc depends on the dihedral angle between the π orbital and the C—X bond direction. Consider the static cases shown in Figure 6.3b. When θ is 0° the C—Mu bond is aligned along the axis of the π orbital; this geometry results in the greatest overlap between the π orbital and the σ bond and the greatest amount of hyperconjugation. For a system with $\rho^{\pi} = 1$ the value of A'_X is then $L + M$. Rotation of the methyl group to $\theta = 30°$ results in a decrease in the overlap and therefore a smaller hfc; the value of A'_X is reduced to $L + 0.75M$. Further rotation of the group to $\theta = 90°$ results in the C—Mu bond lying in the nodal plane of the π orbital, so that there is no hyperconjugation and only spin polarization remains. A'_X then has a value of L.

In the high-temperature limit, where the system has an energy far above the potential barriers that hinder internal rotation, all angles θ contribute equally to A'_X, so that A'_X approaches $L + M/2$ (assuming that $\rho^{\pi}_S = 1$). This is one way that the spin density on an α carbon can be determined.

6.2 Isotope Effects on Muoniated Radicals

6.2.1 Bond Length

Within the Born–Oppenheimer approximation, the potential energy curve corresponding to the stretching of a C—X bond is independent of the isotope. However, isotopic substitution can affect the *vibrationally averaged* bond length. This arises because the potential energy curve for bond stretching is generally anharmonic.

To illustrate this we assume that the potential energy curve can be modelled by a Morse potential (Eq. 6.8) and that only the ground vibrational state is significantly occupied, e.g. low-to-moderate temperatures and a large vibrational frequency.

$$V(x) = D_e\{1 - \exp[-ax]\}^2, \tag{6.8}$$

where $x = (r - r_{eq})$ is the deviation from the potential minimum and $a = (k_e/2D_e)^{1/2}$. D_e is the dissociation energy of the C—X bond measured from the minimum of the potential energy curve, and k_e is the bond force constant, equal to the curvature of the potential at $x = 0$. Since D_e and k_e are properties of the potential they are isotope-independent. The energy of the ground vibrational state, i.e. the zero-point energy (ZPE), is given by

$$\text{ZPE} = \frac{1}{2}hc\overline{\nu} - \frac{[hc\overline{\nu}]^2}{16D_e}. \tag{6.9}$$

The vibrational frequency in wavenumbers, $\overline{\nu}$, is given by

$$\overline{\nu} = \frac{1}{2\pi c}\left(\frac{k_e}{\mu_R}\right)^{1/2} = \frac{a}{2\pi c}\left(\frac{2D_e}{\mu_R}\right)^{1/2}, \tag{6.10}$$

where c is the speed of light in cm s^{-1} and μ_R is the reduced mass of the C—X fragment.

Typical values for a C—H bond are $\overline{\nu} = 3000\ \text{cm}^{-1}$, $D_e = 408$ kJ mol^{-1} and $r_{eq} = 1.09$ Å. These result in a ZPE for the C—H bond of 17.8 kJ mol^{-1}. The lower mass of Mu results in a lower value of μ_R, hence a larger vibrational frequency (8598 cm^{-1}) and a larger ZPE of 49.8 kJ mol^{-1} (Figure 6.4).

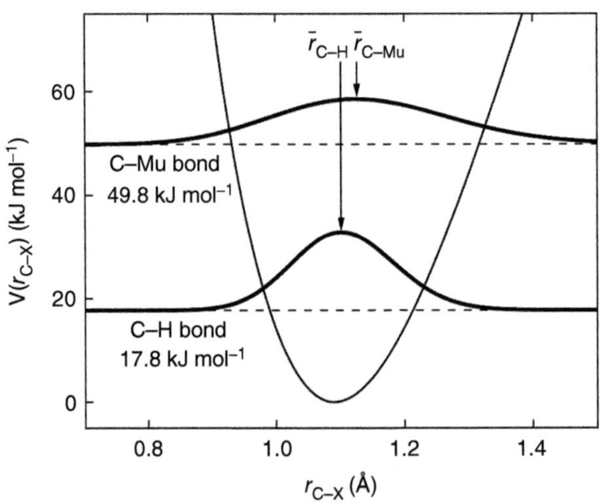

Figure 6.4 The zero-point energies (dashed lines) and probability distributions (ψ^2) (bold curves) of the ground vibration states of C—H and C—Mu in an anharmonic potential corresponding to a typical C—H bond stretching vibration. Because the probability distributions are skewed to longer separations of the nuclei, the vibrationally-averaged bond length for C—Mu is slightly longer than that of C—H.

The larger ZPE of the C—Mu bond combined with the asymmetry of the stretching potential has three notable effects:

- The C—Mu bond is weaker than the corresponding C—H bond; a consequence of the larger ZPE.
- The vibrationally averaged C—Mu bond is longer than the corresponding C—H bond. This was first estimated for a C—X fragment using a Morse potential with realistic parameters, the result being that the C—Mu bond length should be 4.9% longer than the corresponding C—H [15, 16].
- The amplitude of the stretching vibration is larger in the C—Mu bond compared with the corresponding C—H bond. This is relevant to the hyperfine isotope effect discussed next.

6.2.2 Hyperfine Constants

There is a non-linear relationship between the hfc of X and the C—X nuclear separation, as indicated by the lower panel of Figure 6.5. The increase of A'_X with C—X separation is clear when one recognizes that the limit of this bond stretching corresponds to an isolated X atom. Both the larger vibrationally averaged C—Mu bond (for the anharmonic case) and the larger amplitude vibration result in A'_μ being larger than A_p. The magnitude of the isotope effect depends crucially on the structure of the radical and can be temperature dependent. For example, in the C_6H_6Mu radical [6], A'_μ is 20% larger than A_p for the equivalent proton in C_6H_7, while for $C_{60}Mu$ the effect is 12% [17]. Nor is the isotope effect limited to stretching vibrations.

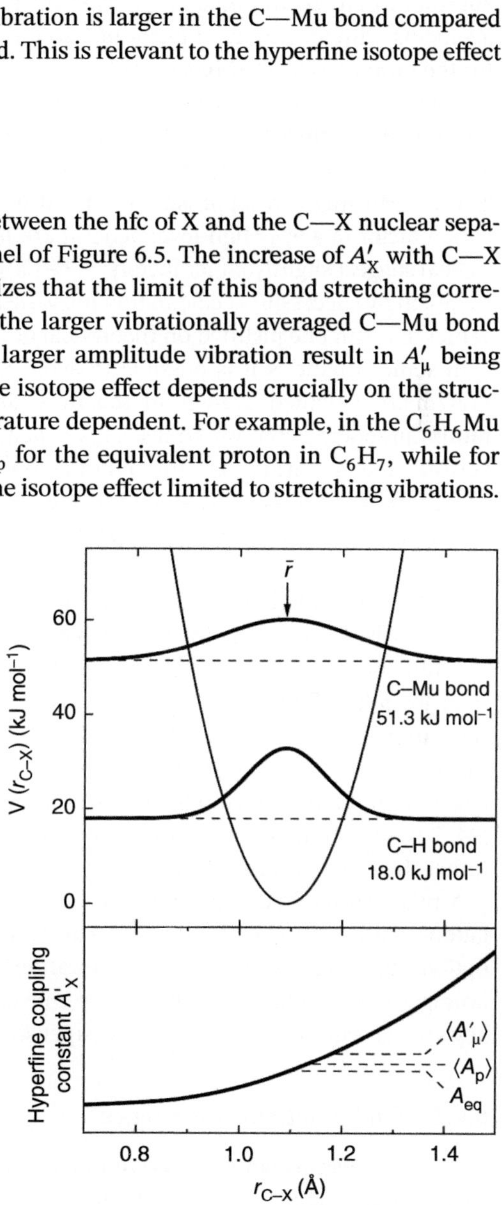

Figure 6.5 Variation of hyperfine coupling constant with nuclear separation for the C–X bond (X = Mu, H). The larger extent of motion for C–Mu results in a larger value for the vibrationally averaged hfc, $\langle A'_\mu \rangle$, than $\langle A_p \rangle$ which is also larger than A_{eq}, the value at the potential minimum. This applies both to a harmonic potential (shown here) and to the anharmonic case (Figure 6.4).

Internal rotations are often important (Section 6.2.3) as are other bending motions, including inversion at the radical centre (Section 6.3.1); in the CH_2Mu radical A'_μ is only 3% larger than A_p [4].

Although not commonly implemented, quantum calculations on muoniated radicals need to include the effects of vibrational averaging in order to accurately calculate the hfcs. Sometimes it is sufficient to apply the isotope effect known for a closely related structure. This strategy has been widely followed for muoniated radicals related to cyclohexadienyl. For example, calculations on radicals formed by Mu addition to the polyaromatic hydrocarbon fluoranthene accounted for vibrational averaging with an empirical factor of 1.20 for A'_μ and a corresponding factor of 0.94 for the proton at the CHMu site [18].

Another strategy is to optimize the structure of the radical and then simulate the isotopic perturbation by increasing the length of the bond involving Mu. Other geometric parameters may then be optimized while freezing the bond to Mu [19]. The C—Mu fragment approach discussed in Section 6.2.1 predicted a bond elongation factor of 4.9%. High-level benchmark investigations on the muoniated ethyl radical suggest slightly larger factors: 5.7% [20] and 7.6% [21]. This method has been surprisingly successful at calculating hfc values, usually within 10% of experimental values, but can't be justified on theoretical grounds [22].

In some situations it is possible to avoid full vibrational averaging by taking a Boltzmann-weighted average of the hyperfine constants corresponding to local minimum-energy conformations. This is justifiable if the radical oscillates rapidly between the conformations and only the lowest vibrational level is occupied. This procedure was followed by Roduner and Reid to model flipping between CHMu conformations in muoniated cyclopentyl [16]. This work included bond elongation to distinguish C—Mu from C—H. In contrast, McKenzie et al. [23] distinguished the three conformations of CH_2Mu in the 4-muonomethyl-oxetan-2-on-4-yl radical on the basis of the zero-point energy in each conformation.

A full, ab initio, approach has been applied in a few cases, but this is laborious, even for relatively small radicals [22, 24, 25]. However, these studies established some important principles. First, vibrational modes involving Mu are usually so high that a local mode approximation can be used [24]. Second, although bond stretching has the dominant effect, motion in the other two dimensions should also be included [24, 25]. Finally, it is not so much the change in bond length that is important but rather the extent of the motion [22].

A practical method applicable even to large radicals is to carry out DFT calculations with commercial software. The Gaussian package of programs includes options to calculate anharmonic vibrational frequencies and apply vibrational averaging to structures and hyperfine constants [26]. Unfortunately, the relevant keyword (Fermi) is computationally expensive and poorly documented at present.

6.2.3 Conformational Preference

It has long been recognized that muoniated radicals favour certain conformations. At first this was ascribed to the effective size of Mu in a molecule, due to the enhanced vibrational amplitudes of C—Mu modes compared to C—H. However, it is now widely accepted that this is a consequence of hyperconjugation: the longer, weaker C—Mu bond facilitates interaction with the p_z orbital on C [27]. Thus, for intramolecular rotation of a β-CH_2Mu group the eclipsed conformation ($\theta = 0°$ in Figure 6.3b) has the lowest energy. Other types of radical geometry can show similar conformational effects, for example the preference for axial C—Mu orientation over equatorial in non-planar cycloalkyl radicals, such as cyclopentyl [16]. However, the discussion here will focus on the internal rotation of methyl groups, illustrated by the muoniated *tert*-butyl radical, which contains both CH_2Mu and CH_3 groups [14].

The potential energy for CH_3 rotation cycles every 60°:

$$V(\theta) = \frac{1}{2}V_6[1 - \cos(6\theta)], \tag{6.11}$$

where the peak-to-peak amplitude V_6 corresponds to the energy difference between the eclipsed ($\theta = 0°$) and staggered ($\theta = \pm 30°$) conformations. This high degree of symmetry results in a constant value for the *average* proton hfc. From Eq. (6.7) and noting that the average value of $\cos^2\theta = 0.5$, we find that the proton hfc has the value $L + M/2$, the free rotation limit mentioned in Section 6.1.2. In contrast, the torsional potential for CH_2Mu,

$$V(\theta) = \frac{1}{2}V_2(1 - \cos(2\theta)) + \frac{1}{2}V_6(1 - \cos(6\theta)), \tag{6.12}$$

has global minima at $\theta = 0°$ and 180° in addition to local minima where C—H is in the eclipsed orientation, as indicated in Figure 6.6.

In the thermodynamically preferred conformation (i.e. at very low temperature) C—Mu is aligned along the axis of the SOMO ($\theta = 0°$ and 180°) and the muon hfc is

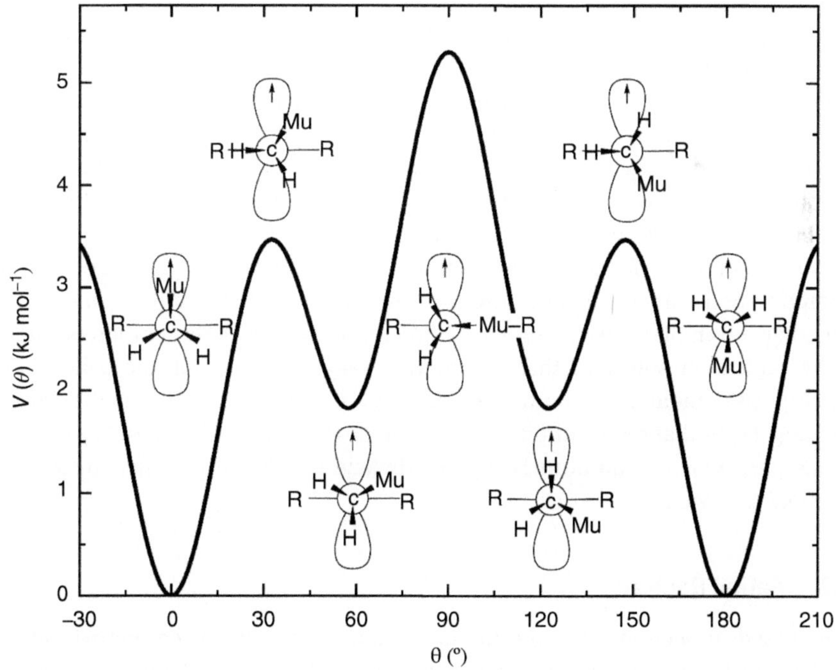

Figure 6.6 Rotational potential for the CH_2Mu group calculated using Eq. (6.12).

a maximum, equal to $L+M$. Since the two C—H bonds are then at an angle of 60° to the SOMO axis, the corresponding proton hfc equals $L+M/4$. At higher temperatures the higher energy conformers can be populated, resulting in lower muon and higher proton hfcs, i.e. *opposite* behaviour.

Further discussion of the temperature dependence is deferred to Section 6.3.

6.3 Intramolecular Motion of Muoniated Radicals

6.3.1 The Muoniated *tert*-Butyl Radical

Although the eclipsed conformation of the muoniated *tert*-butyl radical is thermodynamically preferred, the barrier to rotation is low and higher energy conformations are increasingly accessed at higher temperatures. The quantum-mechanical description is that the population of higher energy torsional states increases with temperature. These states extend over a wider range of rotation angles until, in the high-temperature limit, there is effectively free rotation. Since β-hyperfine constants depend on average dihedral angles (Section 6.1.2), they are inevitably temperature dependent. Moreover, the effect is different for muon and proton hfcs, as shown in Figure 6.7. Although the muon hfc decreases with temperature, the proton hfc for the CH_2Mu group *increases*. In contrast, there is little to no temperature dependence for the protons of the CH_3 groups. This is also true for the average reduced hfc, defined as

$$\overline{A}(CH_2Mu) = \left(A'_\mu + 2A_p\right)/3, \tag{6.13}$$

Note that it is larger than $A(CH_3)$, consistent with the isotope effect on the bond length (Section 6.2.1).

The full temperature dependence can be modelled by taking a Boltzmann-weighted average over torsional energy states. Percival et al. [14] followed a procedure developed by Ramos et al. [29] for the ethyl radical, but modified to use the potential described by Eq. (6.12). Fleming et al. [28] used only the first (V_2) term, but found similar values for the torsional barrier, ~2 kJ mol^{-1} for liquid-phase data and ~3 kJ mol^{-1} for the frozen solution. Such analyses are predicated on the assumption of a planar radical centre, and that the torsional motion is uncoupled from other intramolecular vibrations. Although the average configuration is planar for the radical in solution, that assumption may not hold for the solid state. A slightly pyramidal configuration could explain the jump in A'_μ in the frozen solution [14]. Similar discontinuities have been observed for cis- and trans-sec-butyl radicals [30], where muonium is in a methylene (–CHMu–) position instead of a terminal CH_2Mu group.

6.3.2 Methyl Radicals

As discussed in Section 6.1.1, the hfc of a muon or proton in the methyl radical is a delicate balance between two effects: spin polarization of the C—H bond,

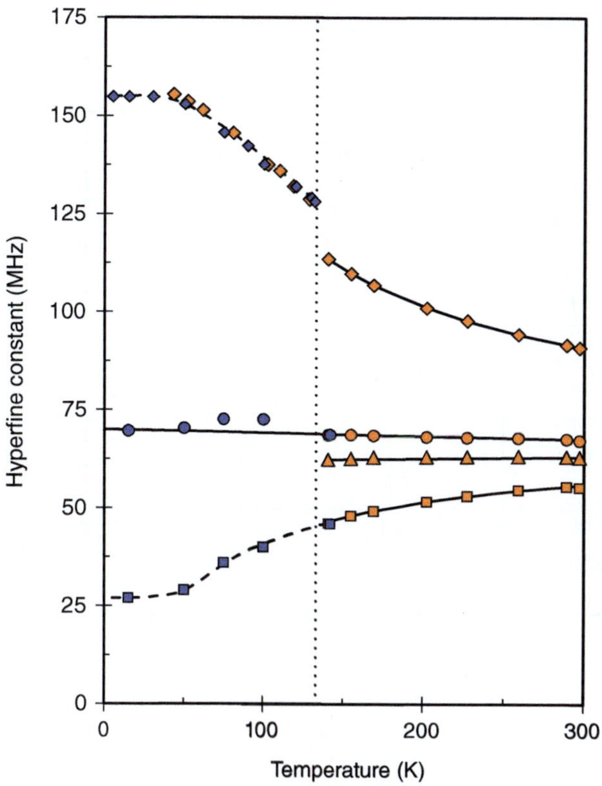

Figure 6.7 Hyperfine coupling constants of $(CH_3)_2C^{\cdot}-CH_2Mu$ in isobutene. Reduced muon hfc values are denoted by diamonds, protons in CH_3 by triangles, protons in CH_2Mu by squares. The circles represent the average hfc for CH_2Mu, defined by Eq. (6.13). The solid lines correspond to fits for the liquid solution data and the dashed lines for solid isobutene. The melting point of isobutene is shown by the dotted vertical line. Experimental data taken from Percival et al. [14] (orange symbols) and Fleming et al. [28] (blue symbols).

which contributes negative spin density, and direct overlap with the SOMO, which contributes positive spin density. The negative sign of the hfc indicates that spin polarization dominates, but the balance of effects means that the proton hfc is very sensitive to the configuration at the radical centre. Out-of-plane bending brings the substituents out of the nodal plane of the SOMO, increasing the positive contribution of spin density and resulting in a smaller magnitude (less negative) coupling constant. The magnitude of the hfc depends on the amplitude of out-of-plane bending, which in turn depends on the ZPE (Figure 6.8a). The out-of-plane vibrational frequency ($\overline{\nu}$) is 606 cm^{-1} for CH_3 and 945 cm^{-1} for CH_2Mu. The larger amplitude motion results in more contributions from positive spin density, and this results in a decrease in $|A'_X|$. It was found that $|A'_X|$ is proportional to $\overline{\nu}$ for methyl radical isotopomers (Figure 6.8c). The lines for Mu, H and D are offset from each other in keeping with the different extents of vibrational motion and different bond lengths, as described in Section 6.2.2.

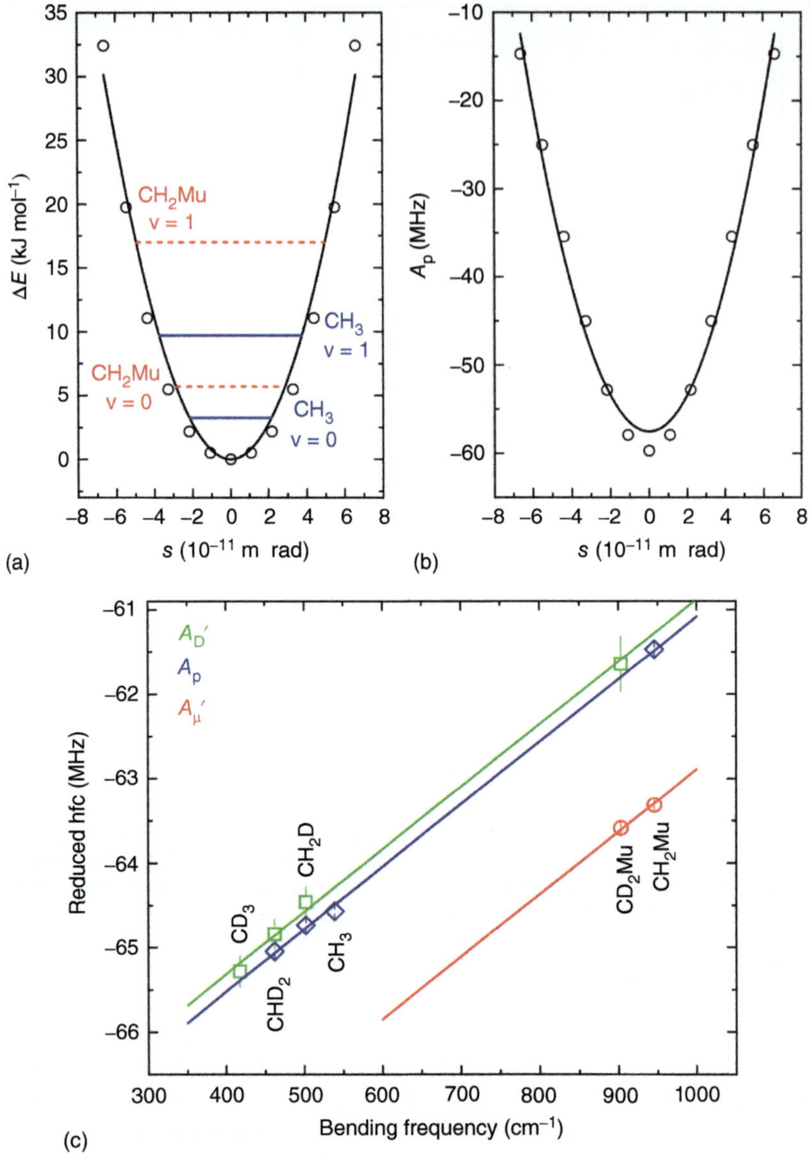

Figure 6.8 Computed dependence of (a) energy; (b) proton hyperfine coupling constant on the out-of-plane displacement s for CH_3 and CH_2Mu. The horizontal lines show the ground vibrational state ($v = 0$) and first excited vibrational state ($v = 1$). (c) Reduced hyperfine coupling constants of the muon (red), proton (blue) and deuteron (green) plotted versus the calculated harmonic out-of-plane bending frequencies for methyl radical isotopomers under the stated conditions: CD_3 (methane, 96 K), CHD_2 (krypton, 86 K), CH_2D (krypton, 86 K), CH_3 (methane, 96 K), CD_2Mu (liquid ketene-d2; Mu hfc at 180 K, D hfc averaged over liquid range) and CH_2Mu (liquid ketene, 184 K). Source: Adapted from figures in McKenzie et al. [4], with permission of the American Chemical Society. Note: The out-of-plane displacement is defined as $s = r(\theta_1 + \theta_2 + \theta_3)/\sqrt{3}$, where r is the C—H bond length and each θ refers to the angle between the bond and the plane defined by the three H atoms. [4].

6.3.3 Other Alkyl Radicals

Similar to the data for *tert*-butyl shown in Figure 6.7, muon and proton hfcs have been measured for other muoniated alkyl radicals, including ethyl [29, 31–33], isopropyl [34] and sec-butyl [30]. Unlike *tert*-butyl these radicals all have α-protons, and like CH_2Mu the proton hfc is almost temperature-independent, with a value close to −62 MHz, consistent with a localized radical centre. Analysis of the muon and β-proton hfcs shows that in all cases the torsional barrier for CH_2Mu rotation is ~3 kJ mol^{-1}. Slightly smaller barriers were reported for CH_2Mu rotation in various chloroalkyl radicals [35]. Not surprisingly the barrier for CH_2Cl rotation is considerably higher (12 kJ mol^{-1} [36]), since, like Mu, Cl has a thermodynamic preference for the eclipsed ($\theta = 0°$) conformation.

When Mu is attached to a ring carbon of a cycloalkyl radical the resulting CHMu group cannot rotate. If the ring is relatively rigid, as in cyclohexadienyl, there are concerted wagging motions of the C—Mu and C—H bonds [6]. A different type of motion arises when the ring is floppy, as in cyclopentyl, C_5H_8Mu. The C_5 ring is non-planar (puckered) and rapidly inverts between two twisted forms. In one conformation the C—Mu bond is more axial and in the other it lies closer to the equator. Analysis of the temperature-dependent hfcs revealed that the more axial C—Mu conformation is 1.2 kJ mol^{-1} lower in energy than the other [16].

6.3.4 Mu Adducts of Carbonyls

Mu adds to the oxygen of a carbonyl to give a carbon-centred radical with an OMu (or muoxy) substituent. These radicals typically have low muon hfcs, e.g. 25 MHz for the 2-muoxy-2-propyl radical in acetone at 25 °C [37]. This is because conformations with the O—Mu bond lying in the nodal plane of the SOMO (Figure 6.9a) have lower energy than those with the O—Mu bond eclipsing the SOMO (Figure 6.9b). The low muon hfc is consistent with restricted librational motion about the minimum-energy geometry (Figure 6.9c). As was explained for CH_2Mu rotation (Section 6.3.1), such

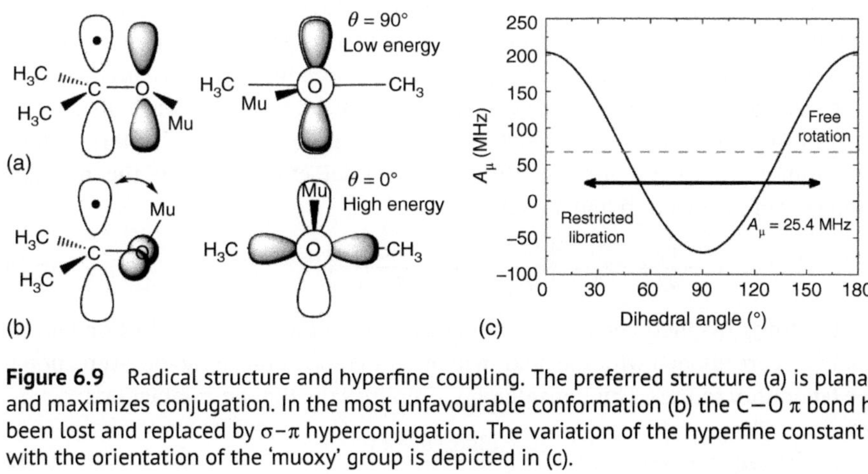

Figure 6.9 Radical structure and hyperfine coupling. The preferred structure (a) is planar and maximizes conjugation. In the most unfavourable conformation (b) the C—O π bond has been lost and replaced by σ–π hyperconjugation. The variation of the hyperfine constant with the orientation of the 'muoxy' group is depicted in (c).

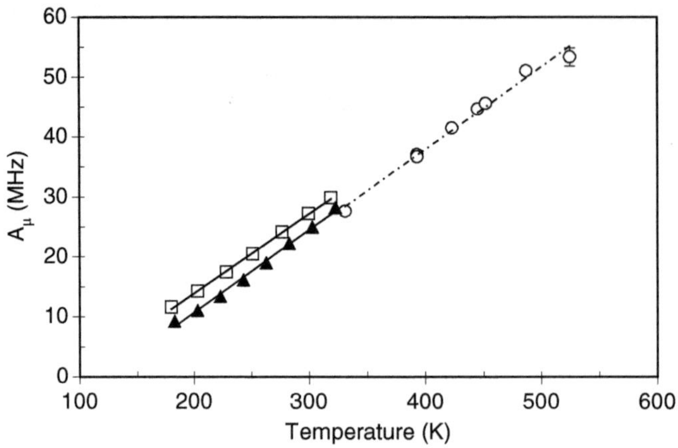

Figure 6.10 Muon hyperfine constants for the muoxy-prop-2-yl radical in pure acetone (squares, Buttar et al. [38]); a 20:1 mixture of acetone:water (filled triangles, Buttar et al. [38]); and hot pressurized water (circles, Ghandi et al. [39]).

motion results in temperature dependence of A_μ, as shown in Figure 6.10. There is also a solvent effect, as hydrogen-bonding to the OMu group provides a drag on the rotation, resulting in an elevated torsional barrier [38].

The muon hfc depends on the substituents at the radical centre. It has been shown that A_μ is proportional to the Raman carbonyl bond stretching frequency of the parent compound, this being a proxy for the strength of the C—O π bond in the radical [40].

6.4 Reorientational Dynamics of Muoniated Radicals

6.4.1 Dipolar Hyperfine Coupling Constants

In addition to the isotropic or Fermi contact interaction, there is a dipole–dipole interaction between the magnetic moments of the electron and nucleus. This gives rise to the anisotropic or dipolar hyperfine coupling. This interaction is time-averaged to zero in low viscosity liquids but is important for radicals in solids or anisotropic environments, which will be covered in detail in Chapters 8 and 9. The spin Hamiltonian for the dipole-dipole interaction between the magnetic moments of the electron and a nucleus X is

$$\hat{H}_{\text{dipolar}} = -\frac{\mu_0}{4\pi} g_e \beta g_X \beta_X \left[\frac{I \cdot S}{r^3} - \frac{3(I \cdot r)(S \cdot r)}{r^5} \right] \tag{6.14}$$

In a strong magnetic field, the magnetic dipole moments of the electron and the nucleus X are aligned with the field (which is by convention along the z-direction in the laboratory frame). This considerably simplifies the dipolar interaction:

$$\hat{H}_{\text{dipolar}} = -\frac{\mu_0}{4\pi} g_e \beta g_X \beta_X M_S M_I \frac{(3\cos^2\theta - 1)}{r^3}. \tag{6.15}$$

M_S and M_I are the spin quantum numbers of the unpaired electron and the nucleus, respectively. The interaction of two classical magnetic point-dipole moments which are aligned in a sufficiently high external magnetic field is rotationally symmetric about the field axis and scales as $(3\cos^2\theta - 1)/r^3$. This interaction disappears for $\theta = 54.7°$ (the *magic angle*). For angles less than 54.7° the interaction is positive, and for angles greater than 54.7° it is negative. Since the unpaired electron is not a point-dipole but is distributed over an entire atomic or molecular orbital, it is necessary to integrate the interaction (Eq. 6.15) over the entire orbital (i.e. over θ and r).

The dipolar hyperfine interaction is expressed in the spin Hamiltonian as a tensor linking the electron and nuclear spin operators. It can be written as a 3 × 3 matrix whose components describe an ellipsoid whose dimensions correspond to the principal components B_X^{xx}, B_X^{yy} and B_X^{zz}. The dipolar tensor is traceless, so

$$B_X^{xx} + B_X^{yy} + B_X^{zz} = 0. \tag{6.16}$$

The coordinate system that results in a diagonal hyperfine tensor will be different for each nucleus X in the radical and depends on the spin density distribution. We will refer to this as the nuclear coordinate system for a given nucleus.

The sign and magnitude of B_X^{xx}, B_X^{yy} and B_X^{zz} can be predicted by judging whether the unpaired electron is mostly in the positive or the negative sector of a double cone with opening angle of 54.7°. A schematic of an α nucleus in a radical with the unpaired electron in a p-orbital is shown in Figure 6.11. The z axis of the nuclear coordinate system is defined to be parallel with the axis of the SOMO and the x axis is parallel with the C—Mu bond. When the magnetic field is along the nuclear z axis, the SOMO is partly in the positive sector and partly in the negative sector of the double cone. The positive and negative contributions cancel out and leave $B_X^{zz} \sim 0$. When the magnetic field is along the nuclear y axis, the SOMO is entirely in the negative sector of the double cone, which results in B_X^{yy} being negative. When the magnetic field is along the nuclear x axis, the SOMO is almost entirely in the positive sector of the double cone, resulting in B_X^{xx} being positive. The values of B_μ^{xx}, B_μ^{yy} and B_μ^{zz} in the CH_2Mu radical obtained from a DFT calculation (UB3LYP/6-311++G(2df,p)) are +124.05, −126.43 and +2.38 MHz, respectively. The dipolar component along the

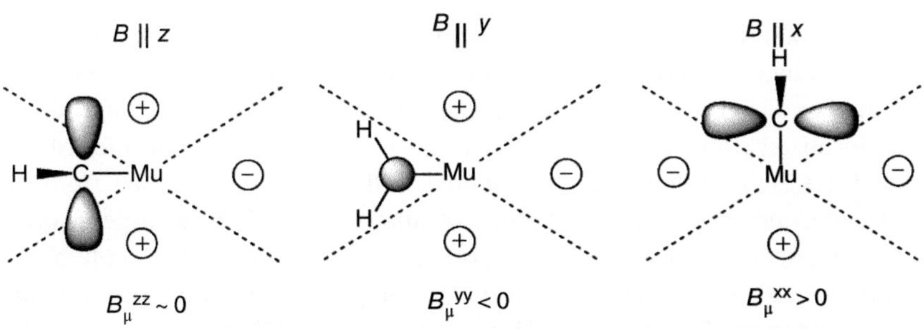

Figure 6.11 Sign of $(3\cos^2\theta - 1)$ for an α nucleus for different orientations of a radical with respect to the magnetic field. The direction of the C—Mu bond defines the x-axis; the z-axis is along the symmetry axis of the SOMO; and the y-axis is in the plane of the molecule.

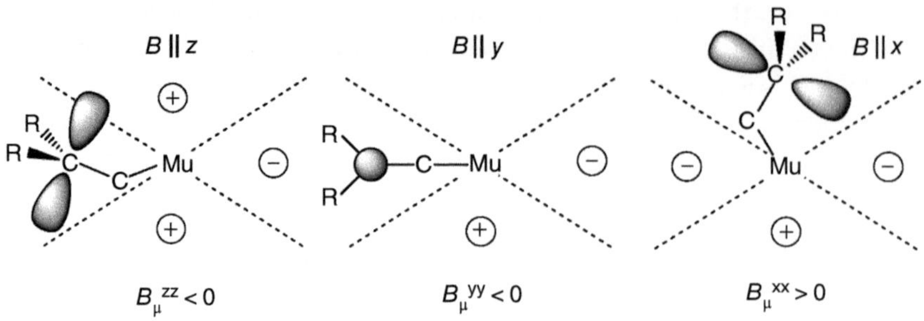

Figure 6.12 Sign of $(3\cos^2\theta - 1)$ for a β nucleus for different orientations of a radical with respect to the magnetic field.

C—Mu bond is positive and approximately 50% of the isotropic muon hfc. This has not been measured experimentally for muoniated radicals, but the dipolar α-proton hfcs of the malonyl radical, $\cdot CH(COOH)_2$, have been determined in a single crystal of irradiated malonic acid and found to be $B_p^{xx} \approx +31 \pm 2$ MHz, $B_p^{yy} \approx -31 \pm 2$ MHz and $B_p^{zz} \approx 0 \pm 2$ MHz [41]. The corresponding values for a muoniated radical would be $+99 \pm 6$ MHz, -99 ± 6 MHz and 0 ± 6 MHz, respectively.

A schematic of a β nucleus in a radical with the unpaired electron in a p-orbital is shown in Figure 6.12. In this case we will define the x axis of the nuclear coordinate system as connecting the carbon of the radical centre to Mu and the z axis lying in the plane defined by Mu–C–C. When the magnetic field is along the nuclear z axis, the SOMO is mostly in the negative sector of the double cone, which results in B_X^{zz} being negative. When the magnetic field is along the nuclear y axis, the SOMO is entirely in the negative sector of the double cone, which results in B_X^{yy} being negative. When the magnetic field is along the nuclear x axis, the SOMO is almost entirely in the positive sector of the double cone, resulting in B_X^{xx} being positive. The values of B_μ^{xx}, B_μ^{yy} and B_μ^{zz} in the muoniated ethyl radical obtained from a DFT calculation (UB3LYP/6-311++G(2df,p)) are +29.35, −17.56 and −11.79 MHz, respectively. The dipolar couplings of β-nuclei are smaller than those of α-nuclei due to the larger distance between the nucleus and the unpaired electron and the r^{-3} dependence. The hyperfine tensor of β-nuclei in methyl groups are often close to axial, with $B_X^{yy} \approx B_X^{zz} \approx -\frac{1}{2}B_X^{xx}$.

Derivatives of the muoniated cyclohexadienyl radical are some of the most commonly studied radicals. The muon and methylene proton are in a β position, but one has to account for the spin density distributed around the ring, rather than the localized spin density considered above. The dipolar muon hfcs of the muoniated 2,3,5,6-tetramethylcyclohexadienyl radical were determined from an orientation-dependent single-crystal study of durene [42]. There is large positive principal component B_μ^{xx} of 8.15 MHz that is approximately along the C—Mu bond, at an angle of 25° with respect to the molecular plane, a negative out-of-plane principal component B_μ^{zz} of −9.05 MHz at an angle of 75° with respect to the molecular plane, and a small in-plane principal component B_μ^{yy} of +0.9 MHz.

Motion of the radical will average the dipolar hyperfine interaction. This effect is easiest to visualize by considering the hyperfine ellipsoid and a rotation axis that

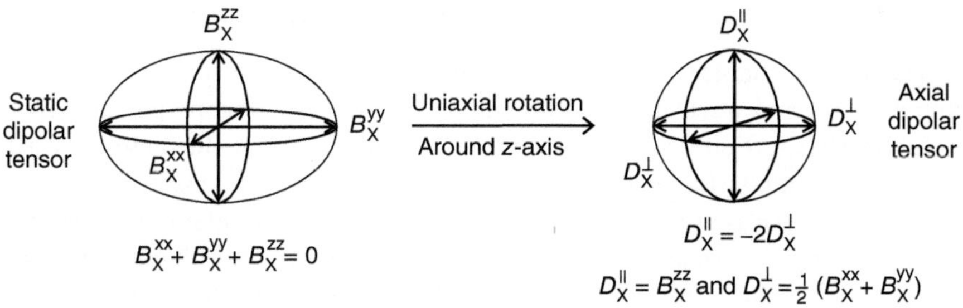

Figure 6.13 Averaging of the principal components of the dipolar hyperfine tensor of nucleus X by fast uniaxial rotation around the *z*-axis.

coincides with one of the principal components of the nuclear coordinate system (Figure 6.13). The component along the axis rotation of the original tensor remains invariant (i.e. not influenced by the averaging process). It becomes the parallel component of the new (axial) hyperfine tensor, which is given the symbol $D_X^{\parallel}$. The two components perpendicular to the rotation axis are averaged and are given the symbol $D_X^{\perp}$. The requirement that the dipolar tensor is traceless results in

$$D_X^{\parallel} = -2D_X^{\perp}. \tag{6.17}$$

Thus, in the case of an axially symmetric hyperfine tensor, its value can be reported just by giving either the $D_X^{\parallel}$ or $D_X^{\perp}$ values.

The situation is the same if the rotation axis does not coincide with one of the principal axes; it just makes the mathematics more difficult. For rapid rotation about an arbitrary axis,

$$\left(A_X^{iso} + D_X^{\parallel}\right)^2 = \boldsymbol{r}^T \cdot \boldsymbol{A}^T \cdot \boldsymbol{A} \cdot \boldsymbol{r}, \tag{6.18}$$

where $\boldsymbol{A}$ is the hyperfine tensor, $\boldsymbol{A}^T$ is its transpose, $\boldsymbol{r}$ is the normalized rotation vector and $\boldsymbol{r}^T$ its transpose [2].

Additional motion on top of uniaxial rotation will further average the dipolar hyperfine couplings. A frequently used approximation is the restricted random walk model where the rotation axis wobbles randomly within a cone of half-angle θ_C [43]. The effective dipolar coupling constant, $\left\langle D_\mu^{\parallel}\right\rangle$, becomes [44]

$$\left\langle D_\mu^{\parallel}\right\rangle = D_\mu^{\parallel}\left(\frac{\cos\theta_C + \cos^2\theta_C}{2}\right). \tag{6.19}$$

Rapid isotropic motion, which corresponds to $\theta_C = 90°$ in Eq. (6.19) or averaging over rotation axes covering the full 2π solid angle, leads to dipolar hfcs with a value of zero. This averaging depends on the timescale of the motion, and this is related to the values of the anisotropic muon hfcs. Perhaps the most interesting case is when the muoniated radical is undergoing anisotropic motion. In that case, the motion partially averages the dipolar hyperfine interaction and leads to dipolar hfcs with non-zero values. By measuring the dipolar hfcs one can then determine the extent of the anisotropic motion of the radical.

6.4.2 Effect of Hyperfine Anisotropy on Δ_1 and Δ_0 Resonances

As described in Chapter 2, Δ_1 resonances arise from mixing of spin states with the same electron and nuclear spin orientations but different muon spin directions. These spin states are only mixed in the presence of anisotropy, so the presence of a Δ_1 resonance can be considered to be diagnostic of a frozen state or anisotropic motion. The Δ_1 resonance is suitable for studying anisotropic reorientational motion of radicals in single crystals as well as in polycrystalline or amorphous states [45]. The lineshape of the Δ_1 resonance is extremely sensitive to and characteristic of the type of reorientational motion that the muoniated radical is undergoing.

Analytical expressions are only available for radicals with an axially symmetric dipolar hyperfine tensor transforming to a state off resonance. The time-integrated polarization as a function of field for a radical with an axially symmetric muon dipolar tensor in this case is given by [46]:

$$P_z(B,\theta) = 1 - \frac{0.5q^2 P_z^0}{(\lambda/2\pi)^2 + q^2 + \left(\nu_\mu - \nu_\mu^0\right)^2} \tag{6.20}$$

where P_z^0 is the muon polarization at time zero, θ is the angle between the unique axis of the hyperfine tensor and the applied magnetic field, λ represents the relaxation of the oscillation,

$$q = \frac{3}{4} D_\mu^{\parallel} \sin\theta\cos\theta, \tag{6.21}$$

and

$$\nu_\mu^0 = \frac{1}{2}\left[A_\mu^{\mathrm{iso}} + \frac{D_\mu^{\parallel}}{2}(3\cos^2\theta - 1)\right]. \tag{6.22}$$

The relaxation rate is given by:

$$\lambda = \lambda_0 + \lambda', \tag{6.23}$$

where $\lambda_0 = 0.4551\ \mu\mathrm{s}^{-1}$ is the inverse muon lifetime and λ' is the rate for a transformation to a state off resonance. The Δ_1 resonance of a single crystal has a Lorentzian shape with a half-width at half-maximum of $q^2 + (\lambda/2\pi)^2$ and a position and amplitude that depends on the orientation of the molecule with respect to the magnetic field (Figure 6.14).

A powder pattern is obtained by integration of Eq. (6.20) over θ. The Δ_1 resonance has a characteristic asymmetric shape depending on the sign of the dipolar coupling constant when λ is small. The asymmetry arises from the greater weight of orientations with $\theta = 90°$ over those with $\theta = 0°$. The cusp is located on the high-field side of the resonance when A_μ^{iso} is positive and $D_\mu^{\parallel}$ is negative, and is on the low-field side when both A_μ^{iso} and $D_\mu^{\parallel}$ are positive (Figure 6.15). The resonance is very sensitive to the magnitude of the dipolar coupling constant. The amplitude of the resonance always decreases with increasing λ. A decrease of $\left|D_\mu^{\parallel}\right|$ leads to the amplitude and width of the Δ_1 resonance decreasing and the lineshape becoming more symmetric

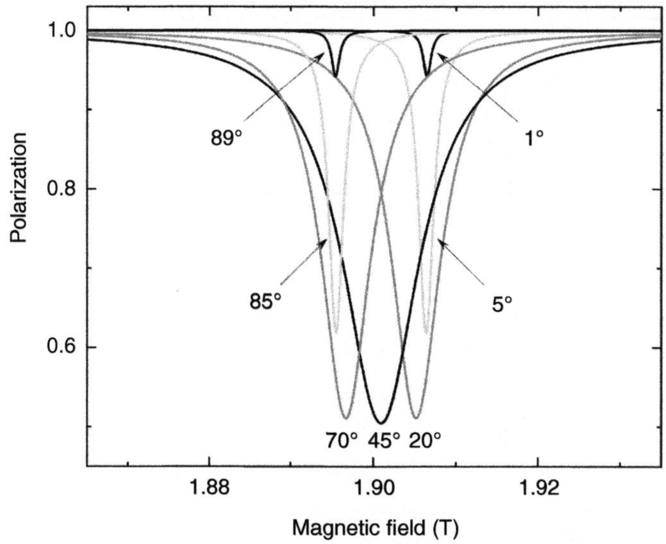

Figure 6.14 Theoretical ALC Δ_1 resonances for a single crystal at different angles between the unique axis and the applied magnetic field. $A_\mu^{iso} = 514.8$ MHz, $D_\mu^{\parallel} = +2$ MHz and $\lambda' = 0\ \mu s^{-1}$.

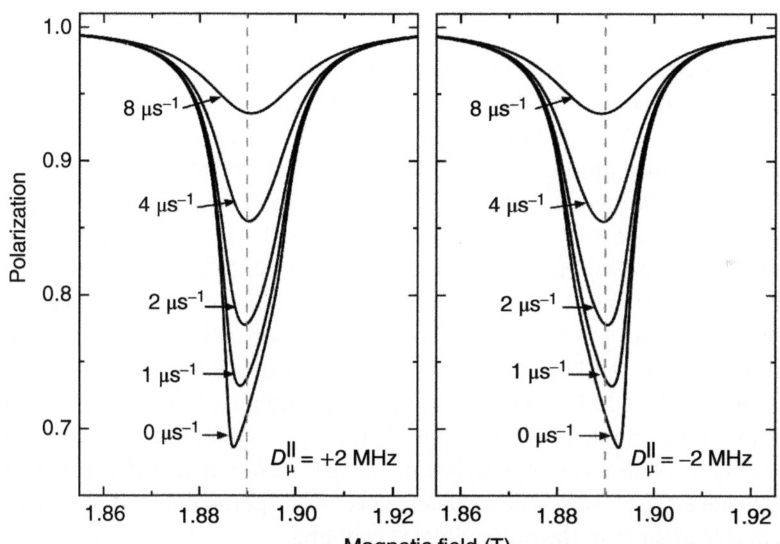

Figure 6.15 Theoretical ALC Δ_1 resonances for a powder with different off-resonance rates (λ') for $A_\mu^{iso} = 514.8$ MHz and $D_\mu^{\parallel} = +2$ MHz and −2.0 MHz. The dotted vertical line corresponds to the position of the Δ_1 resonance calculated using A_μ^{iso}.

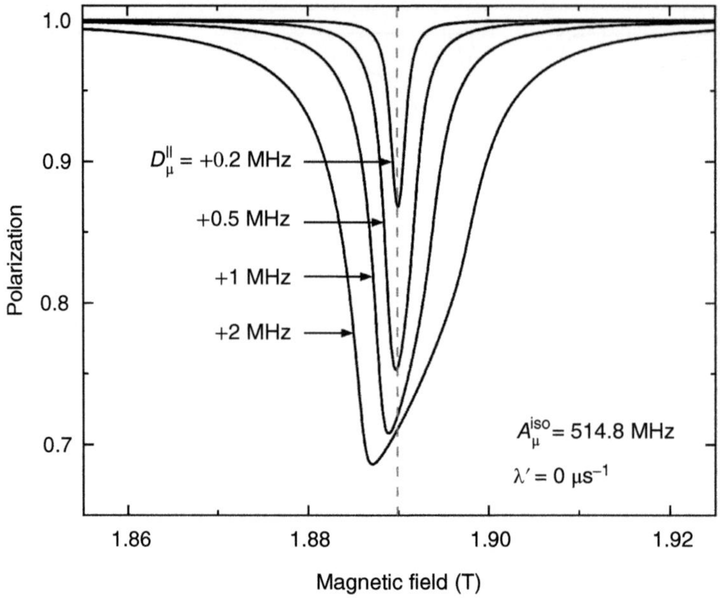

Figure 6.16 Theoretical ALC Δ_1 resonances for a powder with different $D_\mu^{\|}$ values. $A_\mu^{iso} = 514.8$ MHz and $\lambda' = 0\ \mu s^{-1}$. The dotted vertical line corresponds to the position of the Δ_1 resonance calculated usingA_μ^{iso}.

(Figure 6.16). The Δ_1 resonance becomes approximately Lorentzian when the relaxation rate is much larger than the dipolar coupling (i.e. when $\lambda \gg 2\pi q$). An increase of λ' leads to a decrease in amplitude and broadening of the Δ_1 resonance.

Analytical expressions are not available to account for electron spin relaxation but the effect can be simulated using Monte Carlo calculations. Let's consider the example shown in Figure 6.17 with $A_\mu^{iso} = 514.8$ MHz and $D_\mu^{\|} = +2.0$ MHz. At low electron relaxation rates, λ_e, the amplitude and width of the Δ_1 resonance increase with increasing λ_e. This must be contrasted with the relaxation caused by transition to an off-resonance state where the amplitude of the resonance always decreases. The amplitude of the Δ_1 reaches a maximum around $\lambda_e = 10\ \mu s^{-1}$ and then decreases with increasing λ_e. The width of the Δ_1 resonance continuously broadens. Electron spin relaxation results in much larger amplitude resonances than transformation to a state off resonance. This is due to polarization being transferred from the spin state with the opposite electron spin. The resonance becomes approximately Lorentzian when λ_e is much larger than the muon dipolar coupling.

The Δ_0 resonance is due to mixing between spin states that have the same electron spin but opposite muon and proton spins and is observed for muoniated radicals in the solid, liquid or gas phases. Dipolar interactions broaden Δ_0 resonances, decrease the amplitude and make them more asymmetric (Figure 6.18). The resonance shape will depend on the relative orientations of the two hyperfine tensors. In practice, Δ_0 resonances are often not observed in solids due to their smaller amplitude compared with the Δ_1 resonances.

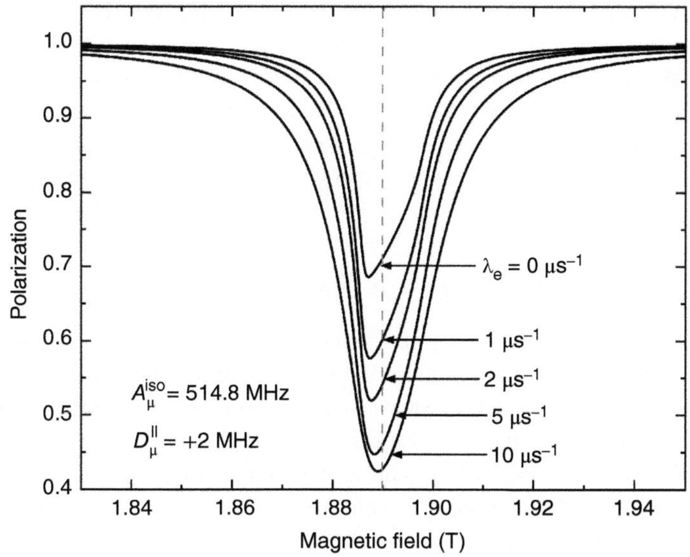

Figure 6.17 Theoretical ALC Δ_1 resonances for a powder with various λ_e values. $A_\mu^{iso} = 514.8\ \mathrm{MHz}$ and $D_\mu^{\parallel} = +2\ \mathrm{MHz}$. The dotted vertical line corresponds to the position of the Δ_1 resonance calculated using A_μ^{iso}.

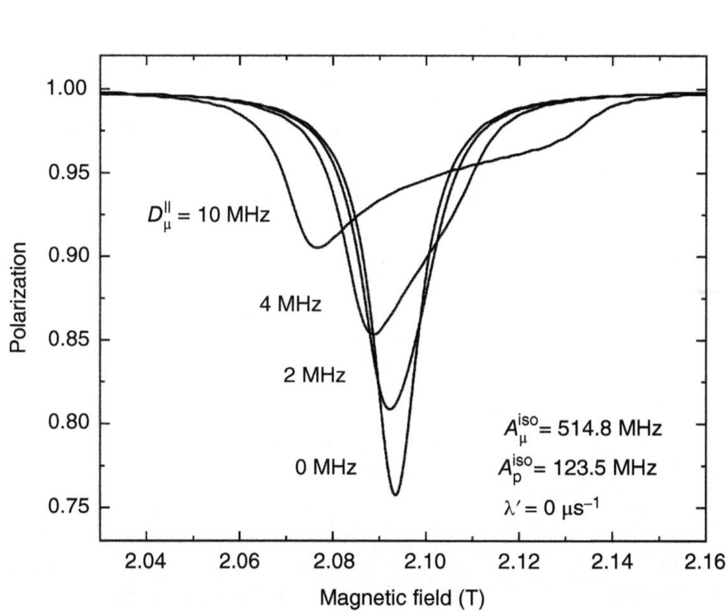

Figure 6.18 Theoretical ALC Δ_0 resonances for a powder with different $D_\mu^{\parallel}$ values. $A_\mu^{iso} = 514.8\ \mathrm{MHz}$, $A_p^{iso} = 123.5\ \mathrm{MHz}$ and $\lambda' = 0\ \mu s^{-1}$.

The above discussion has focused on axially symmetric hyperfine tensors, for which analytical lineshape expressions exist. Predicting the shapes of resonances for radicals with less symmetric hyperfine tensors or including the effects of more complex processes, such as rotational diffusion, chemical reaction and electron spin relaxation, is more difficult and usually requires numerical simulation [47, 48]. ALC-μSR spectra can be simulated using the freely available program 'Quantum' [49].

6.4.3 Anisotropic Motion of Muoniated Radicals in Solids

ALC-μSR can be used to determine the muon dipolar hyperfine coupling constants and this can in turn provide information about the restricted motion of radicals. As an example we consider here how ALC-μSR was used to study the reorientational dynamics of polycrystalline norbornene in its plastic phase, which is between 129 K and the melting point of 320 K [46]. Mu can add to the double bond of norbornene. The Δ_1 resonance of the exo adduct is shown in Figure 6.19a at several temperatures in the plastic phase. The presence of the Δ_1 resonance proves immediately that there is anisotropic motion on a timescale of 10^{-8} s. The Δ_1 resonance becomes narrower and more symmetric with increasing temperature. The asymmetric shape of the resonance indicates that $D_\mu^\parallel$ is negative and the most likely preferred rotation axis is shown in Figure 6.19a. The resonance was fit with Eq. (6.20), and $D_\mu^\parallel$ as a function of temperature is shown in Figure 6.19b. The magnitude of $D_\mu^\parallel$ decreases with increasing temperature due increased wobbling of the preferred rotation axis, which leads to greater motional averaging. The expected value of $D_\mu^\parallel$ for pure uniaxial rotation is −28 MHz, so even at 138 K there is large amplitude wobbling of the main rotation axis, with θ_C calculated using Eq. (6.19) to be ~83°.

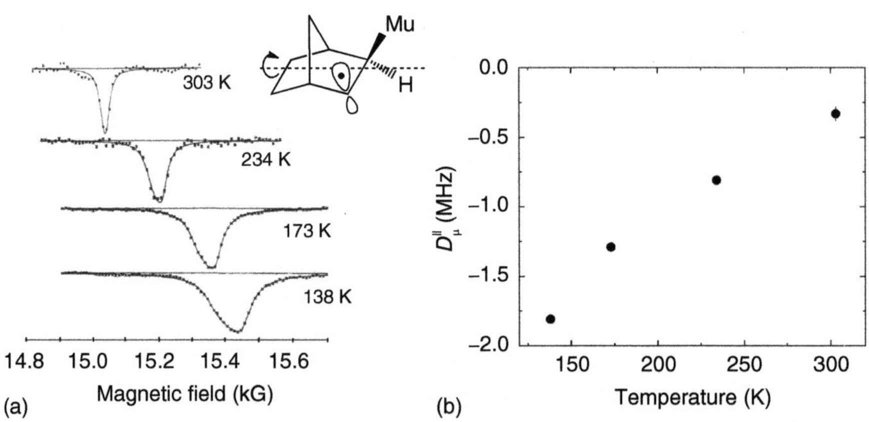

Figure 6.19 (a) The Δ_1 resonance of the exo Mu adduct of norbornene at several temperatures in the plastic phase of polycrystalline norbornene. The preferred reorientation axis of the radical is shown as a dotted line. (b) Temperature dependence of the dipolar muon hfc of the exo Mu adduct of norbornene in the plastic phase of polycrystalline norbornene [46]. Source: (a) Spectra reproduced from Roduner et al. [46] with permission of John Wiley & Sons.

6.5 Solvent Effects on Hyperfine Coupling Constants

The hfcs of a radical are affected by interactions with surrounding solvent molecules. This has been most thoroughly investigated for the C_6H_6Mu radical, where the methylene proton hfc (A_p^{CHMu}) has been measured in a wide range of solvents. The methylene proton Δ_0 resonance shifts to higher fields with increasing polarity of the medium (Figure 6.20) [50]. The variation of the experimental methylene proton and muon hfcs of the radical in several solvents and solvent mixtures of varying polarity can be accounted for by a dipole–dipole reaction field (RF) model that is based on the model of Reddoch and Konishi [51]. This was originally developed to account for the solvent dependence of the ^{14}N hfc in di-*tert*-butyl-nitroxide. The C_6H_6Mu radical has a dipole moment of 0.54 D in vacuum, aligned with the vector connecting the methylene and para carbons. Because the cyclohexadienyl radical is polar it can polarize its environment via dipole–dipole, dipole–induced dipole and dipole–quadrupole interactions. The partial order of the neighbouring solvent molecules results in an electric field along the radical's dipole moment (i.e. the reaction field) that alters the distribution of the unpaired electron and changes the methylene proton hfc. A_p^{CHMu} is linearly dependent on the reaction field, and it has been shown that this results in a linear dependence on the concentration of solvent dipole moments, $[\mu_S]$:

$$A_p^{CHMu} = A + B[\mu_S]. \tag{6.24}$$

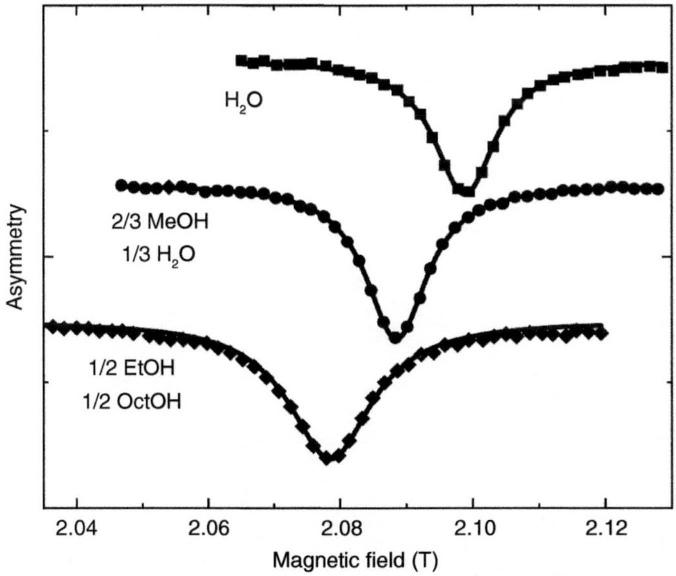

Figure 6.20 ALC-μSR spectra of the methylene proton Δ_0 resonance at 298 K. Abbreviations: MeOH (methanol), EtOH (ethanol) and OctOH (octanol). Source: Data from Reference [50].

A and *B* are empirical constants that depend on the type of radical. For a pure solvent, $[\mu_S]$ is given by

$$[\mu_S] = \frac{\mu_S d_S}{M_S}, \tag{6.25}$$

where μ_S, d_S and M_S are the dipole moment, density and molecular weight of the solvent, respectively. For binary solvent mixtures $[\mu_S]$ depends on the volume fractions of the constituents:

$$[\mu_S] = \frac{\left[\mu_S^{(1)}\right] V_1 + \left[\mu_S^{(2)}\right] V_2}{V_1 + V_2}, \tag{6.26}$$

where $\left[\mu_S^{(1)}\right]$ and $\left[\mu_S^{(2)}\right]$ are the dipole moment concentrations of the pure solvents. With the exception of benzene, A_p^{CHMu} for C_6H_6Mu was found to increase linearly with $[\mu_S]$ in a range of solvents, as shown in Figure 6.21 [50].

Although octadecane and benzene both lack dipole moments their A_p^{CHMu} values are considerably different. This is due to ordering of the quadrupole moments around the radical in benzene [52]. This ordering in turn results in an electric field at the radical, which alters the distribution of the unpaired electron and changes the methylene proton A_p^{CHMu}. The average reaction field, RF, is given by:

$$\mathrm{RF} = -\frac{50 N_A Q^2 d_R}{\pi \varepsilon_0^2 k_B T r_{min}^5} c_Q, \tag{6.27}$$

where Q is the quadrupole moment of the solvent molecules, d_R is the dipole moment of the radical, r_{min} is the minimum separation between the radical and the solvent molecule and c_Q is the concentration of quadrupole moments. A

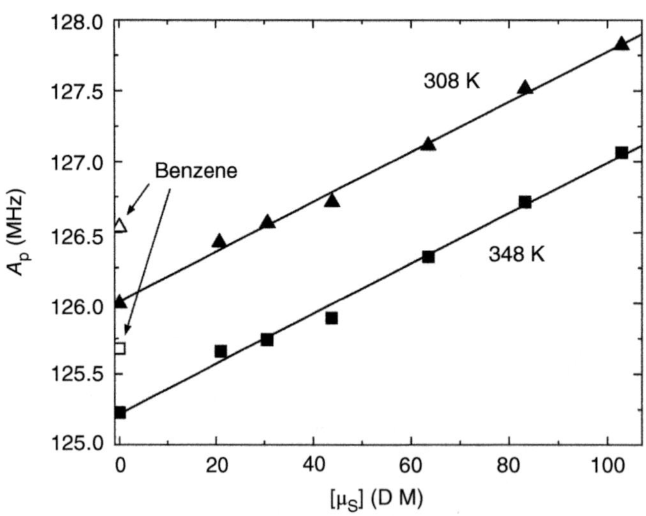

Figure 6.21 Methylene proton hfcs as a function of $[\mu_S]$ for various solvents and solvent mixtures at 308 and 348 K. Source: Data from Reference [50].

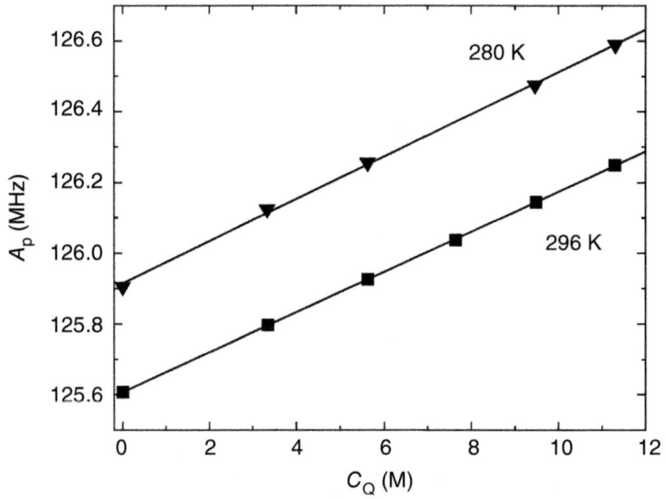

Figure 6.22 The dependence of the methylene proton hyperfine coupling constant on the benzene concentration in mixtures of benzene and cyclohexane at 280 and 296 K. The error bars are smaller than the symbol size. Source: Data from Reference [52].

consequence of this model is that A_p^{CHMu} should increase linearly with c_Q. A_p^{CHMu} of the C_6H_6Mu radical was measured in mixtures of benzene and cyclohexane, a non-polar solvent with no quadrupole moment, with a range of benzene concentrations (Figure 6.22) [52]. A_p^{CHMu} was found to increase linearly with c_Q, which in this case is the concentration of benzene.

6.6 Kinetics of Reactions of Muoniated Radicals

The reaction rates of muoniated radicals can be determined from the broadening of radical signals in both TF-μSR and ALC spectra. In special cases it is also possible to deduce kinetics from the transfer of polarization from one radical to another.

6.6.1 Measuring Chemical Reaction Rates Using TF-μSR

Here we give two examples of measuring reaction rates using TF-μSR. The first involves isomerization of muoniated radicals. Mu adds to the terminal end of the C=C bond of ethyl methacrylate to produces two isomers of the muoniated α-(ethoxycarbonyl)-2-propyl radical [53]. There is no free rotation about the C—C bond, due to its partial double bond character (enol tautomer), and this results in two distinguishable isomers whose structures are shown in Figure 6.23. The cis and trans isomers have A_μ values of 270 and 275 MHz, respectively.

The radical lines in the TF-μSR spectrum broaden at higher temperature as the two isomers interconvert more rapidly. The isomerization is a unimolecular process and so is independent of concentration and exhibits first-order kinetics. The radical

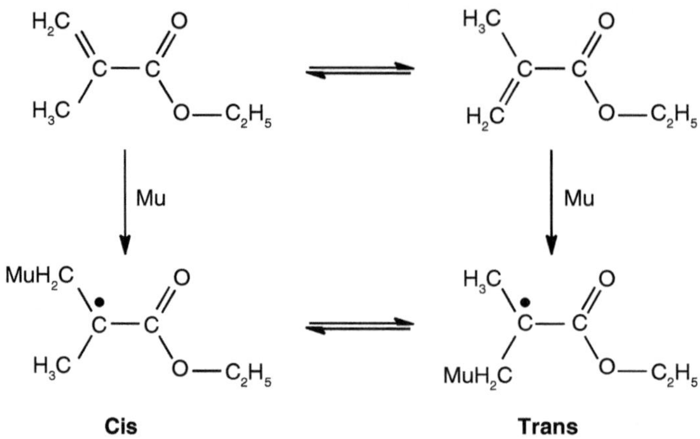

Figure 6.23 Scheme showing the isomerization of ethyl methacrylate and formation of the two isomers of the muoniated α-(ethoxycarbonyl)-2-propyl radical [53].

signals are assumed to decay exponentially in solution, giving a Lorentzian lineshape with width λ. Assuming Arrhenius behaviour for the rate constant, the temperature dependence of the line width parameter is given by:

$$\lambda = \lambda_0 + A\exp(-E_a/RT), \tag{6.28}$$

where λ_0 represents the residual width, which include lifetime broadening due to muon decay. The cis and trans isomers show the same temperature dependence (Figure 6.24), which indicates that the forward and backwards isomerization rates of the radicals are the same.[2] The best fit Arrhenius parameters are $\log(A/\text{s}^{-1}) = 13.2(4)$ and $E_a = 45$ kJ/mol.

The second example involves the reaction between the C_6H_6Mu radical and benzoquinone, which acts as an electron acceptor [54]. One of the TF-μSR precession signals of the radical is shown in Figure 6.25a as a function of the benzoquinone concentration. The exponential decay in the time domain results in the radical line having a Lorentzian shape in the Fourier-transformed spectrum. The damping increases as a function of benzoquinone concentration (Figure 6.25b). As with Mu reactions, the damping rate was fit to

$$\lambda = \lambda_0 + k_R[S] \tag{6.29}$$

where $[S]$ is the concentration of benzoquinone. From the slope the second-order rate constant, k_R, was found to be $(2.8 \pm 0.4) \times 10^8\ \text{M}^{-1}\ \text{s}^{-1}$.

6.6.2 Measuring Chemical Reaction Rates Using ALC-μSR

Chemical reactions also lead to broadening of the resonances in an ALC spectrum, but it is more complicated than in TF-μSR. For reactions that take the radical off

2 In conventional kinetics the forward and backward rates depend on the *sum* of the rate constants, but this does not apply in TF-μSR, where the isomerization product is depolarized.

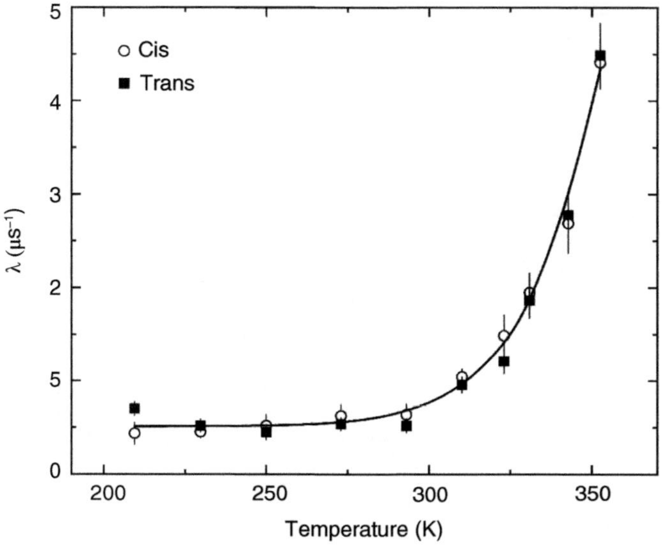

Figure 6.24 Line width parameter λ for the cis and trans isomers of the muoniated α-(ethoxycarbonyl)-2-propyl radical [53].

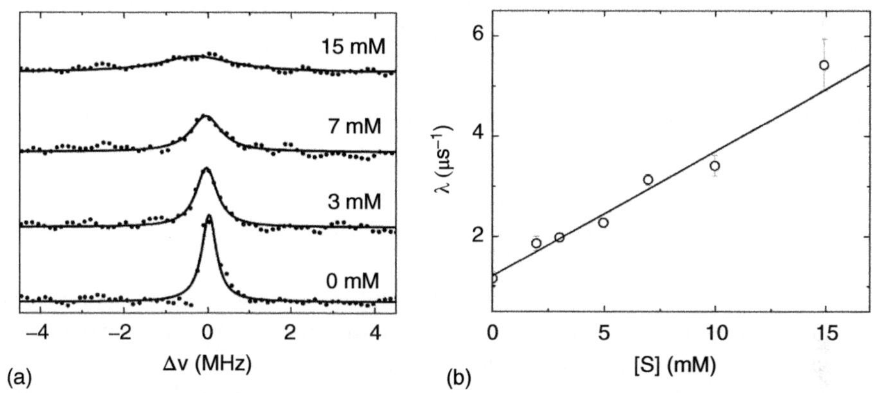

Figure 6.25 (a) One line of the TF-μSR spectrum of C_6H_6Mu in benzene for various concentrations of the electron acceptor benzoquinone ($B = 3$ kG). (b) Damping constants versus benzoquinone concentration. Source: (a) Spectra redrawn from a review article by Fischer [54] with permission of Springer Nature. (b) Data reported by Roduner [55].

resonance (addition, electron transfer, termination and isomerization), the width (FWHM, $\Delta B_{1/2}$) of the Δ_0 resonance is given by [15]:

$$\Delta B_{1/2} = \sqrt{\frac{\lambda_{\text{eff}}^2 + \omega_{\text{ALC}}^2}{\pi^2(\gamma_\mu - \gamma_p)^2}} \tag{6.30}$$

Measuring the width of the Δ_0 resonance will give the effective damping rate λ_{eff}, and if this is done as a function of concentration, one can then determine the rate constant.

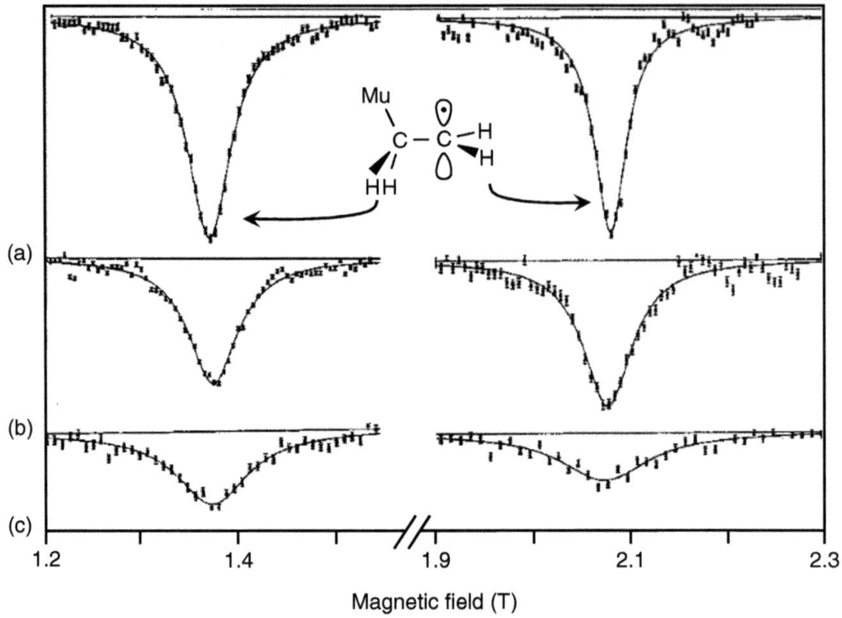

Figure 6.26 Δ_0 resonances of the muoniated ethyl radical at O_2 concentrations of (a) 0, (b) 2.3×10^{16} and (c) 8.2×10^{16} molecules cm^{-3} in ethene at 1.5 bar and 318 K. Source: Adapted with permission from Dilger et al. [57]. Copyright American Chemical Society.

The FWHM of a Δ_0 resonance for reactions involving electron spin relaxation is given by [56]:

$$\Delta B_{1/2} = \frac{(\lambda_e + \lambda_{eff})}{\pi(\gamma_\mu - \gamma_p)}\sqrt{1 + \frac{\omega^2_{ALC}}{2\lambda_e\lambda_{eff} + \lambda^2_{eff}}}. \tag{6.31}$$

λ_e represents the effect of electron spin relaxation at the muon. $\lambda_{eff} = \tau_\mu^{-1} + \lambda_{rem}$ is the sum of the muon decay rate and of the remaining processes which have the same effect as muon decay, such as chemical reaction or muon relaxation. ω_{ALC} is the energy gap in angular frequency units at the avoided crossing and is typically less than 3.5 μs^{-1} for organic radicals. Figure 6.26 shows the broadening of the Δ_0 resonances of the muoniated ethyl radical in the gas phase with increasing concentration of O_2. In this case the broadening is due to both chemical reaction and spin exchange of the muoniated ethyl radical with the paramagnetic O_2. One can separate spin exchange from chemical reaction by performing time-differential measurements in several longitudinal magnetic fields off resonance [57].

6.6.3 Measuring Chemical Reaction Rates from the Transfer of Polarization from a Primary to a Secondary Radical

Most of the muoniated radicals described in this chapter are formed by muonium addition to an unsaturated molecule. Detection by TF-μSR requires substantial transfer of muon spin polarization from muonium to the radical, as described in

Section 5.4. It seems reasonable to ask if the polarization can be transferred from one radical (R1) to another (R2). Extending the formalism of Eqs. (5.10), we find that the polarization transferred to the precession frequencies of the secondary radical can be expressed by [58]:

$$P_{\perp}(\omega_{12}) = \frac{1}{2}\left\{\left[\frac{C}{1+\Delta_{1212}^2} + \frac{S}{1+\Delta_{1412}^2}\right]^2 + \left[\frac{C\,\Delta_{1212}}{1+\Delta_{1212}^2} + \frac{S\,\Delta_{1412}}{1+\Delta_{1412}^2}\right]^2\right\}^{1/2}$$

$$P_{\perp}(\omega_{43}) = \frac{1}{2}\left\{\left[\frac{C}{1+\Delta_{4343}^2} + \frac{S}{1+\Delta_{2343}^2}\right]^2 + \left[\frac{C\,\Delta_{4343}}{1+\Delta_{4343}^2} + \frac{S\,\Delta_{2343}}{1+\Delta_{2343}^2}\right]^2\right\}^{1/2}, \qquad (6.32)$$

where $C = c_{R1}^2 c_{R2}^2 + c_{R1}s_{R1}c_{R2}s_{R2}$, $S = s_{R1}^2 c_{R2}^2 - c_{R1}s_{R1}c_{R2}s_{R2}$ and $\Delta_{klmn} = (\omega_{kl} - \omega_{mn})/\lambda$.

The mixing coefficients c_i and s_j are analogous to those of muonium (Table A.1 in the Appendix) but with the hyperfine frequencies of radicals R1 and R2.

Under the high-field conditions of radical detection the polarization at the lower radical precession frequency is simply

$$P_{\perp}(\omega_{12}) = \frac{1}{2}\left\{\frac{\lambda^2}{\lambda^2 + [\omega_{12}(R1) - \omega_{12}(R2)]^2}\right\}^{1/2}, \qquad (6.33)$$

which shows that the extent of polarization transfer depends on the relative magnitudes of the first-order rate of reaction, λ, and the difference in precession frequencies of the primary and secondary radicals. Some example plots are shown in Figure 6.27.

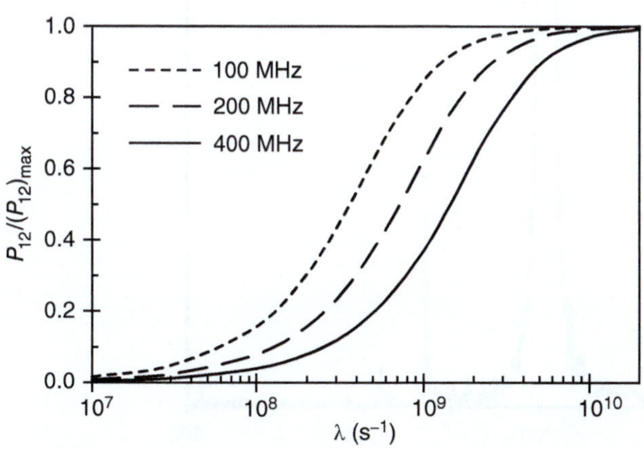

Figure 6.27 Predicted transfer of muon spin polarization from of a muoniated radical reacting at rate λ to form a secondary radical, for cases where the lower precession frequency changes by 100, 200 or 400 MHz.

It is evident from these plots that a secondary radical can only be detected by TF-μSR if it forms on the nanosecond timescale or faster. One such example is the muoniated disilanyl radical described in Section 6.7.1.

6.7 Characterization of Novel Radicals by Muon Spin Spectroscopy

By their very nature muoniated radicals correspond to 'normal' radicals for which a proton has been replaced by a muon. The advent of ALC-μSR extended the range of spin-active nuclei whose hyperfine constants can be measured to D, ^{13}C, ^{14}N, ^{19}F, ^{35}Cl and ^{37}Cl. However almost all radicals studied by muon spin spectroscopy are classified as carbon-centred, i.e. the unpaired electron spin density is localized on one or more carbon atoms. The following sections describe some less common classes of radicals.

6.7.1 Radicals Containing Si or Ge

Silicon and germanium are in the same group of the periodic table as carbon, so it is reasonable to expect some common aspects of chemistry. The earliest detection of a muoniated silicon-centred radical occurred in a study of a silene (a molecule containing —C=Si—). Two radicals were detected (Figure 6.28), and it was found that the carbon in the double bond is more reactive towards Mu addition than the silicon atom by a factor 2.2 [59].

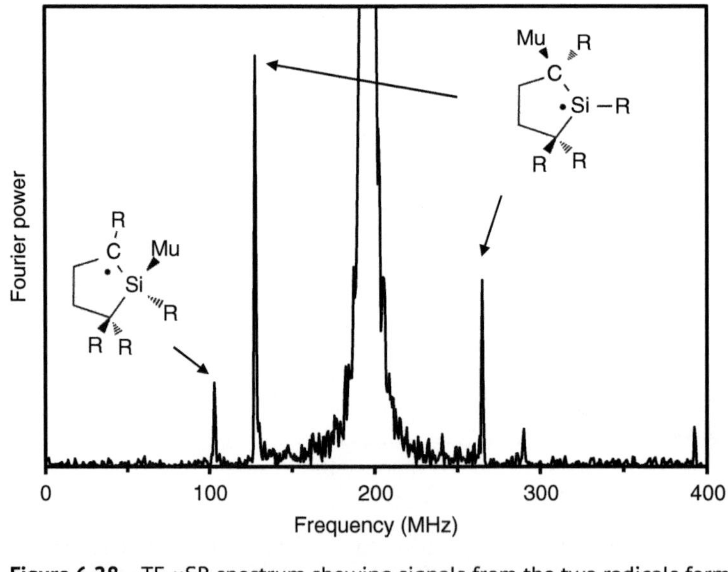

Figure 6.28 TF-μSR spectrum showing signals from the two radicals formed from the addition of Mu to a silene [59].

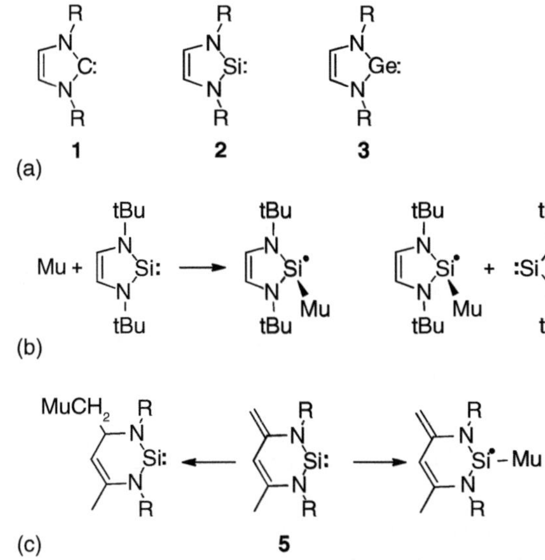

Figure 6.29 (a) Some low-valent molecules with highly reactive dicoordinate atoms: an imidazole ylidene (**1**); a silylene (**2**); a germylene (**3**). R represents a bulky alkyl or aryl group. (b) Mu adds to a carbene to form a radical which then couples with a second carbene molecule. (c) Silylene **5** forms two radicals by Mu addition.

A different class of low-valent silicon-centred molecule has a neutral dicoordinate Si atom and is known as a silylene. This molecule, and the related germylene, are analogous to N-heterocyclic carbenes (imidazole ylidenes), which incorporate dicoordinate carbon (Figure 6.29a).

Bulky substituents (R) are necessary to inhibit dimerization of silylenes and germylenes, and they also control the stability of the radicals formed from Mu addition. Thus, Mu adds to the carbene centre of **1** to form easily-detected radicals [8], but muoniated radicals from silylenes and germylenes can undergo further reaction, as indicated in Figure 6.29b. This radical coupling reaction [60] is inhibited if R is large enough [61]. To detect the disilanyl radical (**4**) by TF-μSR the coupling reaction must occur with a rate $\sim 10^9$ s^{-1} [58], as explained in Section 6.6.3.

It is clear from the hfcs of the product radicals that Mu adds to **1**, **2** and **3** at the dicoordinate atom rather than the C=C double bond in the imidazole ring. However, there is a fine balance of reactivity between the two sites, as evident from the detection of two muoniated radicals formed from **5** (Figure 6.29c) in equal yields [62]. It is significant that **5** contains a diene and that Mu addition at the exo CH_2 promotes unpaired spin delocalization around the ring.

6.7.2 Radicals Containing P

Phosphorus has a nuclear spin ($I = 1/2$) convenient for ALC-μSR, so it is surprising that only a few phosphorus-centred radicals have been reported so far. One area of research concerns phosphaalkenes, molecules containing P=C. It is important to

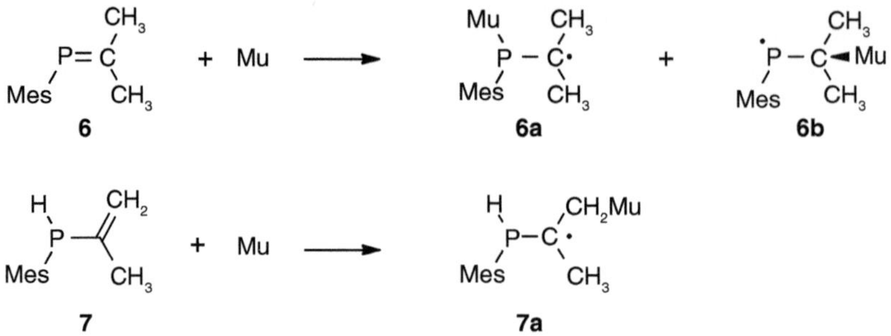

Figure 6.30 Muoniation of phosphaalkene **6** results in radical **6a** (major product) and **6b** (minor product). A different starting material, **7**, gives radical **7a,** an isotopomer of **6a**. Mes = 2,4,6,trimethylphenyl.

establish whether radical attack occurs at the P or C, since the product could determine subsequent radical polymerization. There is even the possibility of modifying the substituents on P and C to steer the reaction in one direction or another [63]. An example of competition for the two reactive sites of a phosphaalkene is shown in Figure 6.30. Both muoniated radicals have been detected, with **6a** more abundant than **6b** [64]. An important clue to identification of the radical signals was a parallel study of the alkene **7**, whose sole product is **7a**. Radicals **6a** and **7a** are isotopomers, differing only in the location of Mu.

The competition between reactive sites has also been explored with phosphasilenes, which contain the Si=P bond. Both Si-centred and P-centred radicals were identified [65].

6.7.3 Radicals Containing a Metal Atom

Mu can react with organometallic molecules. There is a missing fraction, frequently associated with the formation of muoniated radicals, when muons are implanted in diamagnetic metal acetylacetonate compounds [66, 67]. The addition of Mu to the diamagnetic compound results in a paramagnetic molecule with either 17 or 19 valence electrons. The number of valence electrons depends on whether Mu adds to the metal or to a ligand. In contrast, 100% of muons are in a diamagnetic environment when implanted in paramagnetic metal acetylacetonate compounds. This is either due to the termination reaction between Mu and the organometallic radical or rapid spin relaxation.

Muoniated radicals have also been observed from the reaction of Mu with ferrocene, $Fe(C_5H_5)_2$ [68], and ruthenocene, $Ru(C_5H_5)_2$ [69]. In ferrocene, Mu was observed to add to the exterior of the cyclopentadiene ($Cp = C_5H_5$) ligand. In ruthenocene, Mu addition was observed at the ruthenium and at the Cp rings, both from the exterior and interior directions. These structures are shown in Figure 6.31. In both cases, closer agreement between the DFT calculations and the experimental values was obtained if it is assumed that the structures of the Mu adducts of the Cp rings are distorted due to interactions with neighbouring molecules. The TF-μSR and ALC-μSR spectra of are shown in Figure 6.32. The disappearance of the ligand

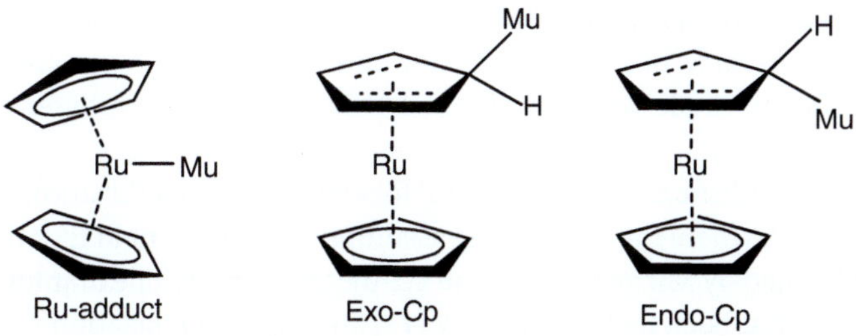

Figure 6.31 Structures of the possible muoniated ruthenocenyl radicals formed by Mu addition to ruthenocene.

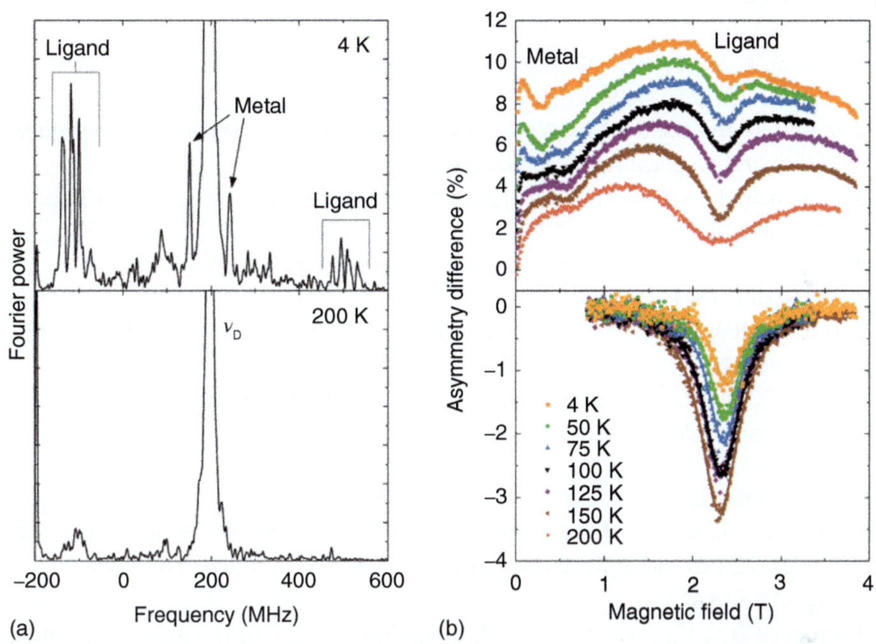

Figure 6.32 (a) TF-μSR spectra of ruthenocene powder at (top) 4 K and (bottom) 200 K. (b) ALC-μSR spectra of ruthenocene powder displayed (top) as the difference between the spectrum at the given temperature and that obtained at 280 K; and (bottom) as the high-field resonance after additional background subtraction. Source: Adapted from spectra in McKenzie et al. [69] with permission of Canadian Science Publishing.

signals in the TF-μSR spectra with increasing temperature and the increasing Δ_1 resonance due to the ligand in the ALC-μSR spectra indicate that the electron spin relaxation rate of the Cp adducts increases with temperature.

6.7.4 Muoniated Radicals Containing No Other Nuclear Moments

Muoniated radicals with no additional nuclear spins have been produced by the addition of Mu to fullerenes (specifically ^{12}C-enriched C_{60} [70]) as well as the

1,2-dicarboxyacetylene dianion [71]. Since these radicals have a two spin-1/2 configuration, the low transverse-field μSR spectrum looks like that of Mu, except for the larger splitting due to the lower muon hfc. Further discussion of fullerenes is deferred to Section 8.2.

Having a radical with a single nuclear spin is useful for exploring spin relaxation in low and zero magnetic field, where the simplifying assumptions of high-field ESR do not apply. The radical formed by Mu addition to the 1,2-dicarboxyacetylene dianion, $^{-}OOC^{\cdot}$–C=C(Mu)–COO^{-}, has been used in this way [71] and the spin relaxation rate in zero magnetic field explained using theory expounded by Fedin et al. [72].

References

1 Gerson, F. and Huber, W. (2003). *Electron Spin Resonance Spectroscopy of Organic Radicals*. Wiley-VCH.
2 Weil, J.A. and Bolton, J.R. (2007). *Electron Paramagnetic Resonance: Elementary Theory and Practical Applications*, 2ee. Wiley.
3 McKenzie, I., Addison-Jones, B., Brodovitch, J.-C. et al. (2002). *J. Phys. Chem. A* 106: 7083–7085.
4 McKenzie, I., Brodovitch, J.-C., Ghandi, K. et al. (2007). *J. Phys. Chem. A* 111: 10625–10634.
5 Percival, P.W., Kiefl, R.F., Kreitzman, S.R. et al. (1987). *Chem. Phys. Lett.* 133: 465–470.
6 Yu, D., Percival, P.W., Brodovitch, J.-C. et al. (1990). *Chem. Phys.* 142: 229–236.
7 McConnell, H.M. and Chesnut, D.B. (1958). *J. Chem. Phys.* 28: 107–117.
8 McKenzie, I., Brodovitch, J.-C., Percival, P.W. et al. (2003). *J. Am. Chem. Soc.* 125: 11565–11570.
9 Rhodes, C.J., Morris, H., and Reid, I.D. (2001). *Magn. Reson. Chem.* 39: 438–442.
10 Kiefl, R.F., Percival, P.W., Brodovitch, J.-C. et al. (1988). *Chem. Phys. Lett.* 143: 613–618.
11 Brodovitch, J.-C., Addison-Jones, B., Ghandi, K. et al. (2015). *Phys. Chem. Chem. Phys.* 17: 1755–1762.
12 Karplus, M. and Fraenkel, G.K. (1961). *J. Chem. Phys.* 35: 1312–1323.
13 Heller, C. and McConnell, H.M. (1960). *J. Chem. Phys.* 32: 1535–1539.
14 Percival, P.W., Brodovitch, J.-C., Leung, S.K. et al. (1988). *Chem. Phys.* 127: 137–147.
15 Roduner, E. (1988). *The Positive Muon as a Probe in Free Radical Chemistry, Chap. 4*, Lecture Notes in Chemistry, vol. 49. Springer-Verlag.
16 Roduner, E. and Reid, I.D. (1989). *Isr. J. Chem.* 29: 3–11.
17 Percival, P.W., Addison-Jones, B., Brodovitch, J.-C. et al. (1995). *Chem. Phys. Lett.* 245: 90–94.
18 Brodovitch, J.-C., Addison-Jones, B., Ghandi, K. et al. (2003). *Can. J. Chem.* 81: 1–6.
19 Lopes de Magalhães, A. and Ramos, M.J. (1990). *Chem. Phys. Lett.* 165: 528–534.
20 Yamada, K., Kawashima, Y., and Tachikawa, M. (2014). *J. Chem. Theory Comput.* 10: 2005–2015.

21 Bohm, M.C., Ramirez, R., and Schulte, J. (2005). *Mol. Phys.* 103: 2407–2436.

22 Hudson, B.S. and Chafetz, S.K. (2013). *Molecules* 18: 4906–4916.

23 McKenzie, I., Addison-Jones, B., Brodovitch, J.-C. et al. (2001). *PhysChemComm* 4: 136–138.

24 Claxton, T.A., Graham, A.M., Cox, S.F.J. et al. (1991). *Hyperfine Interact.* 65: 913–926.

25 Boxwell, M.A., Claxton, T.A., and Cox, S.F.J. (1993). *J. Chem. Soc. Faraday Trans.* 89: 2957–2960.

26 Gaussian, Inc. (2016). *Gaussian 16 and Earlier Versions of this Software Package.* Wallingford CT, USA: Gaussian, Inc.

27 Rhodes, C.J. and Symons, M.C.R. (1988). *J. Chem. Soc. Faraday Trans. I* 84: 1187–1194.

28 Fleming, D.G., Arseneau, D.J., Bridges, M.D. et al. (2013). *J. Phys. Chem. C* 117: 16523–16539.

29 Ramos, M.J., McKenna, D., Webster, B.C., and Roduner, E. (1984). *J. Chem. Soc. Faraday Trans. I* 80: 255–265.

30 Fleming, D.G., Bridges, M.D., Arseneau, D.J. et al. (2011). *J. Phys. Chem. A* 115: 2778–2793.

31 Yu, D. (1989) *Muonium kinetics and intramolecular motion of muonium-substituted free radicals*, Ph.D. Thesis. Simon Fraser University.

32 Percival, P.W., Brodovitch, J.-C., Leung, S.K. et al. (1989). *Chem. Phys. Lett.* 163: 241–245.

33 Bridges, M.D., Arseneau, D.J., Fleming, D.G., and Ghandi, K. (2007). *J. Phys. Chem. C* 111: 9779–9793.

34 Sun-Mack, S. (1994). *Delayed formation and intramolecular motion of muonium-substituted free radicals*, Ph.D. Thesis. Simon Fraser University.

35 Brodovitch, J.-C., Percival, P.W., Addison-Jones, B., and Ji, F. (1997). *Hyperfine Interact.* 106: 163–167.

36 Ji, F., Percival, P.W., Brodovitch, J.-C. et al. (1994). *Hyperfine Interact.* 87: 853–858.

37 Hill, A., Symons, M.C.R., Cox, S.F.J. et al. (1985). *J. Chem. Soc. Faraday Trans. I* 81: 433–448.

38 Buttar, D., Macrae, R.M., Webster, B.C., and Roduner, E. (1990). *Hyperfine Interact.* 65: 927–936.

39 Ghandi, K., Addison-Jones, B., Brodovitch, J.-C. et al. (2003). *J. Am. Chem. Soc.* 125: 9594–9595.

40 Stride, J.A., Jayasooriya, U.A., and Reid, I.D. (2002). *J. Phys. Chem. A* 106: 244–250.

41 McConnell, H.M., Heller, C., Cole, T., and Fessenden, R.W. (1960). *J. Am. Chem. Soc.* 82: 766–775.

42 Roduner, E. (1981). *Chem. Phys. Lett.* 81: 191–194.

43 Marsh, D. (2019). *Spin-Label Electron Paramagnetic Resonance Spectroscopy.* CRC Press.

44 Israelachvili, J., Sjösten, J., Eriksson, L.E.G. et al. (1975). *Biochim. Biophys. Acta* 382: 125–141.

45 Roduner, E. (1993). *Chem. Soc. Rev.* 22: 337–346.

46 Roduner, E., Reid, I.D., Riccò, M., and de Renzi, R. (1989). *Ber. Bunsenges. Phys. Chem.* 93: 1194–1197.

47 Kreitzman, S.R. and Roduner, E. (1995). *Chem. Phys.* 192: 189–230.

48 Tregenna-Piggott, P.L.W., Roduner, E., and Santos, S. (1996). *Chem. Phys.* 203: 317–337.

49 Lord, J.S. (2006). *Physica B* 374–375: 472–474.

50 Vujŏsevìc, D., Dilger, H., McKenzie, I. et al. (2007). *J. Phys. Chem. B* 111: 199–208.

51 Reddoch, A.H. and Konishi, S. (1979). *J. Chem. Phys.* 70: 2121–2130.

52 Dilger, H., Hess, S., Scheuermann, R. et al. (2009). *Physica B* 404: 927–929.

53 Strub, W., Roduner, E., and Fischer, H. (1987). *J. Phys. Chem.* 91: 4379–4383.

54 Fischer, H. (1984). *Hyperfine Interact.* 19: 751–770.

55 Roduner, E. (1981). *Hyperfine Interact.* 8: 561–570.

56 Dilger, H., Roduner, E., Stolmar, M. et al. (1997). *Hyperfine Interact.* 106: 137–142.

57 Dilger, H., Schwager, M., Tregenna-Piggott, P.L.W. et al. (1996). *J. Phys. Chem.* 100: 6561–6571.

58 McCollum, B.M., Brodovitch, J.-C., Clyburne, J.A.C. et al. (2009). *Physica B* 404: 940–942.

59 McCollum, B.M., Abe, T., Brodovitch, J.-C. et al. (2008). *Angew. Chem. Int. Ed.* 47: 9772–9774.

60 McCollum, B.M., Brodovitch, J.-C., Clyburne, J.A.C. et al. (2009). *Chem. Eur. J.* 15: 8409–8412.

61 Mitra, A., Brodovitch, J.-C., Krempner, C. et al. (2010). *Angew. Chem. Int. Ed.* 49: 2893–2895.

62 Percival, P.W., McCollum, B.M., Brodovitch, J.-C. et al. (2012). *Organometallics* 31: 2709–2714.

63 Chandrasena, L. (2019). *Exploring main group radicals using an isotope of hydrogen*, Ph.D. Thesis. Simon Fraser University.

64 Chandrasena, L., Samedov, K., McKenzie, I. et al. (2019). *Angew. Chem. Int. Ed.* 58: 297–301.

65 Samedov, K., Heider, Y., Cai, Y. et al. (2020). *Angew. Chem. Int. Ed.* 59: 16007–16012.

66 Kubo, M.K. and Nishiyama, K. (1999). *J. Radioanal. Nucl. Chem.* 239: 165–167.

67 Kubo, M.K. (2001). *Anal. Sci.* 17: i653–i656.

68 McKenzie, I. (2014). *Phys. Chem. Chem. Phys.* 16: 10600–10606.

69 McKenzie, I. (2018). *Can. J. Chem.* 96: 358–362.

70 Addison-Jones, B. (2000). *Muon spin resonance studies of fullerenyl radicals and associated organic radicals*, Ph.D. Thesis. Simon Fraser University.

71 McKenzie, I. (2011). *Phys. Chem. Chem. Phys.* 13: 1168–1173.

72 Fedin, M.V., Shakirov, S.R., Purtov, P.A., and Bagryanskaya, E.G. (2006). *Russ. Chem. Bull.* 55: 1703–1716.

7

Spin Relaxation Studies

7.1 Probing Spin Relaxation with Muons

Chapters 4 and 5 concern chemical kinetics, but signal decay can also arise from spin relaxation. Some examples of relaxing signals are shown in Figure 7.1, from solutions of manganous nitrate in water, with and without the addition of methanol. The precession frequency is 2.7 MHz in an applied transverse field of 200 G, so these signals are clearly identified as arising from muons in diamagnetic environments. The upper signal is from a highly concentrated solution in water and can be fit with a single, exponentially decaying oscillation; the lower one is from a mixture with methanol and has two components, one relaxing and one not.

Relaxation of the diamagnetic muon signal in pure water is negligible, so the fast relaxation is clearly due to the presence of the paramagnetic ion (Mn^{2+}). Details of the relaxation mechanism will be given in Section 7.3; here, it is only the qualitative observation that is of interest. The two-component relaxation for the mixed liquids implies that muons do not freely exchange between all possible sites in water and methanol. Specifically, in contrast to MuOH in water, muons in the form of $MuCH_2OH$ and MuH in methanol do not exchange with water in the hydration sphere of the paramagnetic ion [1].

As well as qualitatively indicating the role of dynamics in spin relaxation, this example demonstrates the practical aspect of distinguishing diamagnetic signals arising from different environments. NMR spectroscopy can differentiate such signals by their frequencies (chemical shift), but the limited frequency resolution of μSR precludes this option. Thus spin relaxation is frequently employed to determine the fraction of muon stops in the sample container, for example when thick-walled vessels are needed for high-pressure work [2]. The long-lived signal from muon stops in a metal container is easily distinguished from the fast relaxing signal of a suitable test sample.

7.2 Molecular Dynamics from Spin Relaxation

The principles of muon spin relaxation are the same as those well known in other forms of magnetic resonance spectroscopy. Changes of spin population between

Muon Spin Spectroscopy: Methods and Applications in Chemistry and Materials Science, First Edition.
Donald G. Fleming, Iain McKenzie, and Paul W. Percival.

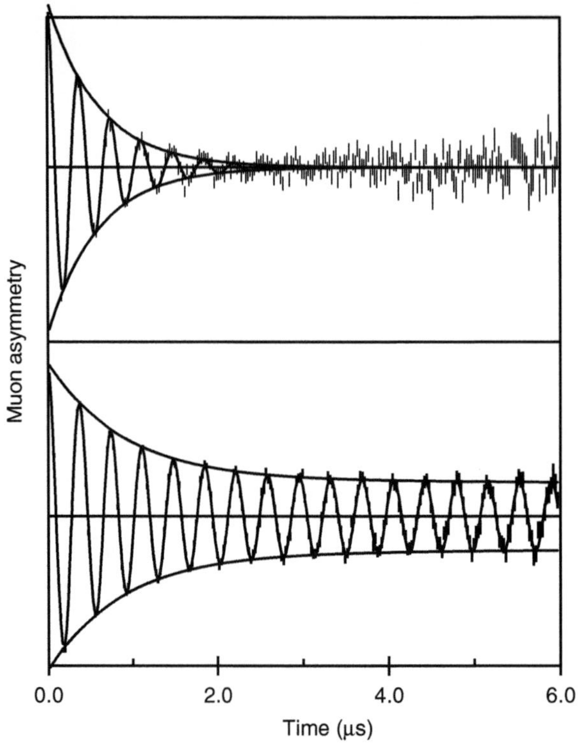

Figure 7.1 Comparison of one-component and two-component relaxation of diamagnetic signals. The upper signal arises from muons in a highly concentrated solution of manganous nitrate in water; the lower signal comes from a mixture of the same aqueous solution with methanol [1].

energy levels are characterized by the *spin-lattice* relaxation time, T_1:

$$\frac{dn}{dt} = -\frac{(n - n_{eq})}{T_1} \tag{7.1}$$

where n is the population difference between the lower and upper states, and n_{eq} is its value at thermal equilibrium. T_1 is the relaxation time appropriate for *longitudinal* relaxation, where spin transitions involve a change in energy. However, there are also interactions between spins that result in relaxation of precession signals in transverse fields, without any overall change in energy. These are characterized by the *transverse* relaxation time, T_2. The equations of motion for an ensemble of spins can be expressed in terms of the components of magnetization:

$$\begin{aligned} \frac{dM_x}{dt} &= \omega_B M_y - \frac{M_x}{T_2} \\ \frac{dM_y}{dt} &= -\omega_B M_x - \frac{M_y}{T_2} \\ \frac{dM_z}{dt} &= -\frac{(M_z - M_{eq})}{T_1} \end{aligned} \tag{7.2}$$

where ω_B is the resonance frequency corresponding to the energy difference between two states of different spin projections in a magnetic field, i.e. $\omega_B = \Delta E/\hbar$.

In magnetic resonance this set of equations is known as the Bloch equations. Although μSR deals with individual spins, the Ergodic Principle (Section 4.1) means that these equations still apply, although it is now more appropriate to refer to components of muon spin polarization, P_i, rather than magnetization. Also, M_{eq} can be ignored as negligible compared to the beam polarization. In terms of individual spins, the decay of transverse magnetization (or polarization) results from a spreading out (dephasing) of the spin vectors in the precession plane. This happens if there is a static distribution of local magnetic fields, or random changes in the muon environment, so that the ensemble of spins has a distribution of precession frequencies.

It is fluctuating magnetic interactions that drive spin relaxation processes. By making certain assumptions about their time dependence (an exponentially decaying autocorrelation function), it is possible to show that the relaxation times take the general form [3]:

$$\frac{1}{T_1} = \gamma_N^2 \frac{1}{2} \overline{(H_x^2 + H_y^2)} \frac{2\tau_c}{1 + \omega_0^2 \tau_c^2}$$

$$\frac{1}{T_2} = \gamma_N^2 \left[\overline{H_z^2} \tau_c + \frac{1}{2} \overline{(H_x^2 + H_y^2)} \frac{\tau_c}{1 + \omega_0^2 \tau_c^2} \right] \tag{7.3}$$

where τ_c is the *correlation time* that characterizes the fluctuating local magnetic field with components (H_x, H_y, H_z). If the motion is isotropic, such that $\overline{H_x^2} = \overline{H_y^2} = \overline{H_z^2}$, and the motion is fast ($\omega_0^2 \tau_c^2 \ll 1$; the *motional narrowing* limit), then $T_1 = T_2$.

In general the correlation functions include different factors, depending on the dynamic process that modulates the specific magnetic interaction. However, the principles remain the same, and information about dynamics can be extracted from detailed study of the relaxation times. Examples of more complex cases are considered in Sections 7.3 and 7.5.

7.3 Muon Spin Relaxation Studies of Aqueous Solutions of Manganese(II) Ions

The fast relaxation evident in Figure 7.1 arises from fluctuating interactions of the muon spin with the electron spin of the manganous ion (S = 5/2). There are two classes of interaction: dipolar (K_D) and scalar hyperfine (K_S). The fluctuations arise from a variety of dynamical processes: the electron spin relaxation of the Mn^{2+} ion (characterized by times T_{1e} and T_{2e}); exchange of MuOH between the hexaaqua paramagnetic complex and bulk water (characterized by τ_m) and tumbling of the complex (characterized by τ_r). For the purpose of this discussion it is assumed that all muons replace protons in water molecules, i.e. the small fraction that form MuH is ignored. Furthermore, the residence time of an individual water molecule in the complex (τ_m) is ~50 ns, long compared with the other correlation times, but short enough that muons can be considered to be distributed statistically between bound and free water molecules.

By applying the well-established theory of proton spin relaxation in aqueous solutions of paramagnetic ions [4] it is predicted that [5, 6]

$$\frac{1}{T_1} = xK_D\left[\frac{6\tau_{c1}}{1+\omega_\mu^2\tau_{c1}^2} + \frac{14\tau_{c2}}{1+\omega_e^2\tau_{c2}^2}\right] + xK_S\left[\frac{2\tau_{e2}}{1+\omega_e^2\tau_{e2}^2}\right] \tag{7.4}$$

$$\frac{1}{T_2} = xK_D\left[4\tau_{c1} + \frac{3\tau_{c1}}{1+\omega_\mu^2\tau_{c1}^2} + \frac{13\tau_{c2}}{1+\omega_e^2\tau_{c2}^2}\right] + xK_S\left[\tau_{e1} + \frac{\tau_{e2}}{1+\omega_e^2\tau_{e2}^2}\right] \tag{7.5}$$

where x denotes the fraction of water bound to the ion, in this case as the hexaaqua complex $[Mn(H_2O)_5(MuOH)]^{2+}$;

$$K_D = \frac{1}{15}\gamma_\mu^2\gamma_e^2 S(S+1)(1/r^6) \tag{7.6}$$

and

$$K_S = \frac{1}{3}S(S+1)(A/\hbar)^2 \tag{7.7}$$

The correlation times are given by

$$1/\tau_{c1} = 1/T_{1e} + 1/\tau_r + 1/\tau_m \tag{7.8}$$

$$1/\tau_{c2} = 1/T_{2e} + 1/\tau_r + 1/\tau_m \tag{7.9}$$

$$1/\tau_{e1} = 1/T_{1e} + 1/\tau_m \tag{7.10}$$

$$1/\tau_{e2} = 1/T_{2e} + 1/\tau_m \tag{7.11}$$

Note that tumbling of the complex (correlation time τ_r) affects only the orientation-dependent dipolar interaction. Water exchange affects both spin interactions, but $1/\tau_m$ is negligible compared to the electron spin relaxation rates.

Because the correlation functions in Eqs. (7.4) and (7.5) contain both the muon (ω_μ) and electron (ω_e) Larmor precession frequencies it is possible to extract correlation times by searching for conditions under which $\omega\tau = 1$. If the magnetic field is increased, there should be step decreases in the relaxation rates, as various terms 'switch off' for $\omega\tau > 1$. An example can be seen in Figure 7.2 for the longitudinal muon spin relaxation rates measured for an aqueous solution of manganous nitrate of composition $Mn(NO_3)_2 \cdot 48.9H_2O$ [5].

The step at about 100 G occurs because $\omega_e\tau_e \sim 1$, implying $\tau_e \sim 1$ ns in agreement with published electron spin relaxation times [7]. The drop above 1600 G occurs for $\omega_e\tau_{c2} \sim 1$ and demonstrates the near dominance of the $1/\tau_r$ term in Eq. (7.9). The detailed fit shows $\tau_r \sim 20$ ps, again in agreement with published values [7]. τ_r increases at lower temperature, and as it approaches τ_e the distinction between the two steps is less clear (see the blue curve in Figure 7.2). This also explains the single step observed in the field dependence of transverse muon spin relaxation rates measured for higher concentrations of $Mn(NO_3)_2$ [6].

The ability to vary the magnetic field over a wide range is a valuable feature of μSR. It is demonstrated again in Section 7.5.

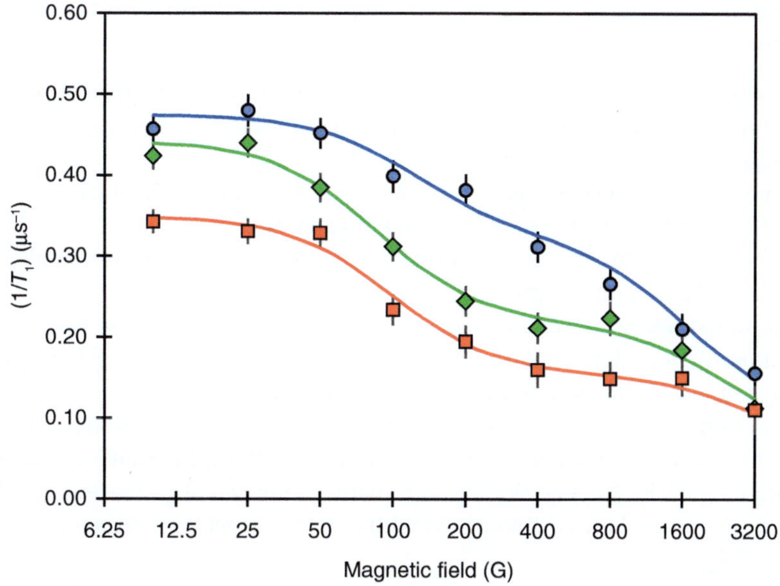

Figure 7.2 Longitudinal muon spin relaxation rates for a solution of composition $Mn(NO_3)_2 \cdot 48.9H_2O$ at temperatures 15 °C (blue circles), 35 °C (green diamonds) and 60 °C (red squares) [5]. The lines through the points represent a global fit to the model described by Eq. (7.4). The change in temperature affects the correlation times and hence the shape of the curves.

7.4 Muonium Spin Exchange with Paramagnetic Species

Electron spin relaxation rates are generally much stronger than nuclear spin relaxation rates, simply because the magnetic interactions are stronger due to the larger magnetic moment of the electron. It is therefore not surprising that low concentrations of paramagnetic ions are effective at relaxing muonium precession signals, many orders of magnitude lower than those discussed for *muon* spin relaxation in Section 7.3. However, it is important to remember that μSR monitors the muon spin polarization, so fast electron spin relaxation is only apparent via the muon–electron hyperfine interaction. In discussing muonium and muoniated radicals we often draw parallels to electron spin resonance, but a more apt comparison would be to ENDOR, electron-nuclear double resonance [8], in which a nuclear spin is used to probe paramagnetic systems. The effect of electron spin relaxation on the nuclear resonance signal is manifest as the Overhauser effect [3].

The general effect of electron spin relaxation on muon spin polarization has been investigated theoretically by several authors, notably Nosov and Yakovleva [9], Ivanter and Smilga [10] and Turner et al. [11]. Patterson's seminal review of muonium in semiconductors [12] includes a useful summary of limiting cases and cautions the reader about the different notation conventions used in early papers from the various Russian groups. A series of papers by Senba focuses on situations more specific to chemistry, in particular the effect of electron spin exchange in encounters between muonium atoms and paramagnetic atoms or molecules [13, 14]. The key findings are:

- In low transverse field, where ν_{12} and ν_{23} are indistinguishable, for spin ½ encounters the muonium signal relaxes at a rate one half of the electron spin-flip rate, i.e. $\lambda_{Mu} = \lambda_{SF}/2$.
- At intermediate transverse fields (~50 G), $\lambda_{Mu} = (3/4)\lambda_{SF}$, i.e. 1.5 times faster than the low field limit, independent of the spin of the partner.
- In low transverse field, encounters between Mu and a spin-1 molecule result in relaxation of the muonium signal at a rate $\lambda_{Mu} = (16/27)\lambda_{SF.}$
- In longitudinal fields the spin statistics are the same as in transverse field, but the non-oscillating signal relaxes with a rate $\lambda_{LF} = (\lambda_{SF}/2)/\left(1 + x_B^2\right)$, where $x_B = (\omega_e + \omega_\mu)/\omega_0$ is the dimensionless field parameter.
- If the electron spin relaxation rate is much faster than the hyperfine frequency, so that the muon spin cannot follow the electron spin flips, the transverse-field precession signal looks very similar to a diamagnetic signal (there is a small shift from the muon Larmor frequency) with a field-independent relaxation rate $\lambda_D = \omega_0^2/(2\lambda_{SF})$.

The field dependence of relaxation rates was already mentioned in Section 4.6, as a practical means to separate signal decay rates into the two components of spin relaxation and chemical reaction. Thus, the longitudinal-field data for Mu + NO shown in Figure 4.16 can be analysed with Eq. (7.12) [15]:

$$\lambda_{LF} = \lambda_{ch} + \lambda_{se} = \left(k_{ch} + \frac{1}{1+x^2}k_{se}\right)[\mathrm{NO}] \tag{7.12}$$

where the subscript 'ch' refers to chemical reaction and 'se' refers to spin-flip exchange. It should be noted that in this formulation the experimentally determined muonium spin-exchange rate constant k_{se} is one-half of the *electron* spin-flip rate, so that in terms of the average relative velocity and spin-exchange cross section

$$k_{se} = \bar{v}\,\sigma_{se} = \frac{1}{2}\bar{v}\,\sigma_{SF} \tag{7.13}$$

This point is emphasized because an alternative definition is used elsewhere [16], in which the factor ½ is incorporated in Eq. (7.12) instead of (7.13).

The factor ½ arises because only half of the spin-exchange encounters result in a spin flip of the muonium electron. This is illustrated here by considering the effect of spin-exchange encounters of like and unlike electron spins on a muon-spin-polarized state at zero field:

$$\begin{aligned}\mathrm{Mu}(\alpha_e\alpha_\mu) + \mathrm{NO}(\alpha_e) &\rightarrow \mathrm{Mu}(\alpha_e\alpha_\mu) + \mathrm{NO}(\alpha_e)\\ \mathrm{Mu}(\alpha_e\alpha_\mu) + \mathrm{NO}(\beta_e) &\rightarrow \mathrm{Mu}(\beta_e\alpha_\mu) + \mathrm{NO}(\alpha_e)\end{aligned} \tag{7.14}$$

The upper line shows that there is no effect on the muon spin in the case of identical electron spins, in contrast to the second line, where exchange of unlike electron spin orientations results in a muonium spin combination which is not an eigenstate. Because of the strong hyperfine interaction the $|\beta_e\alpha_\mu\rangle$ state evolves cyclically into the $|\alpha_e\beta_\mu\rangle$ state, resulting in muon spin depolarization. This simple picture applies

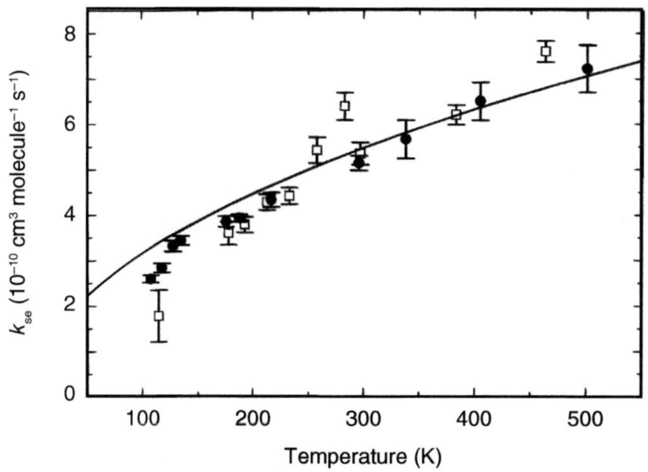

Figure 7.3 Spin-exchange rate constants for Mu + O_2 in the gas phase determined by longitudinal-field (open squares [16]) and transverse-field techniques (filled circles [17]). The line through the points is a fit of $T^{1/2}$ dependence. Source: Adapted from a figure by Himmer et al. [16] with permission from the American Chemical Society.

only to Mu spin exchange with $S = ½$ partners. A formal theoretical analysis shows that the spin-statistical factor depends on the spin of the exchange partner according to [11]:

$$f = \frac{8S(S+1)}{3(2S+1))} \tag{7.15}$$

Thus $f = ½$ for $S = ½$ (the case described by Eq. (7.13)), but $f = 16/27$ for $S = 1$ (e.g. for O_2).[1]

Since spin exchange is not an activated process, the temperature dependence of the muonium spin-exchange rate constant in the gas phase should match the $T^{1/2}$ factor expected from the thermal velocity. As shown in Figure 7.3, data for the Mu + O_2 reaction [16, 17] are consistent with this expectation over a very wide temperature range. The weak temperature dependence of the spin-exchange cross section is also in agreement with the findings of a study on H + H spin exchange [18].

The electron spin-flip exchange cross section is a measurable quantity related to the fundamental quantum process of electron spin exchange. Ideally one should study this process in isolated atom–atom collisions, as was done for H + H [18]. This was the motivation for a study of muonium spin exchange with Cs metal vapour [19, 20]. Muonium decay rates were determined from precession signals in a low transverse field for various concentrations of Cs atoms in nitrogen moderator gas. As evident from Figure 7.4 there is a good straight-line relationship. Measurements of the Mu decay rates made in low (6 G) and intermediate (35 G) fields were found to differ by the factor 1.5 mentioned above, so it was concluded that there is no significant contribution from chemical reaction. Thus the slope of the line through the

1 The reader is cautioned that the statistical angular momentum weighting factor defined by Turner et al. [11] is the inverse of the factor defined in Eq. (7.15).

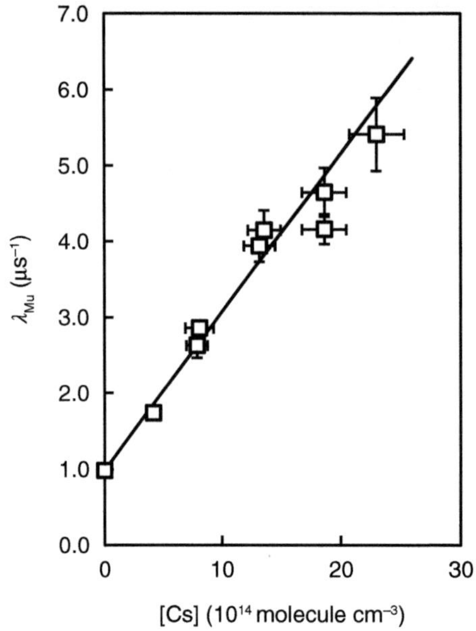

Figure 7.4 Decay rates of the Mu precession signal in a weak transverse field for various concentrations of Cs in 2 atm nitrogen at 566 K [19]. The slope of the line gives k_{se}.

experimental points gives k_{se}, and thence, via Eq. (7.13), the spin-flip exchange cross section σ_{SF}.

There is no direct calculation of the spin-exchange cross section for Mu + Cs, but a calculation for H + Cs [21] predicts a value over twice as large as determined for Mu [19]. In the absence of any direct measurement of the H + Cs cross section it remains an open question whether this is a real isotope effect or not. However, similar comparisons of H and Mu spin exchange with NO and O_2 show that the cross sections for H are roughly three times larger than for Mu [11], suggesting that cross sections are indeed mass dependent. This may be related to the larger number of orbital angular momentum partial waves in the scattering process for the heavier H atom [19].

Muonium spin-exchange reactions also occur in liquids, and it was originally assumed that the measured rate constants represent the diffusion limit [22]. However, this ignores the spin-statistical factor. For example, Eq. (7.15) predicts that $f = 5/8$ for Mu encounters with a spin-3/2 system such as Cr(III) [23, 24]. In liquids the cage effect results in multiple collisions per encounter (Section 5.3), leading to a 50:50% mixture of spin-polarized muonium. By this argument the encounter rate should be twice the observed reaction rate [23, 24].

7.5 Spin Relaxation in Muoniated Radicals

Just as for muonium, we expect that the major source of muon spin relaxation in muoniated radicals is electron spin relaxation communicated via hyperfine coupling. The various relaxation processes include electron-spin exchange, fluctuation of anisotropic g- and hyperfine tensors, modulation of isotropic hyperfine interactions and the spin–rotation interaction [3, 8]. The first of these is an *inter*molecular process and can be discounted for isolated radicals, such as in dilute solution. However, the *intra*molecular interactions are subject to fluctuations induced by molecular collisions (temperature-dependent diffusion in liquids; temperature- and pressure-dependent collisions in gases). Depending on the nature of the specific spin interactions the relaxation processes may result in electron spin flips, coupled muon–electron spin flips, or just muon spin flips. These general considerations have been expressed in a phenomenological model of muon spin relaxation to rationalize the magnetic field dependence of muon spin relaxation rates measured for the muoniated ethyl radical, $MuCH_2CH_2^{\cdot}$ in the gas phase [25]. In the style of Eqs. (7.4) and (7.5), labelling the magnetic interactions Δ_E, Δ_{ME} and Δ_M, the relaxation rates are written as

$$\frac{1}{T_1} = \Delta_E^2\left(\frac{1}{1+x_B^2}\right)\left[\frac{2\tau_c}{1+\omega_e^2\tau_c^2}\right] + \Delta_{ME}^2\left[\frac{2\tau_c}{1+\omega_e^2\tau_c^2}\right] + \Delta_M^2\left[\frac{\tau_c}{1+\omega_{12}^2\tau_c^2} + \frac{\tau_c}{1+\omega_{34}^2\tau_c^2}\right] \tag{7.16}$$

and

$$\frac{1}{T_2} = (\Delta_E^2 + \Delta_{ME}^2)\left[\frac{\tau_c}{1+\omega_e^2\tau_c^2}\right] + \frac{1}{2}\Delta_M^2\left[\frac{\tau_c}{1+\omega_{12}^2\tau_c^2} + \frac{\tau_c}{1+\omega_{34}^2\tau_c^2}\right] + \Delta_S^2\tau_c \tag{7.17}$$

where x_B is the usual dimensionless field parameter and $\omega_{12} = 2\pi\nu_{12}$ and $\omega_{34} = 2\pi\nu_{34}$ are muonium transition frequencies (Table A.2). The last term in Eq. (7.17) accounts for additional, *secular* spin-dephasing processes, which don't involve a change in energy. All other terms depend on magnetic field, via x_B itself, and via the correlation functions which contain the electron Zeeman frequency or the muonium transition frequencies. For simplicity in this treatment a common correlation time is assumed, this being related to the time between collisions in the gas phase, and thus inversely proportional to the pressure. A schematic representation of the different field regions in which the various terms of Eq. (7.16) are effective is displayed in Figure 7.5.

As expected, given the smaller magnetic moment of the muon, there is only a minor contribution from processes involving pure muon spin flips. Also, the field region where the Δ_M term is dominant is not experimentally accessible. However, the other two regions can be distinguished, as shown by the longitudinal-field data in Figure 7.6. Equivalent data were acquired for transverse field, and the full set was analysed by a global fit to Eqs. (7.16) and (7.17) [25]. Similar but less extensive studies were also carried out for isotopic species, $MuCD_2CD_2^{\cdot}$ and $Mu^{13}CH_2{}^{13}CH_2^{\cdot}$, and for the muoniated tert-butyl radical. The best fit parameters for all these data sets are shown in Table 7.1.

Comparison of the Δ_E fit variable with predictions based on g-factor anisotropy (Eq. (7.18)) shows that for the ethyl radicals Δ_E relaxation is dominated by the

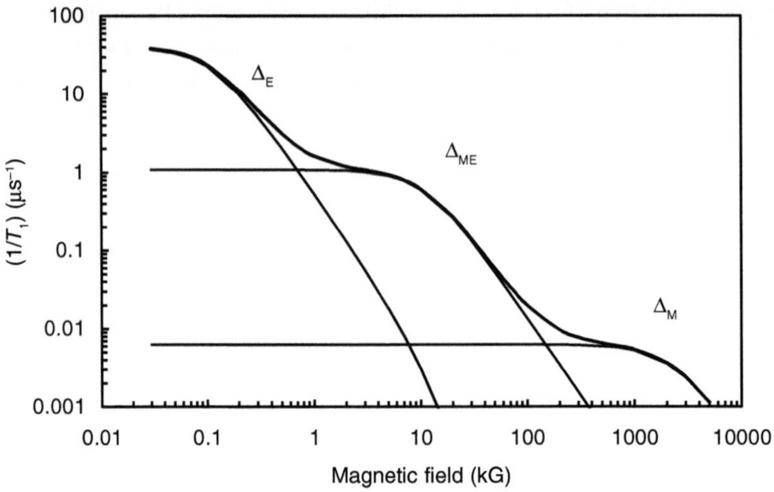

Figure 7.5 Field dependence of the longitudinal relaxation rate as predicted by Eq. (7.16) for a typical set of parameter values appropriate for a muoniated radical [25]. The field regions where successive terms of Eq. (7.16) dominate are denoted by Δ_E, Δ_{ME} and Δ_M.

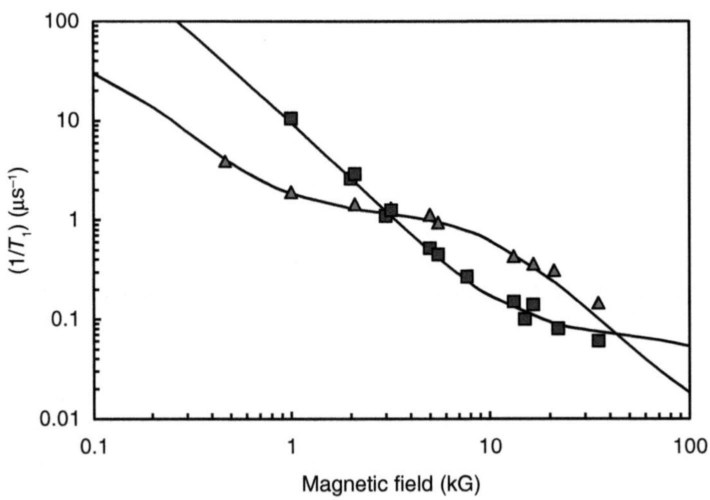

Figure 7.6 Muon relaxation rate as a function of longitudinal magnetic field for the $MuCH_2CH_2^{\cdot}$ radical in ethene at pressures of 1 atm (blue squares) and 11.9 atm (red triangles). The lines through the points represent a global fit of Eq. (7.16) to the data. Source: Selected data taken from the work of Fleming et al. [25].

electron spin-rotation mechanism. The spin-rotation contribution is considerably smaller for the tert-butyl radical, which has a five-fold larger moment of inertia (I_{av}).

$$\Delta_E^2 = \left(\frac{k_B T}{I_{av}}\right) \overline{\Delta g^2} \tag{7.18}$$

Electron–nuclear hyperfine coupling also plays a significant role, both through its effect on electron spin relaxation (affecting the Δ_E term) and through coupled spin

Table 7.1 Isotropic Hyperfine Coupling Constants (A_μ) and Relaxation Parameters for $MuCH_2CH_2^\cdot$, $MuCD_2CD_2^\cdot$, $Mu^{13}CH_2{}^{13}CH_2^\cdot$ and $MuCH_2\dot{C}(CH_3)_2$ at 300 K [25].

Parameter	$MuCH_2CH_2^\cdot$	$MuCD_2CD_2^\cdot$	$Mu^{13}CH_2{}^{13}CH_2^\cdot$	$MuCH_2\dot{C}(CH_3)_2$
A_μ (MHz)	330	340	331	290
Δ_E (μs^{-1})	2124 ± 20	1389 ± 38	2474 ± 55	379 ± 34
Δ_{ME} (μs^{-1})	328 ± 2	339 ± 28	320 ± 10	156 ± 7
Δ_M (μs^{-1})	23 ± 2	30 ± 4	50 ± 11	84 ± 5
Δ_S (μs^{-1})	441 ± 9	859 ± 47		
τ (ps) at 1 atm	65 ± 1	93 ± 2	54 ± 9	128 ± 19

flips (Δ_{ME}). The reduction of Δ_E for $MuCD_2CD_2^\cdot$ compared with $MuCH_2CH_2^\cdot$ and $Mu^{13}CH_2{}^{13}CH_2^\cdot$ is consistent with this.

The best fit value of τ_c is 5.5 ps at 11.9 atm (the data shown in Figure 7.6) and proportionately larger at lower pressures. This correlation time can be interpreted as the average time for reorientation of the radical due to molecular collisions. Assuming a simple hard-sphere collision model the collision rate ($1/\tau_c$) can be translated to a cross section, $\sigma \approx 100$ Å^2, which is close to the geometrical cross section found from the viscosity of ethene.

The first detection of a muoniated radical in the gas phase [26] also concerned the ethyl radical. Higher pressures were necessary, due to the use of a decay muon beam, and relative low fields (1 and 2 kG) were used, leading to high relaxation rates ($\sim 10\ \mu s^{-1}$). The important advance in the work of Fleming et al. [25] detailed above is the use of magnetic-field dependence to separate the classes of relaxation mechanisms (Figure 7.5), and to extract a correlation time from the steps that occur when $\omega_e\tau_c = 1$. As pointed out in Section 7.3, electron spin resonance spectroscopy generally lacks this advantage. Indeed, standard X-band spectrometers display typical signals (organic radicals; $g \sim 2.0$) close to 3.4 kG, where electron relaxation times (T_{1e}, T_{2e}) are predicted to be short and the lines broad. In some circumstances, such as high temperature and pressure so that τ_c is small, polyatomic radicals can be detected in the gas phase [27]. In principle a better strategy would be to use a higher field, so that $\omega_e\tau_c > 1$. However, although high-frequency ESR spectrometers are now available, the correspondingly smaller resonators cannot accommodate large-volume gas samples. Thus, as pointed out by Dilger et al. [28], μSR is superior for gas-phase radical spectroscopy.

7.6 Muon Spin Relaxation During Chemical Reaction

The previous sections of this chapter concern spin relaxation of directly detected muoniated species, either the reactant (Mu) or end-product. We now consider the effects of spin relaxation *during* a reaction, exemplified by the gas-phase addition of muonium to carbon monoxide, to produce the muoniated formyl radical, MuCO.

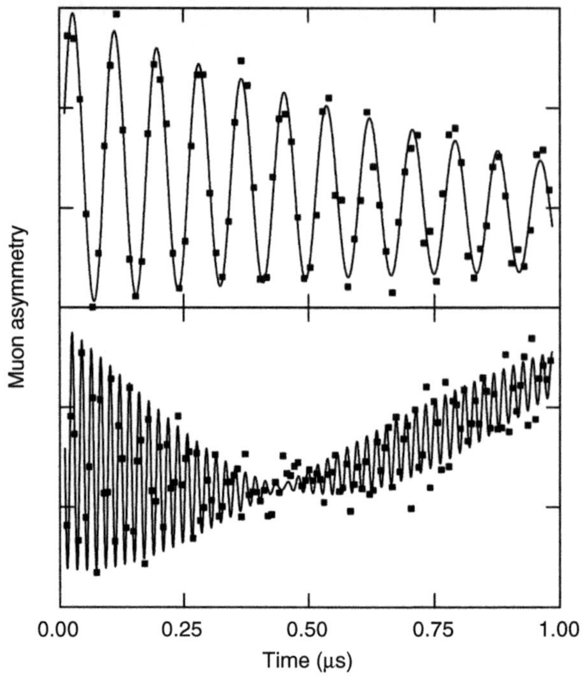

Figure 7.7 Muonium precession at 8 G (top) and 37 G (bottom) in nitrogen gas (5 bar) with added partial pressure of 70 Torr CO. Fits to the data give decay rates of 0.93 ± 0.03 and $1.4 \pm 0.1\ \mu s^{-1}$, respectively. Source: Adapted from Pan et al. [30], with the permission of AIP Publishing.

In terms of the Lindemann mechanism, expressed in general form in Eq. (4.16), the reaction is considered to proceed through an excited intermediate:

$$\text{Mu} + \text{CO} \underset{k_d}{\overset{k_a}{\rightleftharpoons}} \text{MuCO}^* \xrightarrow[\text{M}]{k_s} \text{MuCO} \tag{7.19}$$

Although MuCO is a radical, it has no nuclear moments other than the muon (ignoring the 1.1% natural abundance of ^{13}C). Therefore the transverse-field μSR spectrum looks like that of Mu, apart from the effect of a smaller muon hyperfine constant (estimated to be ~1400 MHz [29]). At the low transverse fields used for standard muonium decay kinetics (Section 4.1) the degenerate ν_{12} and ν_{23} frequencies are the same for Mu and MuCO, so there would be no spin dephasing at the detectable frequency. Nevertheless, the Mu precession signal decays in the presence of CO [30], suggesting that spin relaxation is important. This is confirmed by the relative decay rates of the precession signals at 8 G and 37 G (Figure 7.7).

The only significant mechanism is electron spin relaxation of the MuCO* intermediate caused by spin rotation (Eq. (7.18)), communicated to the muon via the hyperfine interaction. The rate of this process has been estimated to be $\sim 10^{11}\ s^{-1}$ [30], so the MuCO product (stabilized by collisions with moderator gas molecules M) would experience essentially instantaneous depolarization on the microsecond time scale of a μSR experiment, effectively relaxing the Mu signal at the corresponding rate of MuCO formation.

This is not the full story, however, because the effective rate constant for Mu decay was found to depend on magnetic field as well as on moderator gas pressure [30], implying a spin relaxation component in addition to the chemical process. As shown in Figure 4.17 the experimentally determined rate constant is consistent with the form

$$k_{\text{eff}} = fk_{\text{SF}} + k_0[\text{M}] \tag{7.20}$$

where f is the spin-statistical factor (1/2 at low field, 3/4 at intermediate field; Section 7.4), k_{SF} is the component due to electron spin relaxation and $k_0 = k_a k_s / k_d$ is the termolecular rate constant. At low moderator pressures the observed signal decay is due to Mu atoms from dissociation of the excited intermediate MuCO* that has undergone spin relaxation in the same way as described above for the stabilized MuCO product. However, the appropriate correlation time is now the dissociation lifetime of the excited molecule, τ^*, estimated to be $\lesssim 10^{-13}$ s [30], considerably shorter than the time between collisions.

References

1 Newman, K.E., Brodovitch, J.-C., and Percival, P.W. (1985). *Chem. Phys. Lett.* 113: 347–350.

2 Brodovitch, J.-C., Leung, S.K., Percival, P.W. et al. (1988). *Radiat. Phys. Chem.* 32: 105–109.

3 Carrington, A. and McLachlan, A.D. (1967). *Introduction to Magnetic Resonance with Applications to Chemistry and Chemical Physics*. Harper & Row.

4 Connick, R.E. and Fiat, D. (1966). *J. Chem. Phys.* 44: 4103–4107.

5 Newman, K.E., Brodovitch, J.-C., and Percival, P.W. (1986). *Chem. Phys. Lett.* 124: 279–282.

6 Newman, K.E., Adamson-Sharpe, K.M., Brodovitch, J.-C., and Percival, P.W. (1984). *Hyperfine Interact.* 18: 709–713.

7 Gumerov, F.M., Kozyrev, B.M., and Vishnevskaja, G.P. (1975). *Mol. Phys.* 29: 937–951.

8 Atherton, N.M. (1973). *Electron Spin Resonance; Theory and Applications.* Wiley.

9 Nosov, V.G. and Yakovleva, I.V. (1963). *Sov. Phys. JETP* 16: 1236–1246.

10 Ivanter, I.G. and Smilga, V.P. (1969). *Sov. Phys. JETP* 28: 796–801.

11 Turner, R.E., Snider, R.F., and Fleming, D.G. (1990). *Phys. Rev. A* 41: 1505–1516.

12 Patterson, B.D. (1988). *Rev. Mod. Phys.* 60: 69–159.

13 Senba, M. (1990). *J. Phys. B* 23: 4051–4069.

14 Senba, M. (1995). *Phys. Rev. A* 52: 4599–4615.

15 Pan, J.J., Arseneau, D.J., Senba, M. et al. (2000). *Phys. Chem. Chem. Phys.* 2: 621–629.

16 Himmer, U., Dilger, H., Roduner, E. et al. (1999). *J. Phys. Chem. A* 103: 2076–2087.

17 Senba, M., Fleming, D.G., Arseneau, D.J. et al. (1989). *Phys. Rev. A* 39: 3871–3883.

18 Desaintfuscien, M. and Audoin, C. (1976). *Phys. Rev. A* 13: 2070–2073.

19 Pan, J.J., Senba, M., Arseneau, D.J. et al. (1993). *Phys. Rev. A* 48: 1218–1226.

20 Pan, J.J. (1995). A μSR study of spin relaxation of small molecules in the gas phase, Ph.D. Thesis, University of British Columbia.

21 Cole, H.R. and Olson, R.E. (1985). *Phys. Rev. A* 31: 2137–2141.

22 Walker, D.C. (1983). *Muon and Muonium Chemistry*. Cambridge University Press.

23 Lazzarini, E., Stadlbauer, J.M., Venkateswaran, K. et al. (1994). *J. Phys. Chem.* 98: 8050–8052.

24 Stadlbauer, J.M., Venkateswaran, K., Porter, G.B., and Walker, D.C. (1997). *J. Phys. Chem. A* 101: 4741–4744.

25 Fleming, D.G., Pan, J.J., Senba, M. et al. (1996). *J. Chem. Phys.* 105: 7517–7535.

26 Roduner, E. and Garner, D.M. (1986). *Hyperfine Interact.* 32: 733–739.

27 Batchelor, S.N., Henningsen, B., and Fischer, H. (1997). *J. Phys. Chem. A* 101: 2969–2972.

28 Dilger, H., Roduner, E., Stolmar, M. et al. (1997). *Hyperfine Interact.* 106: 137–142.

29 Webster, B. (1997). *J. Chem. Soc., Faraday Trans.* 93: 205–210.

30 Pan, J.J., Arseneau, D.J., Senba, M. et al. (2006). *J. Chem. Phys.* 125: 014307.

8

Aspects of Materials Chemistry

8.1 Muonium in Confined Spaces

Chapters 4–7 of this book deal with the interactions of muonium with matter – spin exchange and chemical reactions – and the properties of free radicals which are products of such reactions. But what if we are interested in muonium itself? The usual approach to learn about an isolated atom is to extrapolate observations made in gases at low pressures. This is how the 'vacuum' value of the muonium hyperfine constant, 4463.302765(53) MHz, was determined [1]. Interest in this quantity stems from tests of quantum electrodynamics (QED). Unlike the proton nucleus of the H atom, the muon is 'point-like', which allows for high-precision calculations of the muonium hyperfine interval and various fundamental constants [1].

In contrast to particle physicists, chemists are interested in the *deviation* of the muonium hfc from its vacuum value. Since the isotropic hyperfine constant (hfc) is determined by the Fermi contact interaction, which is proportional to $\psi(0)^2$ (Section 6.1), it follows that the interaction of muonium with its environment will affect the hfc. It may seem surprising, therefore, that the Mu hfc in liquids is only 1% or so below the vacuum value (A_{vac}) [2]. The reason is that Mu, H and other weakly solvated solutes exist in a cavity, or 'bubble', in the solvent. As discussed in Section 5.2, the configuration of the cavity walls may be dynamic, but on average the solute atom resides at the centre of a spherical void [3].

To explore the behaviour of Mu in a cavity in a controlled manner it is better to have a rigid cage. Suitable structures include fullerenes, clathrate hydrates and zeolites, and each of these will be discussed in more detail later in this chapter. First we examine the case of silicon-oxide cages in the class of compounds known as silsesquioxanes. As indicated in Figure 8.1, the Si_8O_{12} cage is almost cube-shaped and can easily host a small guest atom or ion.

H atoms have been generated in such materials by γ-irradiation, and ESR studies have shown that the encapsulated H atoms are stable for months at room temperature, both in the solid state and in solution [4]. This is remarkable stability for an otherwise highly reactive paramagnetic species and makes such systems potential targets for development as qubit elements of a quantum computer. The hyperfine constant for the caged atoms (H@SiQ) varies with the nature of the host (SiQ) but is

Muon Spin Spectroscopy: Methods and Applications in Chemistry and Materials Science, First Edition.
Donald G. Fleming, Iain McKenzie, and Paul W. Percival.

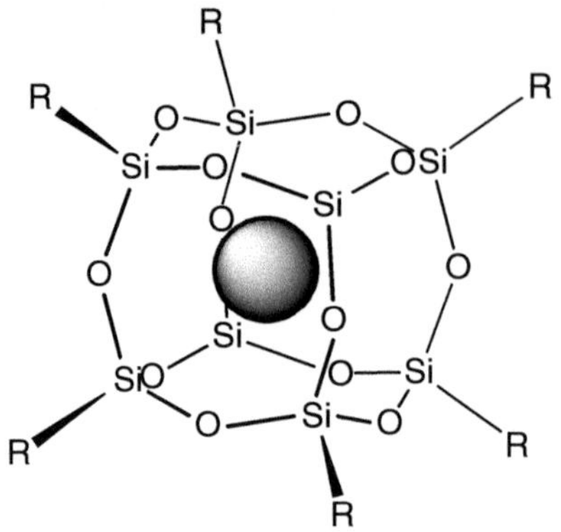

Figure 8.1 Schematic representation of an octasilsesquioxane molecule, where R may be H, methyl, ethyl, propyl or larger substituent such as $-O-Si(CH_3)_3$. The sphere denotes an encapsulated atom.

very close to the vacuum value, typically within 0.5% [5, 6]. The deviation is less for D@SiQ but more so for Mu@SiQ.

The principal ideas governing the interaction of a trapped H atom with the matrix can be traced back to the work of Adrian [7]. He proposed that two opposing effects are in play. On the one hand, non-bonded (van der Waals) interactions transfer spin density from the guest atom to the nearest-neighbour atoms of the matrix. Opposite to this are Pauli exclusion forces, which serve to compress the H atom orbital and thus increase the spin density at the nucleus. In addition to these *static* effects on the guest atom spin distribution are the dynamic effects caused by 'rattling' of the atom in the cage. The dynamic effects are both temperature- and isotope-dependent. To demonstrate the behaviour in different chemical systems it is convenient to express theoretical predictions and experimental data in terms of a normalized hfc, $f = A/A_{\text{vac}}$.

Roduner developed a phenomenological model to explain the temperature and isotope dependence of the hyperfine constants for hydrogen isotopes in liquid water and ice [8], and this has subsequently been applied to other systems, including silsesquioxanes [5, 6]. The model is based on the spherical harmonic oscillator and assumes that the spin density depends linearly on the quadratic displacement of the guest atom from the centre of the cage. Thus

$$f = 1 + c_0 + c_2 \langle R^2 \rangle , \tag{8.1}$$

where c_0 accounts for static effects and c_2 describes the dependence on displacement. The sign for c_0 may be positive or negative, depending on the relative effects of wavefunction compression and spin density transfer under static conditions (zero Kelvin). The dynamic part depends on transfer of spin through van der Waals interactions and results in a negative sign for c_2. The mean displacement is

$$\langle R^2 \rangle = \frac{h\nu}{k} \left(\frac{3}{2} + n_x + n_y + n_z \right), \tag{8.2}$$

where ν, k and n_i are the frequency, force constant and vibrational quantum numbers of the oscillator. Since ν is proportional to $m^{-1/2}$, it is evident that a lighter atom (Mu) will exhibit greater displacement and consequently a lower hfc for a given oscillator state (specified by n_x, n_y and n_z). Similarly, the hfc will fall as higher states become populated ($n_i > 0$) as the temperature increases. To calculate the temperature dependence of f it is necessary to take the Boltzmann average of Eq. (8.2) over all thermally accessible states, which results in

$$f = 1 + c_0 + c_2 \frac{h\nu}{k} \frac{\sum \left(\frac{3}{2} + N\right) \exp\left[-Nh\nu/\left(k_B T\right)\right]}{\sum \exp\left[-Nh\nu/\left(k_B T\right)\right]}, \tag{8.3}$$

where $N = n_x + n_y + n_z$.

This model has been successfully applied to fit temperature-dependent hfcs for D, H and Mu, using only three (isotope-independent) variable parameters: c_0, c_2 and k. An example is given in Figure 8.2, which displays the temperature dependence of the hyperfine constants for Mu, H and D atoms in $Si_8O_{12}(C_2H_5)_8$. The solid line represents the best fit of the model to H atom data (not shown). The two dashed lines are predictions for D and Mu based on the same three fit parameters as H. They do not reproduce the experimental data exactly, but that is not surprising given the simplicity of the model. More important are the principles: lighter isotopes deviate most from their vacuum hyperfine constant, but they have less temperature dependence because it is more difficult to populate vibrational levels above the ground state.

The model parameters used for Figure 8.2 predict 1016 cm^{-1} for the zero-point vibrational frequency of Mu and 0.66 Å for the root mean square displacement $\langle R^2 \rangle_{000}^{1/2}$. The corresponding values for H are 341 cm^{-1} and 0.38 Å. Remarkably, this rattling frequency for H in a rigid cage is not much different from the values estimated for liquid water (252 cm^{-1} [8]) and ice (256 and 209 cm^{-1}, depending on direction [9]).

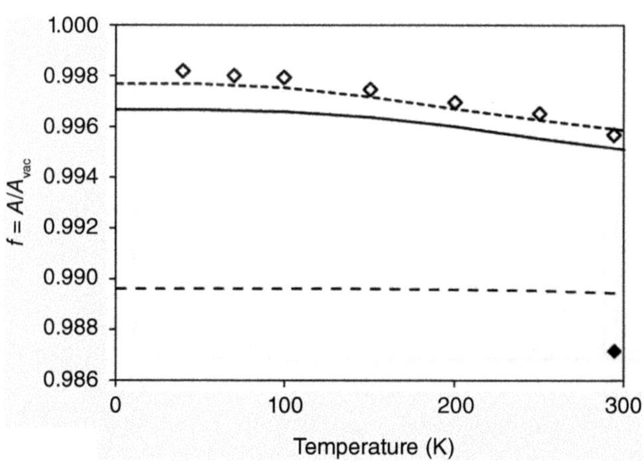

Figure 8.2 Deviation of hyperfine constants from the vacuum value for Mu, H and D atoms in octasilsesquioxane $Si_8O_{12}(C_2H_5)_8$. The diamonds denote experimental data for D (open symbols) and Mu (filled symbol) [5]. The lines represent the predictions of Eq. (8.3) with coefficients c_0, c_2 and k adjusted to fit the H atom data (not shown).

8.2 Muonium and Muoniated Radicals in Fullerenes

Like the silsesquioxanes discussed above, fullerenes have well-defined cage structures. However, they also have unsaturated bonds and should therefore react with muonium to form one or more free radicals. Remarkably, the low-field TF-μSR spectrum of C_{60} exhibits precession signals characteristic of a muoniated radical and *also* those of muonium [10]. C_{60} powder was the first material for which both Mu and radical signals were observed simultaneously. In most circumstances either Mu is long-lived and no radical is seen, or Mu reacts very fast so that there is a coherent transfer of spin polarization to the radical product. As explained in Sections 5.4 and 6.6.3, loss of spin coherence during reaction is manifest as decay of the muon precession signal. Consequently it should be impossible to observe both muonium and its reaction product simultaneously. Yet it happens for C_{60}, as evident from the spectrum displayed in Figure 8.3 [11].

The unbound Mu must be in a protected (chemically inert) environment, as endohedral muonium, Mu@C_{60}. The radical signal results from C_{60}Mu, the radical formed by Mu addition to the *outside* of the cage, as indicated in Figure 8.4. Evidently there is no such reaction on the inside of the cage, which would entail configuration inversion at the point of attachment [12].

Another unusual feature of the muonium–C_{60} system is that the radical can be detected at low field. As explained in Section 2.7.1, muoniated free radicals are usually studied at high field, where the muon precession frequencies collapse to two

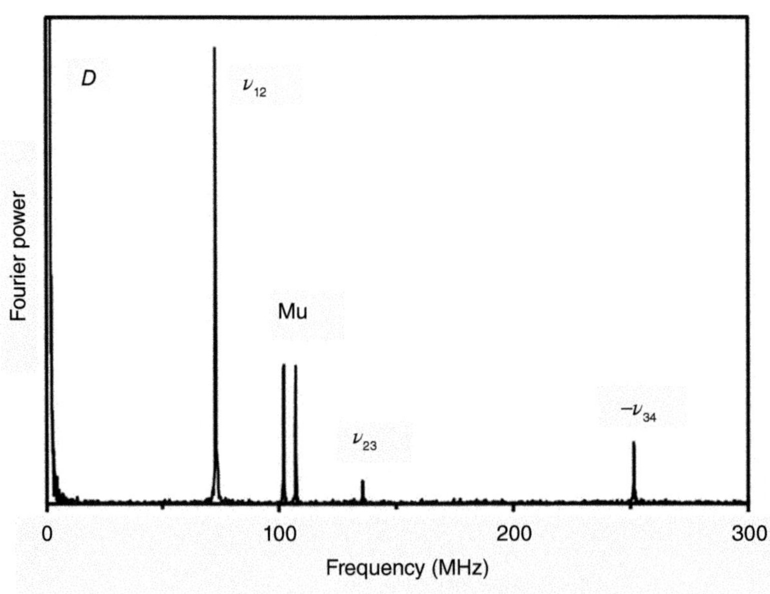

Figure 8.3 Fourier transform μSR spectrum obtained from C_{60} powder in a 75 G field. ν_{12}, ν_{23} and ν_{34} denote precession frequencies of C_{60}Mu; the remaining frequency, ν_{14}, is predicted to be at 460 MHz, off scale. The pair of frequencies near 100 MHz are due to endohedra1 muonium, Mu@C_{60}.

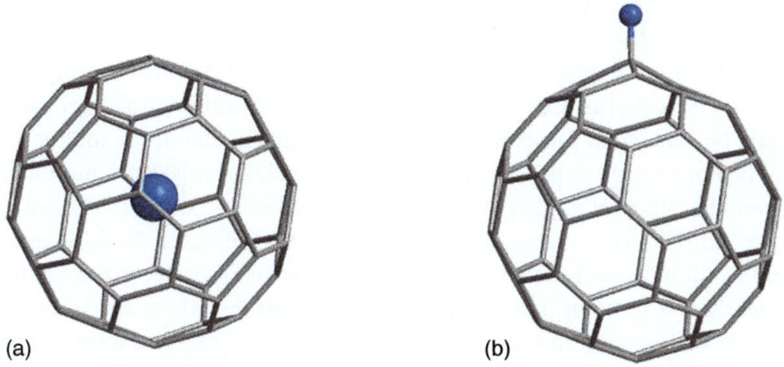

Figure 8.4 Mu@C_{60} (a) and C_{60}Mu (b). The non-bonded endohedral muonium atom is drawn larger than the bound atom in the radical to emphasize its freedom of motion.

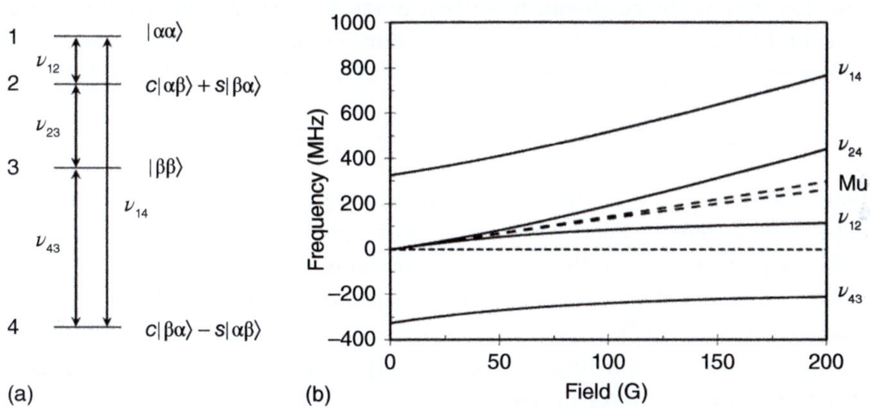

Figure 8.5 Energy levels and transitions for the C_{60}Mu spin system at 75 G (a), where $c = 0.88$ and $s = 0.48$. The field dependence of the spin precession frequencies for C_{60}Mu (solid lines) and Mu@C_{60} (dashed lines) is compared in (b). All except ν_{14} can be seen in the spectrum of Figure 8.3.

values. At low fields there are typically a large number of transition frequencies, and the muon polarization (and hence signal) is dispersed. The simplicity of the spectrum shown in Figure 8.3 thus stems from the lack of protons or other spin-active nuclei in C_{60}Mu[1].

Thus the energy levels and allowed transitions are given by the Breit–Rabi formalism, as displayed for muonium in Figure 2.12, but with a reduced hyperfine frequency. The spin precession frequencies for both Mu@C_{60} and C_{60}Mu are displayed in Figure 8.5.

1 This is not quite accurate. Carbon-13 has a natural abundance of 1.1%, so the probability that an individual C_{60} molecule contains no ^{13}C is only ~50%. However, the probability that a ^{13}C nuclear spin interacts with the unpaired electron in C_{60}Mu is considerably smaller, depending on the extent of the unpaired spin distribution in the radical.

The muon hyperfine constant of endohedral muonium (4341 MHz at 5 K [13]) is close to the vacuum value, consistent with the view that Mu is unbound inside the cage. In contrast the smaller hfc of the C_{60}Mu radical (324.6 MHz at room temperature [11]) is more typical of an alkyl radical (see Section 6.1.2), although a direct comparison would be somewhat misleading. As explained in Chapter 6, the hfc for an alkyl radical is temperature dependent and its 'typical' value results from motional averaging over the dihedral angle between the C—Mu bond and the half-filled *p* orbital on the neighbouring carbon atom. For C_{60}Mu that dihedral angle is zero, which would result in a much larger hfc *if* the π spin density were localized on the α carbon. Clearly it is not; nor is the spin delocalized over the whole structure, as suggested by an early model of C_{60} as an aromatic polyene [14]. The answer lies somewhere between these views, with the magnitude of the hfc depending on spin contributions from all three neighbouring carbons [12].

It is important to realize that the muon hfc is not directly proportional to the sum of π spin densities on the neighbouring carbon atoms. Instead, as pointed out by Whiffen [15] for the case of cyclohexadienyl, one must combine the appropriate orbital coefficients before squaring them. Thus the hfc for a β-H or Mu in C_{60}X is [12]

$$A_\beta = \left(\rho_a^{1/2} + \rho_b^{1/2} + \rho_c^{1/2}\right)^2 \left(L + M\left\langle \cos^2\phi \right\rangle\right), \tag{8.4}$$

where a, b and c label the three carbons adjacent to C—X and the dihedral angle ϕ can be taken as zero for the equilibrium conformation.

While Eq. (8.4) successfully explains the measured muon hfc in C_{60}Mu it concerns only the immediate environment of Mu (or H) attachment. To learn about the extended distribution of unpaired spin it is necessary to examine ^{13}C hyperfine constants, which are related to unpaired spin density via the Karplus–Fraenkel relation (see Eq. (6.6) in Section 6.1.1). This was first done by Klemt et al. [16] for C_{60}H, which was produced photochemically and studied by CW ESR. Later work by Gross et al. [17] confirmed the ESR splittings, although neither study provided signs for the hyperfine constants. Nevertheless, the largest splitting should be positive and correspond to the large unpaired spin density adjacent to the site of H attachment. However, the largest ^{13}C splitting found for C_{60}H is only about half that found for equivalent alkyl adducts C_{60}R [18], and the corresponding spin density is inconsistent with the proton hfc and theoretical predictions [12, 19]. One possibility is that some ^{13}C splittings in the C_{60}H ESR spectrum are obscured by the intensity distortions associated with chemically induced electron polarization (CIDEP). Such effects are clearly manifest in the ESR spectrum of $C_{60}CD_3$ [20].

There are no such complications for C_{60}Mu, which was studied by RF-μSR and ALC-μSR in a dilute solution of 99% enriched $^{13}C_{60}$ in decalin [21]. The ALC spectrum (displayed in Figure 8.6) contains as many as 17 resonances (although not all can be distinguished by eye in this single display). The signals below 13.2 kG correspond to positive ^{13}C hfcs and those on the high-field side to negative values. The intensities depend on the hfcs, as shown in Eq. (2.33), so small coupling constants give rise to weak resonances in the middle region where there are many overlapping signals.

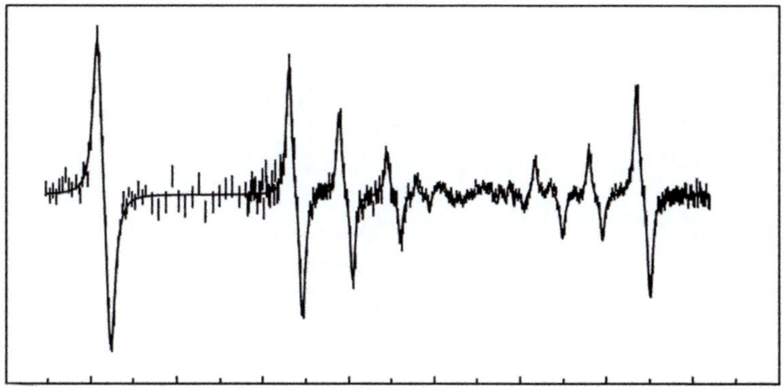

Figure 8.6 ALC-μSR spectrum of the $^{13}C_{60}Mu$ radical in decalin solution at 20 °C [21].

Figure 8.7 Map of unpaired spin density (%) in the $^{13}C_{60}Mu$ radical as determined by assigning ^{13}C hyperfine coupling constants with the aid of theoretical predictions [12].

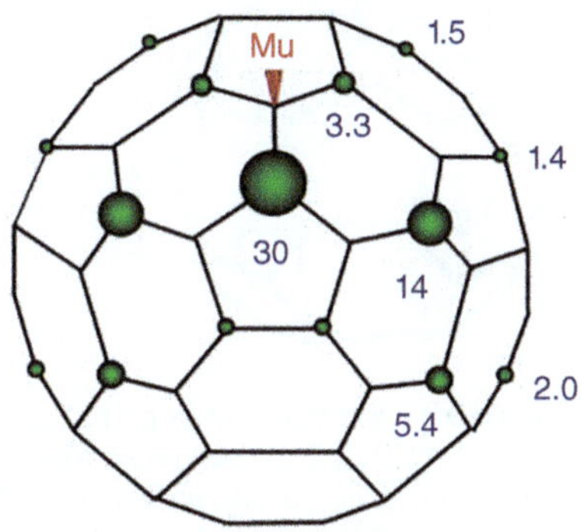

Analysis of the spectrum provides ^{13}C hfcs ranging from +52.6 to −25.4 MHz and the largest value translates to a C_α unpaired spin density of 30%, consistent with both the muon hfc and theoretical predictions. Figure 8.7 is a map of unpaired spin density derived by matching experimental data with predictions. The distribution is consistent with expectations for an alternant hydrocarbon (the network of six-member rings) but tempered by the curvature of the carbon framework.

Investigation of the spin distribution in $C_{70}H$ is more complicated than for $C_{60}H$ because there are multiple isomers. All carbons are identical in C_{60} (being at the intersection of two hexagons and one pentagon), but there are five distinct carbon sites in C_{70}, as shown in Figure 8.8. Those labelled **A**–**D** have similar environments to C_{60}, but carbon type **E** is different, being at the intersection of three hexagons. Its benzenoid character is indicated in Figure 8.8 by the aromatic rings. The curvature of the carbon framework is also smallest at **E**, which could well affect its reactivity, since radical addition would require distortion of the carbon framework to achieve a more tetrahedral configuration.

Mu addition to C_{70} results in either three [22] or four [23, 24] radicals, consistent with ESR studies of H and alkyl radical addition [25, 26]. There is general agreement

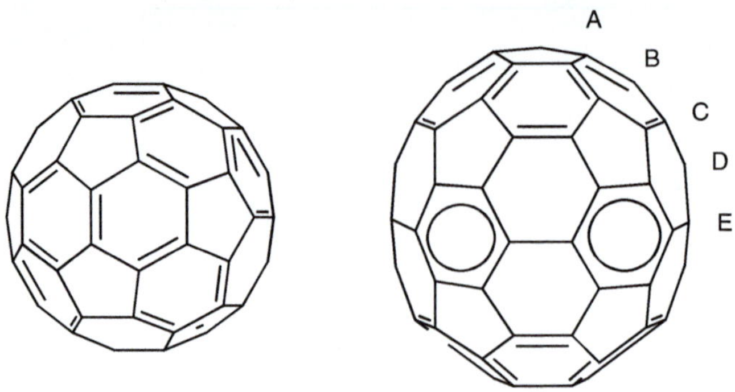

Figure 8.8 Comparison of C_{60} and C_{70} structures. All carbons are equivalent in C_{60}, but there are five different carbon sites in C_{70}.

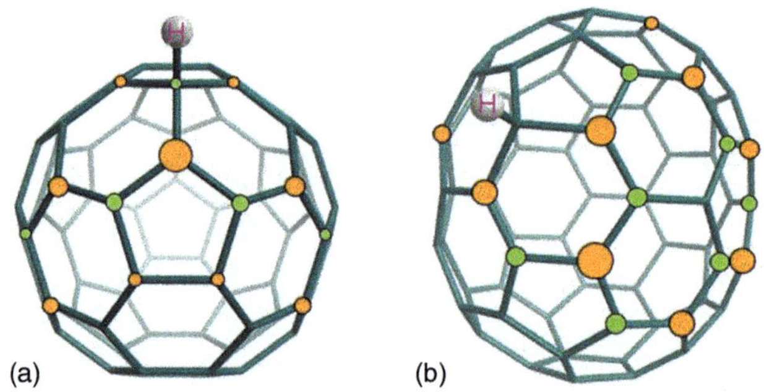

Figure 8.9 Examples of unpaired spin distribution in $C_{70}H$ isomers with the H atom attached to (a) a **B** carbon (view from above); and (b) a **D** carbon (view from the side). The orange and green disks represent positive and negative spin density, respectively, with area proportional to magnitude. Source: Adapted from Brodovitch et al. [24] with permission of the Royal Society of Chemistry.

that addition does not occur at site **E**. Similarly, four $C_{70}Mu$ radicals were detected by ALC-μSR, using a dilute solution of enriched $^{13}C_{70}$ [24]. A total of 36 resonances were recorded and matched with muon hfcs to derive ^{13}C hyperfine couplings. Comparison with DFT predictions then led to the radical assignments. Figure 8.9 shows examples of the resulting maps of unpaired spin density.

8.3 Muonium and Radicals in Clathrates

Gas hydrates, or more generally clathrate hydrates, are another example of porous media which have been investigated with muon spin spectroscopy. They are solid crystalline materials composed of a framework of hydrogen-bonded water molecules arranged to form cages which can contain small guest molecules [27]. The most common crystal structures are cubic and composed of two sizes of cavity: dodecahedrons (5^{12}) and tetradecahedrons ($5^{12}6^2$) for the so-called Structure I; and dodecahedrons (5^{12}) and the larger hexadecahedrons ($5^{12}6^4$) for Structure II. Figure 8.10 shows

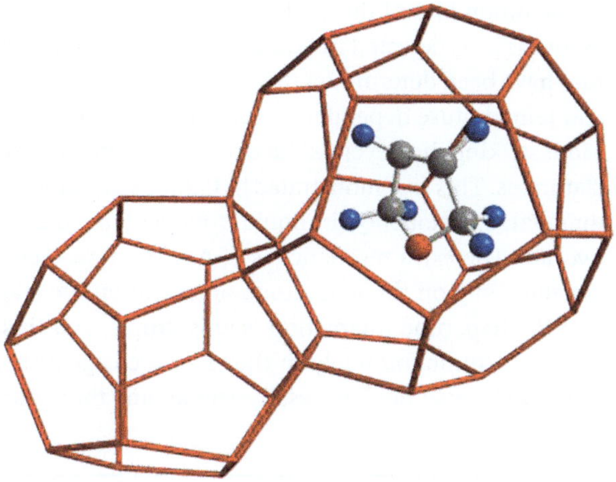

Figure 8.10 Two cages of a structure II clathrate hydrate with a guest molecule (2,5,-dihydrofuran). The red framework represents a lattice of water molecules with oxygen atoms at each vertex and hydrogen atoms between them.

two neighbouring cages of a type II clathrate. The smaller cage can accommodate methane and smaller molecules, but larger molecules (here dihydrofuran) only fit in the larger cavity.

When clathrate hydrates of unreactive molecules are irradiated with muons, muonium is formed. Thus Mu has been detected in the type II clathrates of cyclopentane and tetrahydrofuran [28]. From the splitting of the transverse-field precession frequencies in intermediate fields (see Section 2.6) it was found that the muon hyperfine constant is only 1–2% lower than for Mu in vacuum, indicating that there is only minor interaction between Mu and the host lattice, similar to Mu@SiQ and Mu@C_{60} as described in Sections 8.1 and 8.2.

If the guest molecule is unsaturated (cyclopentene instead of cyclopentane; dihydrofuran instead of tetrahydrofuran), then muoniated free radicals are detected instead of muonium [28–30]. If the guest molecule is symmetric only one radical is formed, but inequivalent reactions sites can lead to multiple radicals [29, 30], as shown in Figure 8.11.

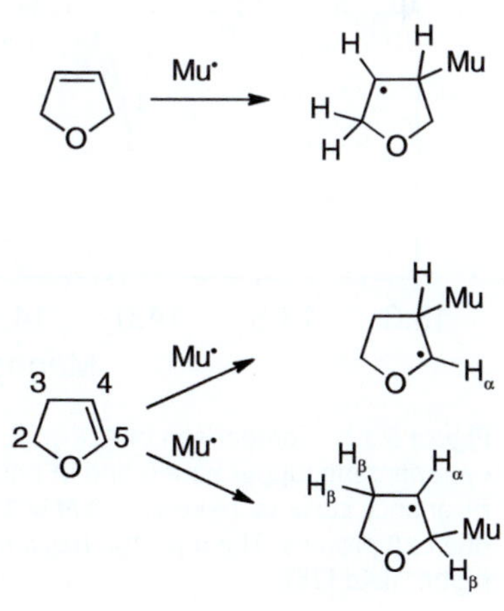

Figure 8.11 Muonium addition to 2,5-dihydrofuran and 2,3-dihydrofuran.

The muon hfcs of the enclathrated radicals are similar to, but not quite the same as for radicals detected in liquid solution at the same temperature. Similarly, proton hfcs have been determined from ALC-μSR spectra, and they have magnitudes, signs and temperature dependence similar to those of the same radicals in liquid phase. More striking, however, is the existence of an additional ALC signal for radicals in clathrates. This is demonstrated in Figure 8.12, which compares ALC-μSR spectra for muoniated cyclopentyl in liquid cyclopentene and in polycrystalline cyclopentene hydrate. The extra resonance is of $\Delta M = 1$ type (Section 2.8), and its field position depends only on the magnitude of the muon hfc (Eq. (2.37)). Its existence reveals that the hyperfine interaction is anisotropic, and this in turn implies that the radical is not tumbling freely in the clathrate cage as it does in the liquid phase. The sharp transverse-field precession signals and the symmetrical line shapes of the ALC

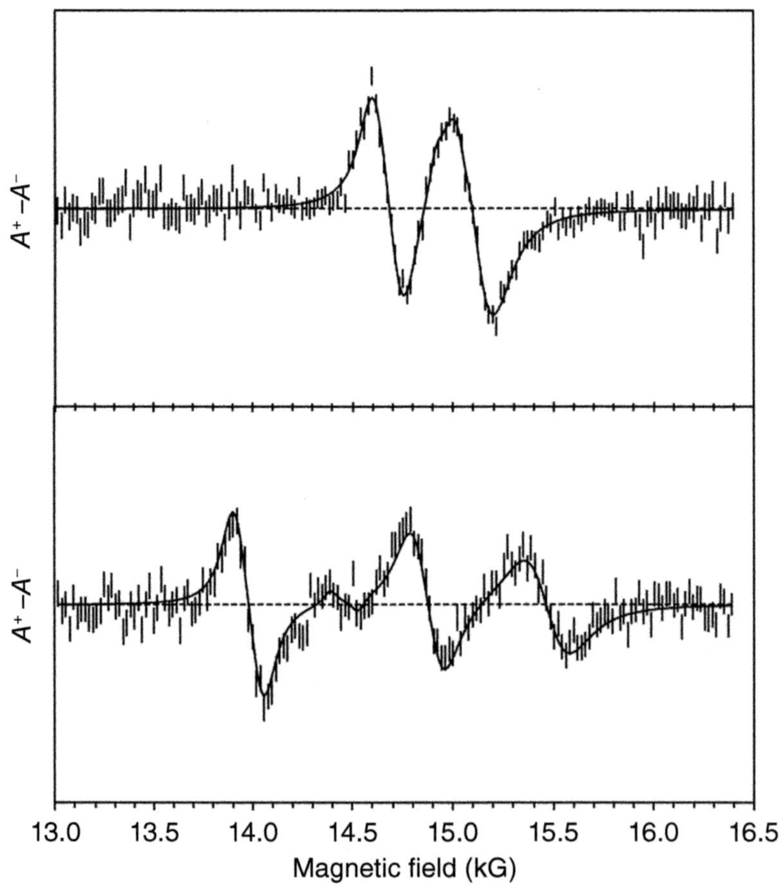

Figure 8.12 Comparison of ALC-μSR spectra for muoniated cyclopentyl in liquid cyclopentene (upper panel) and in polycrystalline cyclopentene hydrate (lower panel). The resonance close to 14 kG is of $\Delta M = 1$ type. Those at higher fields are $\Delta M = 0$ resonances due to β protons. The α proton has a negative hfc and its ALC resonance appears at much higher field [28].

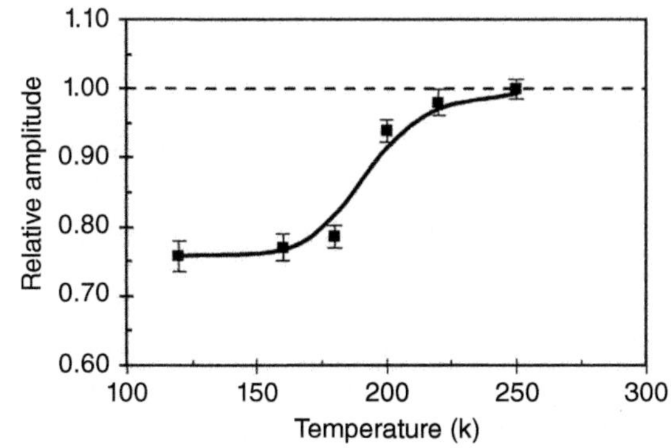

Figure 8.13 Relative radical yield as determined from the intensity of $\Delta M = 1$ ALC signals for the $(CD_3)_2C$-OMu radical in acetone-d6 hydrate. The line through the points represents a simple model for activated diffusion of Mu. The best fit corresponds to an activation energy of about 26 kJ mol^{-1} [31].

spectra suggest that the anisotropy is small, at least at relatively high temperature (as in Figure 8.12) [28]. A possible explanation for the anisotropy is that transient hydrogen bonding provides a 'tether' to restrict the motion of guest molecules and the related radicals.

At low temperatures (~100 K) the radical signals are broad and asymmetric, with powder-pattern line shapes typical of a random distribution of frozen orientations. Furthermore, in some hydrates muonium can be detected in addition to the radical signal. The obvious inference is that Mu is formed when muons stop in empty cages, where they are chemically stable, at least on the microsecond time scale. Muons which stop in occupied cages quickly form radicals, although it is not known if this occurs via muonium or through an ionic route (muon attachment followed by charge neutralization). The Mu signal disappears as the temperature is raised, and the radical yield increases, as shown in Figure 8.13. This behaviour is consistent with activated transport of Mu from empty cages to those containing reactive guest molecules.

8.4 Muoniated Radicals in Zeolites

Microporous materials are generally important in the realm of heterogeneous catalysis, and zeolites in particular have important applications in chemical industry. However there is relatively little known at the microscopic level about the interactions and reactions of guest molecules in different zeolite frameworks, and in particular the involvement of free radicals. Zeolites have alumino-silicate structures incorporating charge-compensating cations, and typically act by facilitating proton transfer from Brønsted acid sites to guest molecules, forming carbenium ion intermediates [32], which may then undergo charge neutralization to form free radicals. Alternatively, neutral free radicals may be formed by direct addition of H atoms to unsaturated guest molecules, as evident from ESR studies of the HZSM-5 zeolite loaded with various hydrocarbons and subjected to γ-radiolysis [33]. Remarkably, such 'H-atom transfer' reactions in zeolites have been more extensively studied by μSR [34–39].

Profiled here are studies of the muoniated tert-butyl radical and the muoniated cyclohexadienyl radical in zeolites, formed by Mu addition to isobutene and benzene, respectively. A large number of different zeolite structures are known [40], all based on silica and aluminosilicate frameworks arranged to form a variety of geometries with different internal spacings and void sizes. Two classes of zeolite materials have been widely employed in μSR studies: ZSM-5, which has intersecting channels that give small void regions of a few Å, and the synthetic 'faujasites' (NaY, HY and USY), which have large internal voids known as supercages, with diameters ~12 Å, connected by windows ~7.5 Å in diameter.

Unlike the fullerenes and clathrate hydrates, there is often significant interaction between a free radical and the host zeolite framework. Thus, TF-μSR spectra of radicals in zeolites typically have broad, anisotropic lines. More useful is the ALC-μSR technique, since the anisotropic muon hyperfine interaction results in a strong Δ_1 resonance (Section 2.8). The resonance position is determined by the isotropic hfc, and the line shape depends on the anisotropy and its modulation by molecular motion (Section 6.4.2). This is well demonstrated by the work of Roduner et al. on the reorientational dynamics of cyclohexadienyl radicals in the high silica ZSM-5 zeolite [34]. Analysis of the temperature-dependent line shapes suggests that the radical performs fast uniaxial rotation about the axis perpendicular to the molecular plane, on a time scale of about 50 ns. It was concluded from the high mobility that the radicals occupy sites at the channel intersections, where there is more free space than inside the channels themselves. Additional signals were noted at high temperatures (above 450 K), consistent with access to the higher energy sites located inside the channels.

A remarkable change in behaviour was noted when benzene was loaded into the H-form of ZSM-5. The ALC signals were found to be much broader in H ZSM-5, with high-temperature shapes similar to those in the high silica sample at 48 K, indicating only limited motion. It was concluded that the radical is locked to the acid site but can still undergo uniaxial rotation, as indicated in Figure 8.14 [35]. The increased broadening and shift of resonances seen at lower temperatures is explained by a dynamic equilibrium between muoniated cyclohexadienyl, C_6H_6Mu, and its protonated form, $C_6H_7Mu^+$:

$$C_6H_6Mu + ZH^+ \rightleftarrows C_6H_7Mu^+ + Z,$$

where Z represents the zeolite acid site. The equilibrium moves to the right at lower temperature, resulting in the 1,3-cyclohexadienyl radical cation, a radical ion detected by ESR after γ-irradiation of 1,3-cyclohexadiene in NaZSM-5 [41].

The interaction between a free radical and the host zeolite framework can also be illustrated by the behaviour of the Mu-*t*-butyl radical in different zeolites. Figure 8.15 compares ALC-μSR spectra for this radical in NaY (blue points) and USY (green points) [39]. The strong signal at ~17.2 kG in the top panel (NaY, 90 K) is due to the muon Δ_1 resonance, which is broadened at higher temperatures and shifts downfield (~14.2 kG at 150 K; ~11.8 kG at 240 K), consistent with expectations for a lower muon hfc at higher temperatures (see Section 6.3.1). There is an additional, weaker signal at ~15.4 kG in the top panel, and this is attributed to the Δ_0 resonance involving the sodium nuclear spin ($I = 3/2$ for ^{23}Na). This resonance is evident at ~13.4 kG at 150 K but only as a shoulder at ~11.4 kG in the scan at 240 K.

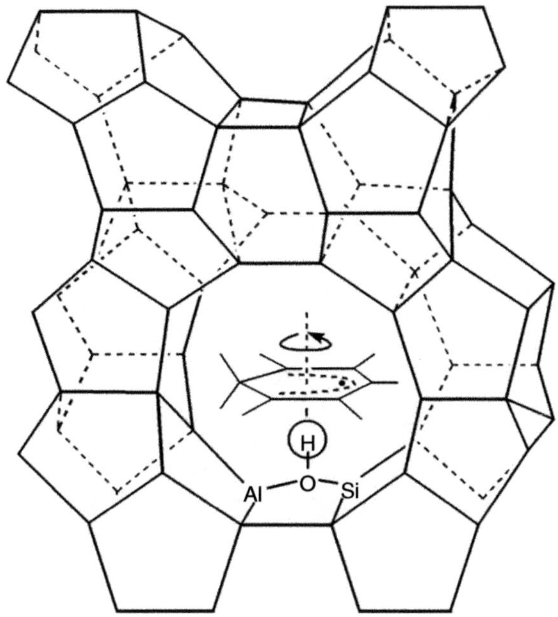

Figure 8.14 Schematic view of a cyclohexadienyl radical docked to an acid site in the HZSM-5 zeolite. Source: Figure taken from Roduner and Dilger [35] with permission from the American Chemical Society.

These signals are absent for the USY zeolite and provide clear evidence for the binding of the Mu-*t*-butyl radical to the cation in NaY. These are the first measurements of sodium hfcs by muon spin spectroscopy; the full data set can be seen in Figure 8.16 [39].

The ALC signals for the radical in USY (green points in Figure 8.15) are completely dominated by the Δ_1 resonances, which are much broader than in NaY. The widths depend on the extent of hyperfine anisotropy sampled by the motion of the radical. This is greater for USY, where the radical is less strongly bound to —OH sites than is the binding to the cation sites in NaY. It is also significant that the broad Δ_1 resonances persist well above the melting point of solid isobutene (133 K), indicating the continued anisotropic environment imposed by the zeolite host.

Other weaker resonances arise from the protons of the $—CH_3$ and $—CH_2Mu$ groups of the radical.

The muon hfcs for the *tert*-butyl radical in NaY and USY zeolites are contrasted in Figure 8.17. Detailed analysis [39] of the temperature dependence is consistent with intramolecular motion of the radical studied in neat isobutene [39, 42], except for lower hfcs at low temperatures in NaY, which can be attributed to some transfer of electron spin density from the radical to the cation in keeping with the observation of Na Δ_0 resonances.

Similar effects have been noted for muoniated ethyl radicals in faujasites [37].

Muoniated cyclohexadienyl radicals have also been studied in the NaY zeolite [38]. There is clear evidence of strong binding of the radical to Na^+ sites, although the Na Δ_0 resonances were not detected. The existence of extra Δ_1 and proton Δ_0 resonances compared to those in bulk benzene suggests significant distortion of the

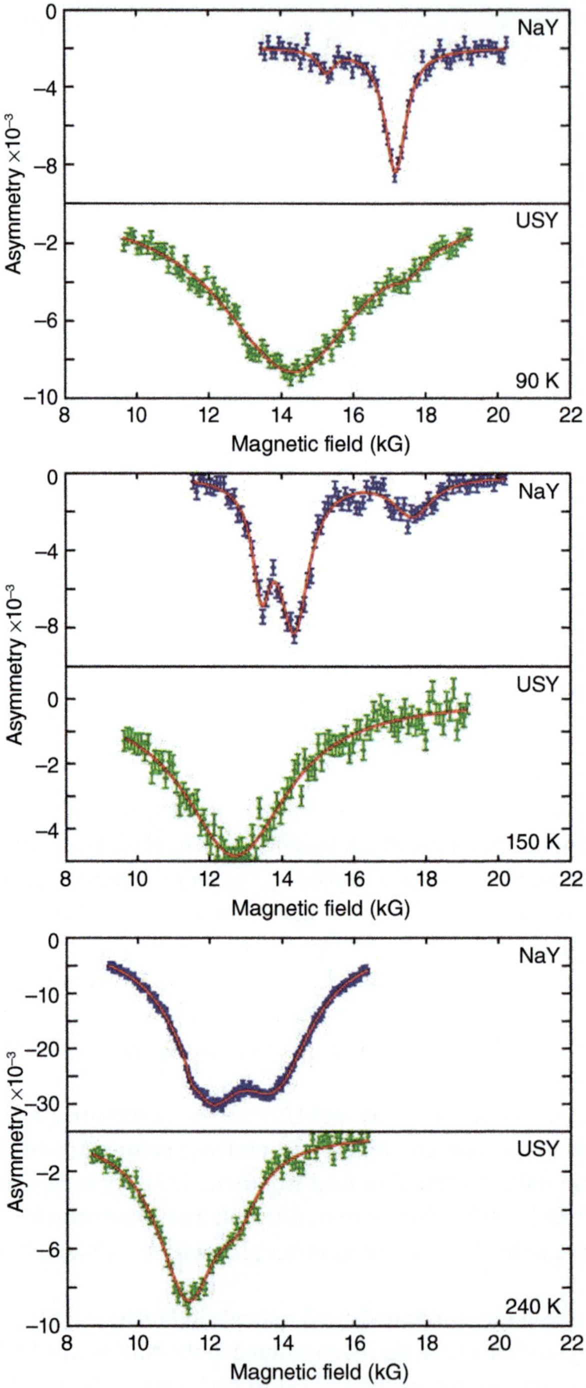

Figure 8.15 Comparison of ALC-μSR spectra for the muoniated *tert*-butyl radical formed from isobutene in NaY and USY zeolites. Source: Adapted from Fleming et al. [39] with permission from the American Chemical Society.

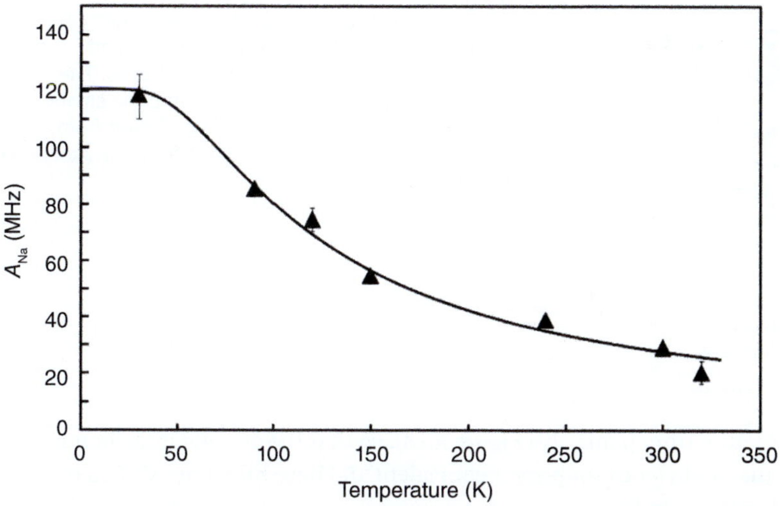

Figure 8.16 Temperature dependence of ^{23}Na hyperfine coupling constants determined from ALC spectra of the muoniated tert-butyl radical in NaY zeolite. The line through the data points is a semiempirical fit consistent with a reduction in electron spin transfer to Na^+ as a result of intramolecular motion of the attached radical [39]. Source: Adapted from Fleming et al. [39] with permission from the American Chemical Society.

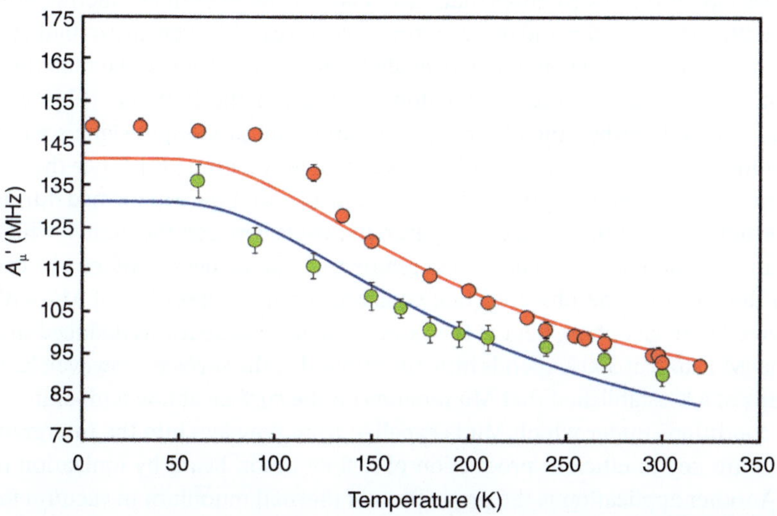

Figure 8.17 Temperature dependence of reduced muon hyperfine coupling constants determined from ALC spectra of the muoniated *tert*-butyl radical in NaY (red data points) and USY (green data points). The lines through the data points represent semiempirical fits consistent with intramolecular motion of the radical [39]. Source: Adapted from Fleming et al. [39] with permission from the American Chemical Society.

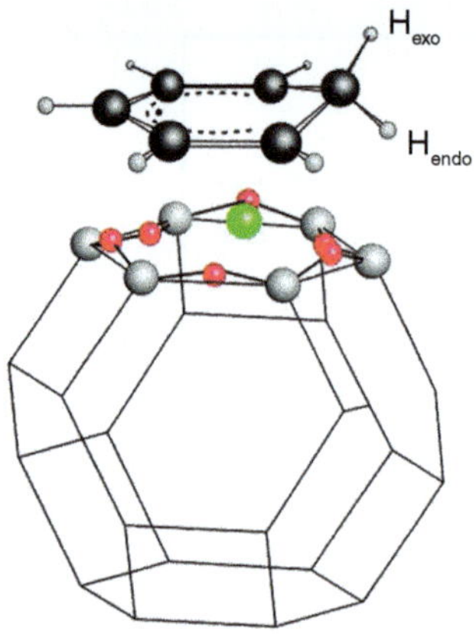

Figure 8.18 Schematic diagram of the cyclohexadienyl radical bound to the sodium cation (green), causing distortion from planarity and inequivalence of the exo and endo H bonds. The light grey circles represent the Si atoms and the smaller red circles the bridging O atoms. Source: Taken from Fleming et al. [38] with permission from the American Chemical Society.

radical structure from planarity (Figure 8.18), so that the exo and endo bond orientations of the methylene group are inequivalent and have hfc values shifted by ~20% compared with the bulk.

8.5 Muonium and Radicals on Surfaces

The previous sections of this chapter have dealt with muonium and muoniated radicals *inside* materials. Now we consider their interactions with external surfaces. There is overlap for those structures that have relatively large cavities: microporous materials with pores 2–50 nm and macroporous structures with even larger diameter pores. The focus here will be on powders made from silica and silica-like materials.

The first evidence for muonium formation in silica was the 1978 report by Marshall et al., which described the observation of muonium precession signals when positive muons were stopped in fine silica powder under vacuum [43]. Since the signal was lost when oxygen was present it was established that the Mu formed during muon thermalization in the silica escapes into the voids between the grains, which were 3–5 nm in diameter. For larger size grains the muons decay before Mu can diffuse to the surface. The observed rate constant for spin relaxation of Mu with oxygen agreed with gas-phase data at the same temperature, so it was deduced that the thermal Mu atom at 300 K spends little time bound to the surface. However, later experiments at 6 K established that Mu remains on the surface at low temperatures [44]. The conditions under which Mu is expelled from powders into the intergranular region are key to efficient production of a slow muon beam by ionization of Mu [45]. Another application is the production of thermal muonium in vacuum for atomic physics experiments, such as measurement of the muonium 1s–2s transition frequency needed for a test of bound-state Quantum Electrodynamics [46].

If reactants with unsaturated bonds are bound to the surface, then free radicals may be formed. The first example of this was the muoniated 1,1,2-trimethylallyl radical detected on silica grains covered with 2,3-dimethylbutadiene [47]:

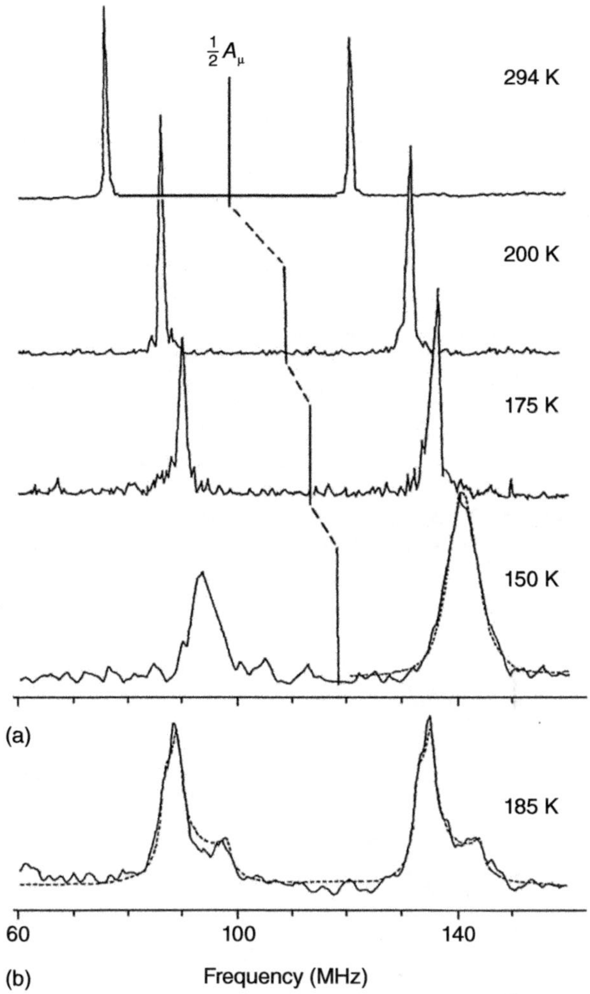

Figure 8.19 TF-μSR spectra of the muoniated radical formed in (a) a monolayer of 2,3-dimethylbutadiene adsorbed on spherical silica grains of 7 nm diameter; (b) the bulk frozen diene. Source: Adapted from Roduner et al. [48] with permission from Springer Nature.

The TF-μSR spectrum was studied over a range of temperatures, as shown in Figure 8.19a. The lines get broader at lower temperatures but remain isotropic down to 150 K, in contrast to the situation for the same radical in bulk frozen material (Figure 8.19b). This implies that the radical on the surface has sufficient translational mobility to average the hyperfine anisotropy.

The muoniated cyclohexadienyl radical has also been studied on silica powder surfaces [48–50]. The TF-μSR spectra suffer from loss of phase coherence due to a slower radical formation rate (see Sections 2.7.3 and 5.4), but that is not a problem for ALC-μSR. The muon Δ_1 resonance is strong and very broad, but the methylene Δ_0 resonance is well resolved (see Figure 8.20). The temperature dependence of its linewidths has been analysed in terms of the motional dynamics of the radical on the surface of spherical grains. The radical is assumed to be physisorbed by hydrogen bonding and lying flat at hydroxyl (—SiOH) sites. As it hops or 'skates' between such sites the angle between the ring symmetry axis changes with respect to the applied magnetic field, leading to loss of muon spin polarization. Detailed line shapes have been calculated with a stochastic Liouville formalism [51] and used to fit

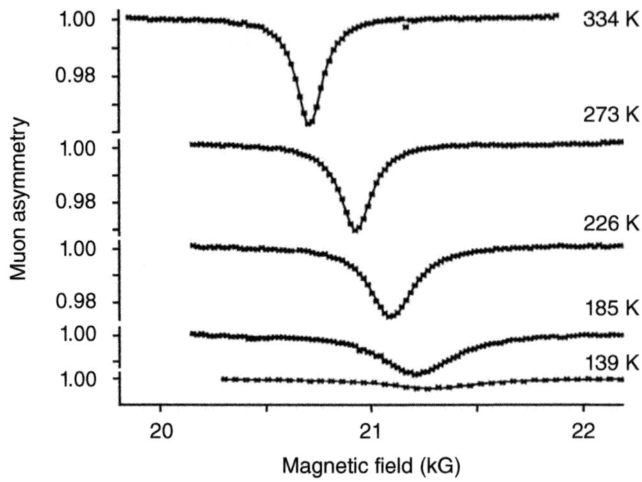

Figure 8.20 Δ_0 ALC resonances due to the methylene proton of the muoniated cyclohexadienyl radical formed in a monolayer of benzene on 7 nm silica grains. Source: Adapted from Reid et al. [49] with permission from Springer Nature.

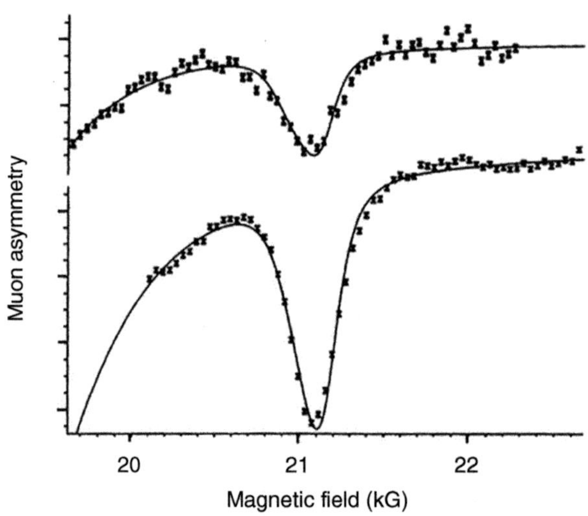

Figure 8.21 Δ_0 ALC resonances due to the methylene proton of the muoniated cyclohexadienyl radical formed in benzene on the palladium catalyst (upper spectrum) compared with a similar coverage of benzene on silica grains (bottom). Source: Adapted from Schwager et al. [50] with permission from Elsevier.

the experimental curves [50]. The resulting diffusion coefficients exhibit an Arrhenius temperature dependence, with activation energy $\sim$10 kJ mol^{-1}, with somewhat higher values for lower surface coverage. Similar results were found from NMR studies of benzene on silica and alumina [52], suggesting that the radical diffuses on the surface in similar manner to benzene. This seems at first surprising, given the open-shell nature of the radical, but it is consistent with the view that van der Waals forces dominate the interaction of the adsorbate with the surface and/or other adsorbed molecules.

An alternative theoretical framework based on the Redfield formalism has also been developed to analyse how motional averaging of anisotropic hyperfine interactions affects muon spin spectra of radicals on a spherical surface [53]. It has been applied to the surface diffusion of the muoniated 2,3,4-trimethylcyclohexadienyl radical, and the results agree with those of the stochastic Liouville approach [53].

Many practical applications of heterogeneous catalysis involve metal surfaces. One way to achieve high surface area is to isolate small grains of metal on the surface of an inert powder or inside a porous material. Such small metallic particles are now often referred to as nanoparticles [54]. An early demonstration of the power of muon spin spectroscopy to investigate such catalysts can be seen in the work of Schwager et al. [50]. As shown in Figure 8.21, there is a dramatic reduction in the signal amplitude of muoniated cyclohexadienyl on Pd-doped silica (8% coverage of benzene) compared with silica (5% coverage). This is consistent with loss of muon spin polarization when the radical is close to the metal; the observed signal simply corresponds to those radicals which did not reach a metal particle before muon

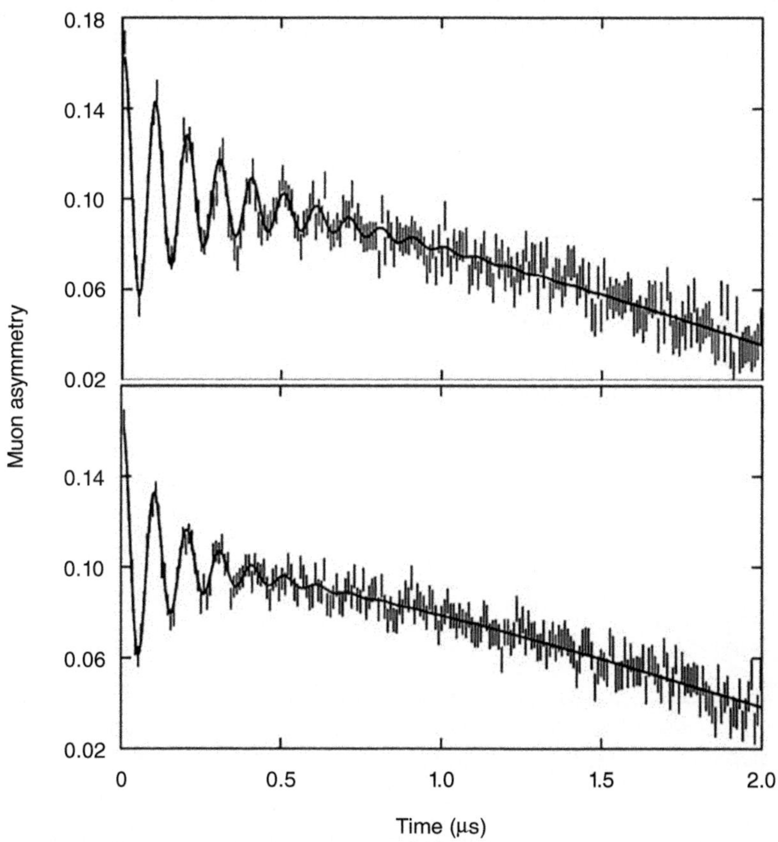

Figure 8.22 Decay of the muonium precession signal in mesoporous silica loaded with 8 nm diameter gold nanoparticles (upper panel; $\lambda_{TF} = 3.8\ \mu s^{-1}$) compared with a similar sample in which the gold particles are covered with ~2 monolayers of benzene (lower panel; $\lambda_{TF} = 6.3\ \mu s^{-1}$). Source: Adapted from Fleming et al. [56] permission from the American Chemical Society.

decay [50]. An alternative explanation is the inhibition of radical formation by a chemisorption reaction of Mu with the Pd nanoparticle, as reported for Mu reacting with gold nanoparticles [55].

Newer results have been reported for Mu interacting with gold nanoparticles embedded in mesoporous silica, which both guarantees isolated nanoparticles and that they are largely uncapped[2], providing an unperturbed reactive metal surface. If Mu thermalizes in a cavity that happens to contain a gold nanoparticle, it chemisorbs almost instantaneously, with a coherent transfer of muonium spin polarization to a diamagnetic state [55]. In contrast, those Mu atoms which thermalize in a different cavity require time to diffuse to the gold particle, and what is observed is the relaxation of the Mu precession signal, as shown in the upper panel of Figure 8.22 [56].

Coating the gold particles with even a monolayer of benzene results in faster Mu decay, as shown for ~2 monolayers in the lower panel of Figure 8.22. The decay rates exhibit a linear dependence on benzene concentration, and this has been studied for gold nanoparticles of various sizes embedded in mesoporous silica (Figure 8.23) [56]. As evident from the slopes of the lines, the metal nanoparticles have a catalytic effect on the reaction rate. The increased slope (and thence rate constant) for the 38 nm particle compared to the smaller particles could indicate a larger number of available sites for the surface-catalysed reaction.

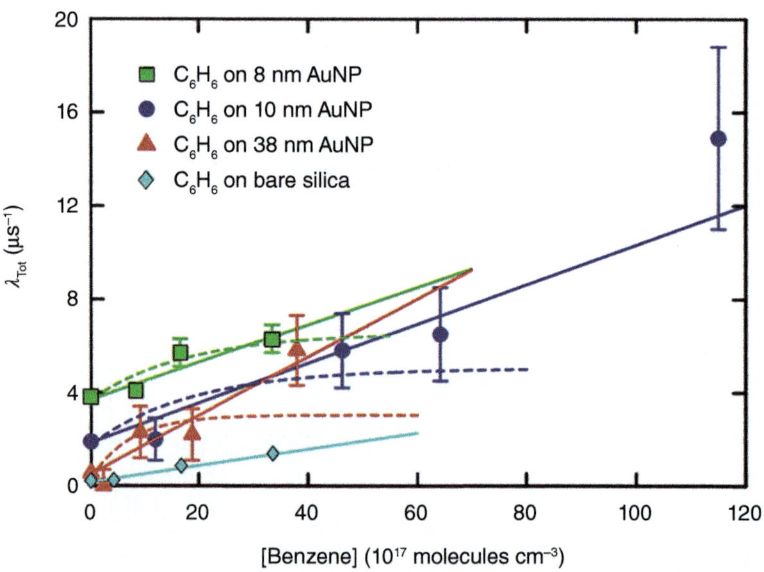

Figure 8.23 Decay rates (λ_{Tot}) for muonium reacting with benzene on bare silica (cyan line) and benzene adsorbed on gold nanoparticles (AuNP) of various sizes embedded in mesoporous silica. Source: Adapted from Fleming et al. [56] with permission from the American Chemical Society.

2 To avoid aggregation, nanoparticles are frequently 'capped' with organic groups, typically thiols.

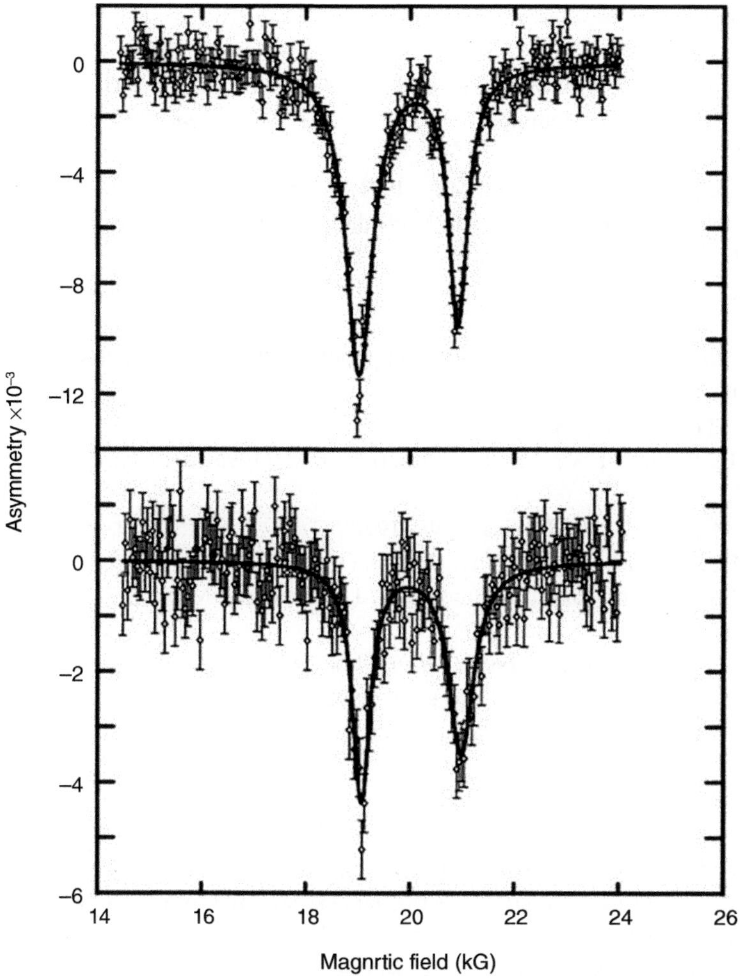

Figure 8.24 Comparison of ALC resonances (Δ_1 at ~19 kG; Δ_0 methylene proton at ~20.8 kG) for the muoniated cyclohexadienyl radical formed in benzene loaded on bare silica (top) and gold-loaded silica (bottom). Note that the amplitudes are much smaller in the latter case, consistent with the chemisorption reaction of Mu with the gold nanoparticles. Source: Adapted from Fleming et al. [58] with permission from the American Chemical Society.

In addition to the dependence on benzene concentration, ALC-μSR studies confirm that the Mu decay results from chemical reaction. Both Δ_0 and Δ_1 resonances of muoniated cyclohexadienyl have been detected in gold-loaded and platinum-loaded silica [57]. Recent results for gold nanoparticles are shown in Figure 8.24. Since the resonance positions are the same as for benzene loaded on bare silica, it can be inferred that there is no direct binding to the metal atoms. Instead, at the high benzene loadings of this study it is likely that benzene preferentially coats the metal nanoparticles and thus shields the radicals from direct interaction with the metal surface.

References

1 Liu, W., Boshier, M.G., Dhawan, S. et al. (1999). *Phys. Rev. Lett.* 82: 711–714.
2 McKenzie, I., Dilger, H., Roduner, E. et al. (2008). *J. Phys. Chem. B* 112: 3070–3076.
3 De Raedt, B., Sprik, M., and Klein, M.L. (1984). *J. Chem. Phys.* 80: 5719–5724.
4 Sasamori, R., Okaue, Y., Isobe, T., and Matsuda, Y. (1994). *Science* 265: 1691–1693.
5 Dilger, H., Roduner, E., Scheuermann, R. et al. (2000). *Physica B* 289–290: 482–486.
6 Gross, B., Dilger, H., Scheuermann, R. et al. (2001). *J. Phys. Chem. A* 105: 10012–10017.
7 Adrian, F.J. (1960). *J. Chem. Phys.* 32: 972–981.
8 Roduner, E., Percival, P.W., Han, P., and Bartels, D.M. (1995). *J. Chem. Phys.* 102: 5989–5997.
9 Bartels, D.M., Han, P., and Percival, P.W. (1992). *Chem. Phys.* 164: 421–437.
10 Ansaldo, E.J., Boyle, J., Niedermayer, C. et al. (1992). *Z. Phys. B* 86: 317–318.
11 Percival, P.W., Addison-Jones, B., Brodovitch, J.-C., and Sun-Mack, S. (1996). *Appl. Magn. Reson.* 11: 315–323.
12 Percival, P.W. and Wlodek, S. (1992). *Chem. Phys. Lett.* 196: 317–320.
13 Kiefl, R.F., Duty, T.L., Schneider, J.W. et al. (1992). *Phys. Rev. Lett.* 69: 2005–2008.
14 Estreicher, S.K., Latham, C.D., Heggie, M.I. et al. (1992). *Chem. Phys. Lett.* 196: 311–316.
15 Whiffen, D.H. (1963). *Mol. Phys.* 6: 223–224.
16 Klemt, R., Roduner, E., and Fischer, H. (1994). *Chem. Phys. Lett.* 229: 524–530.
17 Gross, R. and Dinse, K.P. (1999). *Recent Advances in the Chemistry and Physics of Fullerenes and Related Material*, Electrochemical Society Proceedings Volume 99–12, vol. 7 (ed. P.V. Kamat, D.M. Guldi, and K.M. Kadish), 18–27.
18 Morton, J.R., Preston, K.F., Krusic, P.J. et al. (1992). *J. Phys. Chem.* 96: 3576–3578.
19 Morton, J.R., Negri, F., and Preston, K.F. (1994). *Can. J. Chem.* 72: 776–782.
20 Morton, J.R., Negri, F., and Preston, K.F. (1996). *Appl. Magn. Reson.* 11: 325–333.
21 Percival, P.W., Addison-Jones, B., Brodovitch, J.-C. et al. (1995). *Chem. Phys. Lett.* 245: 90–94.
22 Niedermayer, C., Reid, I.D., Roduner, E. et al. (1993). *Phys. Rev. B* 47: 10923–10926.
23 Addison-Jones, B., Percival, P.W., Brodovitch, J.-C. et al. (1994). *Hyperfine Interact.* 86: 817–824.
24 Brodovitch, J.-C., Addison-Jones, B., Ghandi, K. et al. (2015). *Phys. Chem. Chem. Phys.* 17: 1755–1762.
25 Morton, J.R., Negri, F., and Preston, K.F. (1994). *Chem. Phys. Lett.* 218: 467–474.
26 Borghi, R., Lunazzi, L., Placucci, G. et al. (1996). *JACS* 118: 7608–7617.
27 Sloan, E.D. Jr., and Koh, C.A. (2008). *Clathrate Hydrates of Natural Gases*, 3e. Boca Raton, FL: CRC press.

28 Percival, P.W., Mozafari, M., Brodovitch, J.-C., and Chandrasena, L. (2014). *J. Phys. Chem. A* 118: 1162–1167.

29 Mozafari, M., Brodovitch, J.-C., Chandrasena, L., and Percival, P.W. (2016). *J. Phys. Chem. A* 120: 8521–8528.

30 Mozafari, M., Chandrasena, L., McKenzie, I. et al. (2017). *Can. J. Chem.* 96: 217–225.

31 Mozafari, M., Chandrasena, L., McKenzie, I. et al. (2020). *Radiat. Phys. Chem.* 168: 108532.

32 Haw, J.F. (2002). *Phys. Chem. Chem. Phys.* 4: 5431–5441.

33 Werst, D.W., Han, P., Choure, S.C. et al. (1999). *J. Phys. Chem. B* 103: 9219–9230.

34 Roduner, E., Stolmar, M., Dilger, H., and Reid, I.D. (1998). *J. Phys. Chem. A* 102: 7591–7597.

35 Roduner, E. and Dilger, H. (2001). *JACS* 123: 7717–7718.

36 Fleming, D.G., Shelley, M.Y., Arseneau, D.J. et al. (2002). *J. Phys. Chem. B* 106: 6395–6407.

37 Bridges, M.D., Arseneau, D.J., Fleming, D.G., and Ghandi, K. (2007). *J. Phys. Chem. C* 111: 9779–9793.

38 Fleming, D.G., Arseneau, D.J., Shelley, M.Y. et al. (2011). *J. Phys. Chem. C* 115: 11177–11191.

39 Fleming, D.G., Arseneau, D.J., Bridges, M.D. et al. (2013). *J. Phys. Chem. C* 117: 16523–16539.

40 Baerlocher, C. and McCusker, L.B., Database of Zeolite Structures. http://www.iza-structure.org/databases/.

41 Werst, D.W., Tartakovsky, E.E., Piocos, E.A., and Trifunac, A.D. (1994). *J. Phys. Chem.* 98: 10249–10257.

42 Percival, P.W., Brodovitch, J.-C., Leung, S.K. et al. (1988). *Chem. Phys.* 127: 137–147.

43 Marshall, G.M., Warren, J.B., Garner, D.M. et al. (1978). *Phys. Lett. A* 65: 351–353.

44 Kiefl, R.F., Warren, J.B., Oram, C.J. et al. (1982). *Phys. Rev. B* 26: 2432–2441.

45 Bakule, P., Beer, G.A., Contreras, D. et al. (2013). *Prog. Theor. Exp. Phys.* 2013: 103C01.

46 Khaw, K.S., Antognini, A., Prokscha, T. et al. (2016). *Phys. Rev. A* 94: 022716.

47 Heming, M. and Roduner, E. (1987). *Surf. Sci.* 189–190: 535–542.

48 Roduner, E., Schwager, M., and Shelley, M. (1994). *Mol. Eng.* 4: 259–276.

49 Reid, I.D., Azuma, T., and Roduner, E. (1990). *Nature* 345: 328–330.

50 Schwager, M., Dilger, H., Roduner, E. et al. (1994). *Chem. Phys.* 189: 697–712.

51 Kreitzman, S.R. and Roduner, E. (1995). *Chem. Phys.* 192: 189–230.

52 Boddenberg, B. and Beerwerth, B. (1989). *J. Phys. Chem.* 93: 1440–1447.

53 Schwager, M., Dilger, H., Roduner, E., and Reid, I.D. (1995). *Ber. Bunsenges Phys. Chem.* 99: 142–151.

54 Roduner, E. (2006). *Nanoscopic Materials : Size-Dependent Phenomena*, Chap. 10. Cambridge, UK: Royal Society of Chemistry.

55 Dehn, M.H., Arseneau, D.J., Boni, P. et al. (2016). *J. Chem. Phys.* 145: 181102.

56 Fleming, D.G., MacFarlane, W.A., Xiao, J. et al. (2019). *J. Phys. Chem. C* 123: 27628–27642.

57 Xiao, J., Arseneau, D.J., Bridges, M.D. et al. (2014). *J. Phys. Conf. Ser.* 551: 012044.

58 Fleming, D.G., MacFarlane, W.A., Xiao, J. et al. (2021). *J. Phys. Chem. C* 125: 21410–21423.

9

Soft Matter, Organic Materials and Biological Systems

9.1 Soft Matter

Soft matter refers to materials such as liquid crystals, polymers and biological materials that are deformed or structurally altered by thermal or mechanical stress of the magnitude of thermal fluctuations [1, 2]. These materials can self-organize to form structures much larger than atomic or molecular scale. Subtle changes in conditions, such as a change in temperature or the application of a weak electric or magnetic field, can dramatically change the appearance and properties of the system on a macroscopic scale. The term 'soft matter' was coined around 1970 in the Paris school of physicists around Pierre-Gilles de Gennes, the 1991 Nobel laureate in physics, who is considered by many to be the father of this research field [3]. Rapid expansion of the field is largely due to the wide variety of applications, such as advanced plastics and elastomers, liquid crystal displays, tissue engineering and future 'smart' materials. The category of 'soft matter' encompasses many different types of material, including liquid crystals, polymers, gels, elastomers, colloids and foams.

Studying the dynamics of molecular (or other structural) units is vital to understanding the properties of soft matter [4]. The dynamical processes that take place in soft materials are very complex and the microscopic mechanisms of many of these remain poorly understood. NMR and scattering techniques with X-rays and neutrons [4–6] have provided valuable information about the microscopic behaviour of soft matter; in this chapter we show that muon spin spectroscopy can provide unique and complementary information. Examples have been chosen from studies of liquid crystals, polymers, surfactants and biological systems.

9.2 Thermotropic Liquid Crystals

Liquid crystals are materials that exhibit phases with more order than a liquid, or isotropic (I) phase, but less than the regular structure of crystalline phases [7–9]. Thermotropic liquid crystals form mesophases over a specific temperature range where the molecules show a significant, long-range orientational order due to anisotropic intermolecular forces acting on every molecule. The molecules are composed of rigid and flexible parts; the rigid components impose alignment while the

Muon Spin Spectroscopy: Methods and Applications in Chemistry and Materials Science, First Edition.
Donald G. Fleming, Iain McKenzie, and Paul W. Percival.

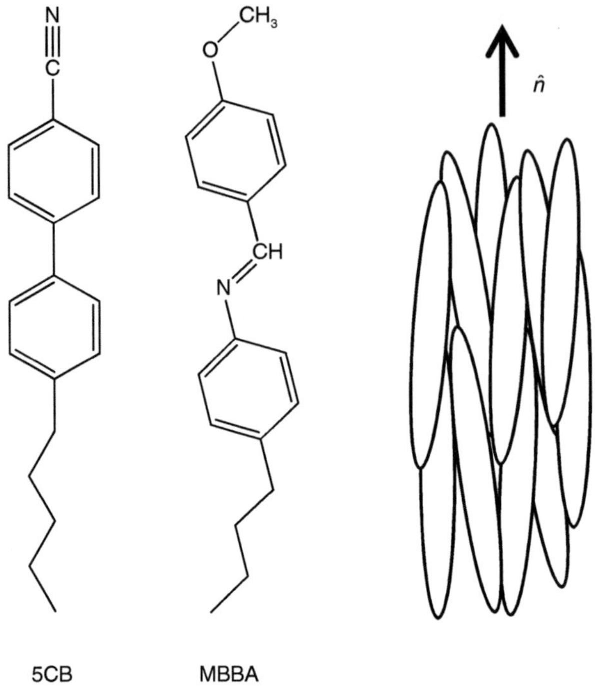

Figure 9.1 The structures of 4-cyano-4′-pentylbiphenyl (5CB) and *N*-(4-methoxybenzylidene)-4-butylaniline (MBBA) and a schematic of the nematic mesophase with the orientation of the director ($\hat{n}$).

flexible segments, typically alkyl chains, provide mobility and hinder crystallization. Liquid crystals are categorized in terms of the shape of the molecules. Calamitic liquid crystals are composed of long rod-like molecules, as shown by the examples in Figure 9.1.

Rod-like molecules can form a wide range of mesophases, but the most common is the nematic (N) phase, where the molecules' positions are disordered but the long molecular axes on average point along a preferred direction known as the director ($\hat{n}$). The orientational order in a liquid crystal is quantified using the order parameter, S, which is defined by

$$S = \int_0^{\pi} \frac{1}{2}(3\cos^2\theta - 1) f(\theta)\, d\Omega, \tag{9.1}$$

where θ is the angle between the rotation axis of a rod-like molecule and the director, and $f(\theta)\, d\Omega$ is the fraction of molecules in a solid angle $d\Omega$ that are oriented at an angle θ. The order parameter decreases with increasing temperature, as shown in Figure 9.2 for MBBA. The application of a large magnetic field can produce an oriented monodomain with $\hat{n}$ aligned along the magnetic field.

The experimental methods commonly used for studying the bulk properties of liquid crystals (e.g. polarizing optical microscopy) give no direct information about molecular motion or ordering. In contrast, magnetic resonance techniques can provide information about liquid crystals at the molecular scale [11]. NMR is frequently used for studying the ordering of liquid crystals, but often requires site-specific isotopic labeling, which can be both difficult and expensive, or requires the inclusion of probes like 1,3,5-trichlorobenzene or propane [12]. ESR is used to study stable spin

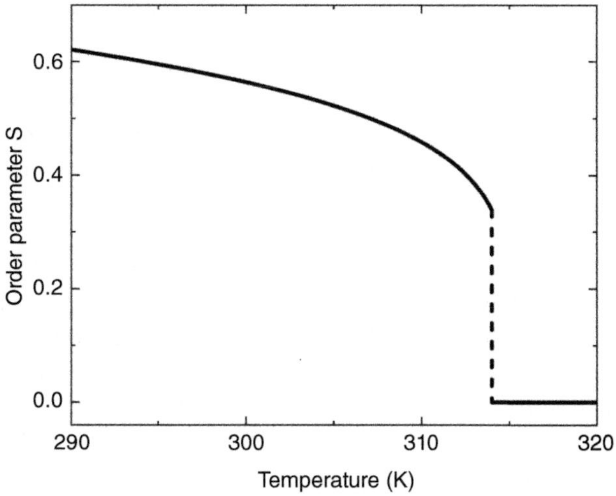

Figure 9.2 The temperature dependence of the order parameter in MBBA determined from the anisotropy of the diamagnetic susceptibility. The order parameter drops from ~0.3 to 0 at the nematic to isotropic phase transition [10].

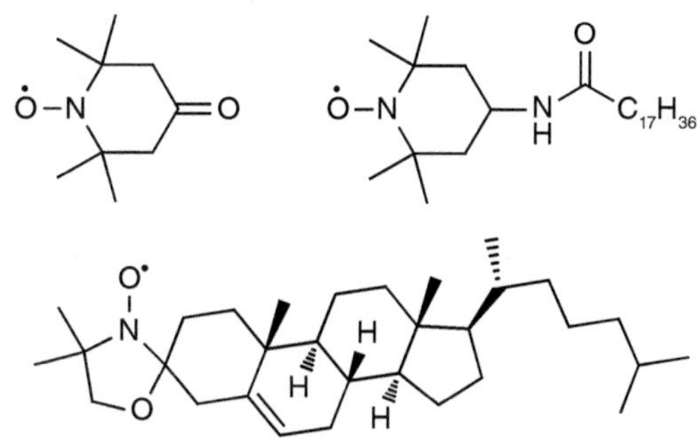

Figure 9.3 The structures of some spin probes commonly used in ESR studies of liquid crystals.

probes dissolved in liquid crystal materials, and this provides useful information about ordering and dynamics of liquid crystal phases [13–16]. The most commonly used spin probes are stable nitroxide radicals, where the unpaired electron is localized in an N–O group, with the nitrogen attached to two alkyl substituents (often in a ring structure) [17]. Nitroxides can be either covalently or non-covalently introduced into materials, and used to monitor molecular-level properties of a material, such as local dynamics and polarity. One issue with the nitroxide spin labels used in ESR studies is that their structures (Figure 9.3) typically differ from the liquid crystal molecules, so that the spin probes reflect the perturbed local environment rather than the intrinsic properties of the material.

In contrast, muon spin spectroscopy employs spin probes made by adding Mu to unsaturated bonds of liquid crystal molecules. This 'internal' spin probe gives a much closer view of the dynamics of the liquid crystal molecules. The timescale to which μSR is sensitive is faster than NMR but slower than neutron scattering. We will consider the results of muon spin spectroscopic studies of two liquid crystals, MBBA and 5CB.

9.2.1 Orientational Ordering of MBBA

MBBA is a rod-like liquid crystal containing two phenyl rings with an imine linkage. There is a nematic-to-isotropic phase transition at T_{NI} ~314–320 K and a nematic-to-crystalline phase transition at ~290–295 K, with the phase transition temperatures depending on the heating and cooling rates. The order parameter of MBBA in the nematic phase is shown in Figure 9.2 [10].

There are two prominent resonances in the ALC-μSR spectra of the nematic phase, at ~1.31 and 1.43 T (Figure 9.4) [18]. The resonance at the lower magnetic field is absent in the isotropic phase, which indicates that it is a Δ_1 resonance. The observed radical was produced by Mu addition to the bridging imine carbon (Figure 9.5).

The muon and methylene proton hfcs in the nematic phase were calculated directly from the Δ_1 and Δ_0 resonance fields. However, for the isotropic phase there is no Δ_1 resonance, so it is necessary to assume that A_μ and the methylene proton A_p are proportional to each other, i.e.

$$A_\mu = KA_p. \tag{9.2}$$

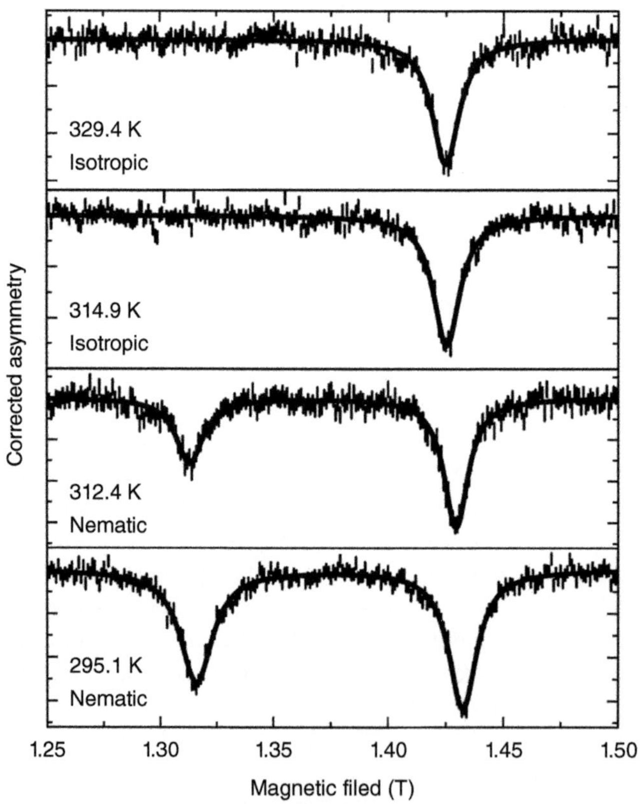

Figure 9.4 ALC-μSR spectra of the Mu adduct of MBBA in the isotropic and nematic phases. The solid lines represent the best fits to the data. Source: Adapted with permission from a figure by McKenzie et al. [18]; copyright 2011 American Chemical Society.

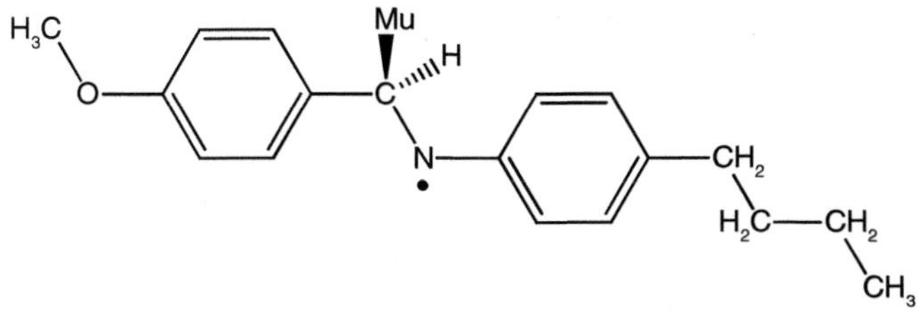

Figure 9.5 Structure of the muoniated radical formed by the addition of muonium to the bridging carbon of MBBA.

In the nematic phase the Mu adduct of MBBA has a K value of 3.956, quite close to the muoniated cyclohexadienyl radical, which has $K = 4.08$ in liquid benzene at room temperature [19]. Inserting Eq. (9.2) into Eq. (2.32) leads to Eq. (9.3), which can be used to determine the proton hfc of the methylene proton in the isotropic phase of MBBA [20]:

$$A_{\mathrm{p}} = \frac{2B_{\mathrm{res}}^{\Delta_0}}{(K-1)/(\gamma_\mu - \gamma_{\mathrm{p}}) - (K+1)/\gamma_{\mathrm{e}}}. \tag{9.3}$$

The values of A_μ and A_{p} as a function of temperature are shown in Figure 9.6. There is a sudden jump in the hfcs at T_{NI}, below which there is a roughly linear increase to lower temperatures. The shift of the hfcs in the nematic phase is due to ordering of the MBBA molecules, with the director aligned along the applied

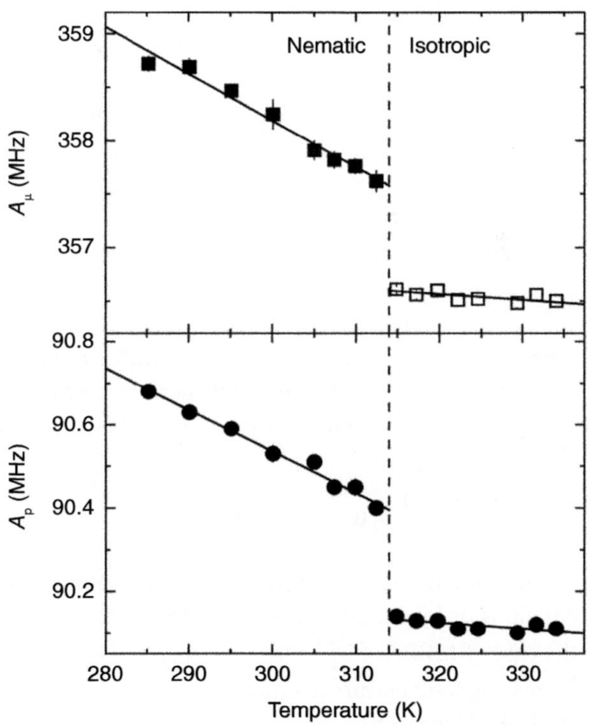

Figure 9.6 Muon and proton hyperfine coupling constants of the radical formed by Mu addition to the bridging carbon of MBBA as a function of temperature in the isotropic and nematic phases. The open squares denote muon hfcs in the isotropic phase, obtained from the Δ_0 resonance assuming $A_\mu/A_p = 3.956$, which is the value in the nematic phase. The dotted vertical line marks the position of the nematic–isotropic phase transition. Source: McKenzie et al. [18].

magnetic field. It is also affected by rapid fluctuations of the molecule about its mean orientation, which average the dipolar hfcs. Both of these effects are explained in detail below.

The muoniated spin probes have axially symmetric muon and methylene proton hyperfine tensors due to rapid rotation around the 'long' molecular axis. One component ($D_k^{\parallel}$) is parallel to the axis of rotation and two components are perpendicular ($D_k^{\perp}$). The dipolar tensor is by definition traceless, so

$$D_k^{\parallel} = -2D_k^{\perp}. \tag{9.4}$$

The values of A_μ and A_p depend on the angle, θ, between the rotation axis of the molecule and the applied magnetic field:

$$A_k = A_k^{\text{iso}} + \frac{1}{2}D_k^{\parallel}(3\cos^2\theta - 1). \tag{9.5}$$

Fluctuations of the molecules about the average orientation will further average the dipolar hfcs, so $D_k^{\parallel}$ in the preceding equation should be replaced by $\left\langle D_k^{\parallel} \right\rangle$. The averaged hfc, $\langle A_k \rangle$, is obtained by integrating over θ, accounting for the angular distribution of the mean orientation of the molecules, $f(\theta)$:

$$\langle A_k \rangle = \int_0^{\pi} \left[A_k^{\text{iso}} + \frac{1}{2}\left\langle D_k^{\parallel} \right\rangle (3\cos^2\theta - 1)\right] f(\theta) 2\pi \sin\theta \, \mathrm{d}\theta. \tag{9.6}$$

In the case of a macroscopically aligned liquid crystal with $\hat{n}$ oriented along the magnetic field, this gives

$$\langle A_k \rangle = A_k^{\text{iso}} + \left\langle D_k^{\parallel} \right\rangle S. \tag{9.7}$$

The increase of the muon and methylene proton hfcs in the nematic phase of MBBA is because both $\left\langle D_k^{\parallel} \right\rangle$ and S are positive. $\left\langle D_k^{\parallel} \right\rangle$ can be determined from Eq. (9.7) since $\langle A_k \rangle$ in the nematic phase is measured by μSR, $A_{\text{k}}{}^{\text{iso}}$ can be obtained from extrapolating the hfcs measured in the isotropic phase and S is known from other techniques, such as NMR, optical birefringence and diamagnetic susceptibility.

The motionally-averaged dipolar hfcs can be related to the fluctuation amplitude if one uses a simple restricted random walk or 'wobbling-in-a-cone' model. In this model the long axis of the molecule performs a random walk over a sector of spherical surface that is bounded by a cone with a vertex angle of $2\theta_c$ (Figure 9.7). This leads to [21]:

$$\left\langle D_k^{\parallel} \right\rangle = \frac{1}{2}D_k^{\parallel}(\cos\theta_c + \cos^2\theta_c). \tag{9.8}$$

The value of $D_p^{\parallel}$, i.e. for pure uniaxial rotation, can be estimated from DFT calculations or by extrapolating to the temperature where the sample crystallizes. In the case of MBBA, $\left\langle D_p^{\parallel} \right\rangle$ was estimated to be 0.82 MHz [18]. $\left\langle D_p^{\parallel} \right\rangle$ decreases with increasing temperature due to the cone angle increasing from ~11° at 285 K to ~31° at 312 K (Figure 9.8).

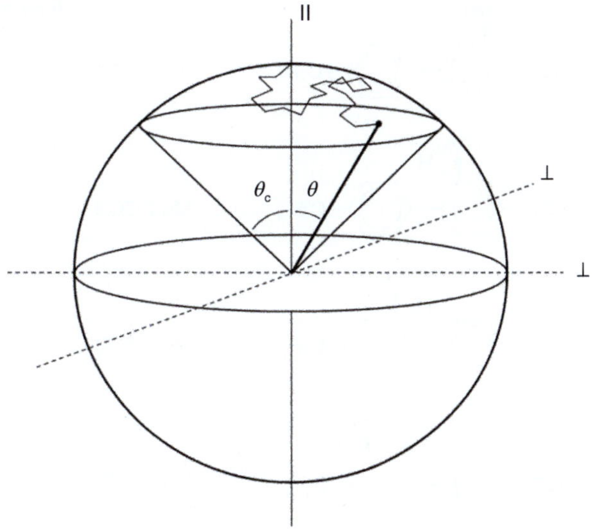

Figure 9.7 Representation of the restricted-random-walk (or wobbling-in-a-cone) model of orientational ordering. The *z*-axis of the muoniated radical performs a random walk over the sector of spherical surface that is bounded by a cone with a vertex angle of θ_c.

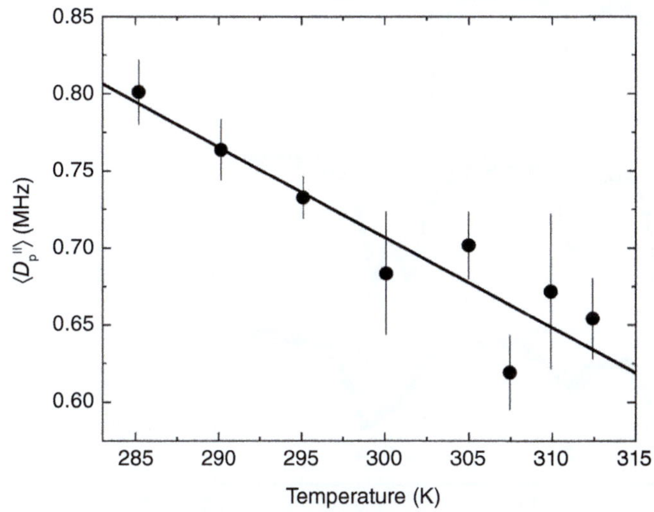

Figure 9.8 Temperature dependence of $\left\langle D_p^{\parallel} \right\rangle$ for the Mu adduct of MBBA. Source: McKenzie et al. [18].

9.2.2 Fluctuations of 5CB

5CB exhibits a nematic phase between 297 and 308 K. It contains a biphenyl group, so four different cyclohexadienyl-type radicals are formed by Mu addition (Figure 9.9). This is clear from the four Δ_0 resonances in the ALC-μSR spectrum (Figure 9.10) [22]. The muoniated spin probes rotate rapidly about their long axis, resulting in an axially symmetric hyperfine tensor, and they are aligned in the nematic phase by the applied magnetic field. No Δ_1 resonances are observed in

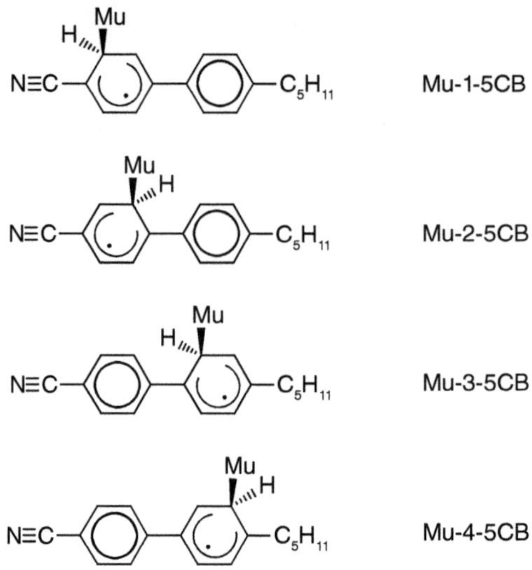

Figure 9.9 The four possible cyclohexadienyl-type Mu adducts of 5CB.

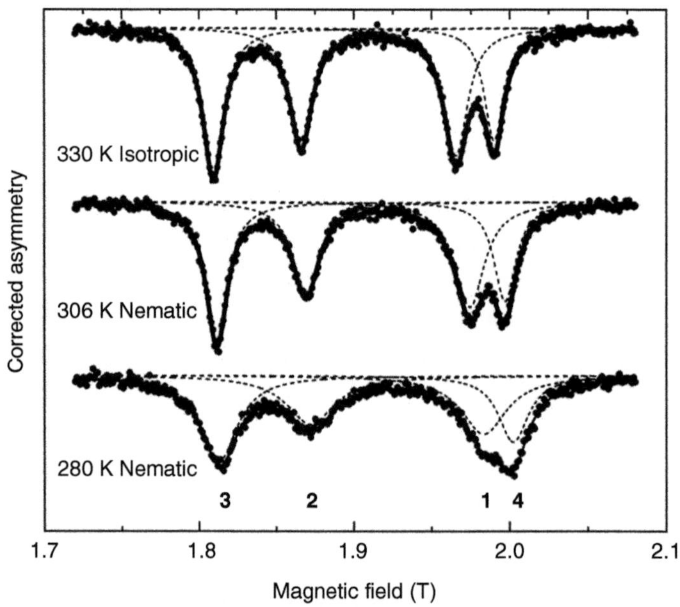

Figure 9.10 ALC-μSR spectra of the Mu adducts of 5CB in the nematic and isotropic phases. The resonances are numbered according to the labels of the radical structures shown in Figure 9.9. Source: McKenzie et al. [22].

the nematic phase of 5CB, even though the molecule is not undergoing isotropic reorientation, because the angle between the main component of the static hyperfine tensor and the principal rotation axis (aligned along the magnetic field) is approximately 60°, resulting in what can loosely be called an intramolecular version of magic-angle spinning. In two of the isomers, Mu-1-5CB and Mu-4-5CB, there is further averaging of the dipolar coupling due to internal rotation of the biphenyl rings. There is a larger barrier to torsional motion of the biphenyl rings in the other two isomers (Mu-2-5CB and Mu-3-5CB), which prevents this further averaging and results in a non-zero hyperfine anisotropy.

The linewidth of the Δ_0 resonance also provides information about the fluctuations of the orientation of the liquid crystal molecules about $\hat{n}$. $\Delta B_{1/2}$ increases significantly with temperature in the nematic phase, due to a decrease in motional averaging, and depends on both the spin relaxation rate, λ, and the dipolar hfc. If the latter is assumed to be very small, λ can be determined from $\Delta B_{1/2}$ using the isotropic formula

$$\Delta B_{1/2} = \frac{\sqrt{\omega_r^2 + \lambda^2}}{\pi(\gamma_\mu - \gamma_p)}, \tag{9.9}$$

where ω_r is the ALC transition frequency on resonance.

The analysis of the spin relaxation in 5CB was based on the model for spin relaxation in muoniated radicals developed by Cox and Sivia [23]. In this model they consider only the four spin states of the muon–electron system and use classical rate equations to compute the evolution of spin-state populations. At most longitudinal fields the analysis of spin relaxation can be complicated, as there are several transitions that are important, but at level crossings the analysis is considerably simpler because there is one dominant transition between the mixing spin states ($1 \Leftrightarrow 2$), so $\lambda = W_{12}$, the transition rate between spin states 1 and 2.

If one type of motion dominates and the radicals can each be characterized by a single correlation time, the relaxation rate is given by

$$\lambda = M_{12}^2 \frac{\tau}{1 + \omega_r^2 \tau^2}, \tag{9.10}$$

where M_{12} is the matrix element responsible for the transition, and τ is the motional correlation time. Muon spin relaxation is predominantly due to the modulation of the dipolar hfc by molecular reorientation, in which case $M_{12}^2 \sim (\gamma_\mu B_D)^2$, where B_D is the effective fluctuating field at the muon and equals $(3/2)^{1/2} D_\mu^{\parallel}/\gamma_\mu$ [24]. Further simplifications are possible because the absence of Δ_1 resonances implies that the radicals are in the fast motional regime ($\omega_r^2\tau^2 \ll 1$). Under these circumstances Eq. (9.10) can be rewritten as

$$\tau = \frac{2\lambda}{3\left(D_\mu^{\parallel}\right)^2}. \tag{9.11}$$

The fluctuation of the 5CB molecules is assumed to be an activated process:

$$\tau = \tau_0 \exp\left(\frac{\Delta E}{k_B T}\right), \tag{9.12}$$

where ΔE is the activation energy for fluctuations about the director. Analysis of the data (Figure 9.11) results in an activation energy of $17.1 \pm 0.9\,\text{kJ mol}^{-1}$ for 5CB molecules in the nematic phase.

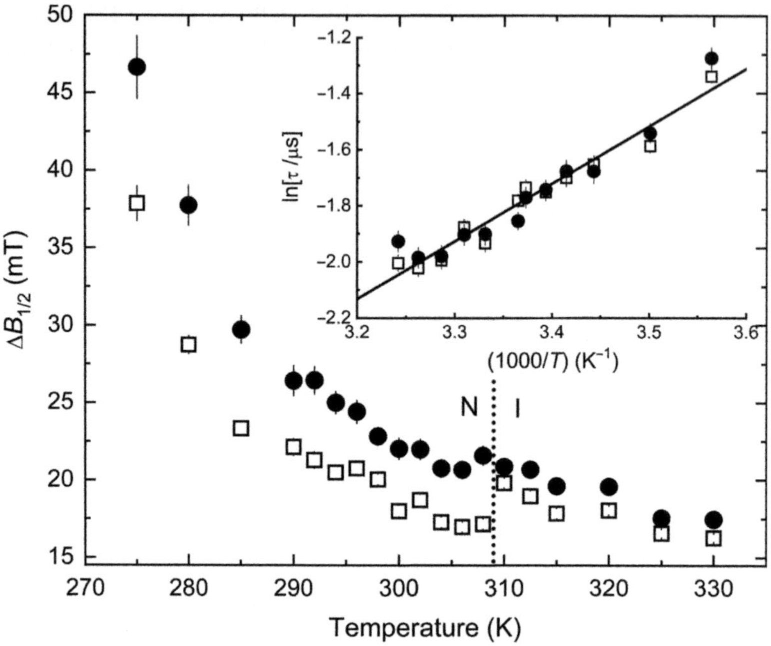

Figure 9.11 Temperature dependence of the width ($\Delta B_{1/2}$) of the Δ_0 resonances for two of the Mu adducts of 5CB: Mu-2-5CB (•) and Mu-3-5CB (□). The dotted vertical line indicates the isotropic-to-nematic phase transition. The inset shows an Arrhenius plot of the nematic phase data, assuming $D_\mu^\parallel = 0.47$ MHz for Mu-2-5CB and $D_\mu^\parallel = 0.42$ MHz for Mu-3-5CB. Source: McKenzie et al. [22].

9.3 Cosurfactants in Bilayers and Micelles

Surfactants are molecules that have hydrophobic and hydrophilic functional groups and which can self-assemble in aqueous solution to form a number of remarkable structures. Examples are 'lamellar' liquid crystalline phases, where the surfactant molecules form arrays of regularly separated bilayer 'sheets' separated by solvent [1]. Subtle changes in the microstructure of the bilayer can have dramatic effects on macroscopic behaviour, which can be brought about by changes in surfactant and electrolyte concentrations or by the addition of a cosurfactant. The cosurfactant may have additional roles, such as being a fragrance, a dye or a pharmaceutical. The cosurfactants are typically present only in very low concentrations, which makes them difficult to study using traditional spectroscopic techniques.

ALC-μSR is an excellent tool for studying cosurfactant molecules with aromatic functional groups because it is sensitive only to the muoniated radicals that are formed from Mu addition, and it is unaffected by the bulk of diamagnetic molecules. It is necessary for Mu to add preferentially with the cosurfactant rather than the surfactant, and this will depend on the relative concentrations of the surfactant and cosurfactant and the reaction rates. Cosurfactant concentrations as low as 1 mM

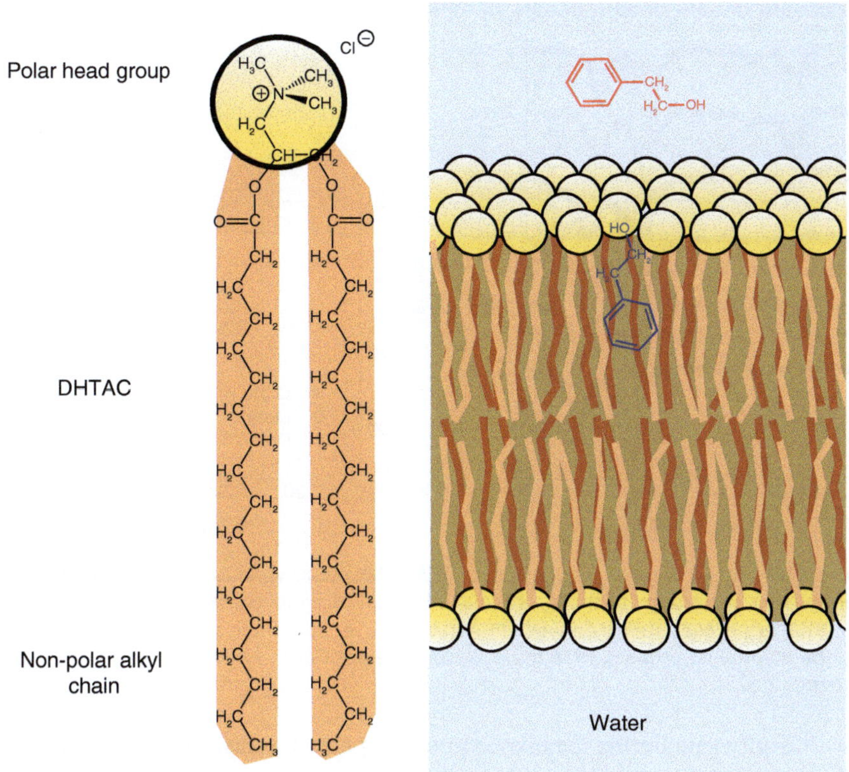

Figure 9.12 Structure of DHTAC showing the polar head group and the non-polar tails, and a schematic representation of a bilayer and locations of the PEA cosurfactant in the low-temperature L_β phase (red) and in the high-temperature L_α phase (blue).

can be studied by this means. For example, ALC-μSR has been applied to study the partitioning of the cosurfactant 2-phenylethanol (PEA) in bilayers of the dichain cationic surfactant 2,3-diheptadecyl ester ethoxypropyl-1,1,1-trimethylammonium chloride (DHTAC) [25–28] and in micelles of the nonionic surfactant tetra(ethylene glycol) n-dodecyl ether ($C_{12}E_4$) [29].

DHTAC forms bilayers (Figure 9.12) that undergo a structural change between 328 and 330 K; below this temperature the system is in the L_β phase, where the motion of the alkyl chains is limited, and above this temperature it is in the L_α phase, where the alkyl chains have considerably more mobility. In the L_β phase the ALC-μSR spectrum shows three Δ_0 methylene proton resonances but no Δ_1 resonance (Figure 9.13a), indicating that the motion of the radical is isotropic [25, 26]. The magnitude of the hfcs can be used to determine the polarity of the local environment of the muoniated radicals. A polarity scale has been defined for each isomer based on the 'aqueous character' of the environment, by calibration with a range of solvents from water (100% aqueous) to $C_{18}H_{38}$ (0% aqueous) [26]. The % aqueous character of the three isomers in the L_β phase (Figure 9.13b) indicates that the radicals are in an aqueous environment. The PEA-Mu radicals (and by inference the PEA cosurfactants)

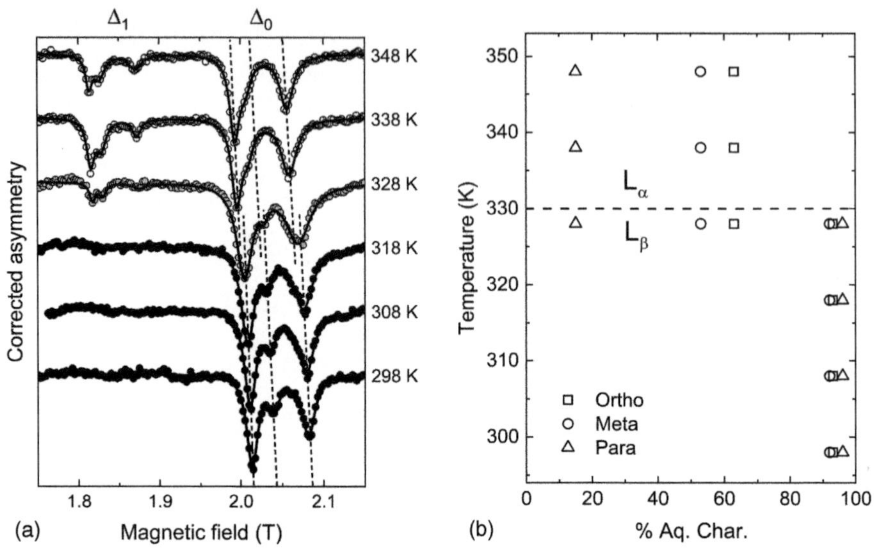

Figure 9.13 ALC-μSR resonance spectra of PEA in DHTAC at various temperatures. The open symbols correspond to the L_α phase and the closed symbols are for the L_β phase. At 328 K there is coexistence between PEA in the bilayer and in the aqueous layers [28]. (b) Local polarity determined from the Δ_0 resonance positions as a function of temperature [28].

have difficulty penetrating the bilayer in the L_β phase because of the tight packing of the alkyl chains.

Δ_1 resonances are present in the L_α phase, indicating that the motion of the radical is anisotropic, likely around the long axis of the molecule. In addition, the position of the Δ_0 resonances indicates a lower % aqueous character around PEA. The results indicate that PEA resides within the DHTAC bilayer in the L_α phase. The penetration of PEA into the DHTAC bilayer is a consequence of the higher mobility of the alkyl chains in the bilayer. The differences in polarity sensed by the ortho, meta and para spin probes could be evidence of a water gradient, and hence a polarity gradient, within the bilayer. At 328 K, which is near the L_α–L_β transition, there is coexistence between radicals in the bilayer and in the aqueous region.

The location of the muoniated radicals was confirmed by studying the reaction between the radicals and paramagnetic Ni^{2+} cations, which do not partition into the bilayer [30]. The spin-exchange rate, k_{ex}, was determined from the width of the Δ_0 resonances as a function of Ni^{2+} concentration (Figure 9.14). Measurements were also performed on aqueous solutions of PEA and $NiCl_2$. At 298 K, the average value of k_{ex} for the three isomers was found to be approximately the same in the DHTAC solution and in the $NiCl_2$ solution. This indicates that the muoniated radicals are in close proximity to the Ni^{2+} ions and therefore PEA resides in the aqueous layers and not in the DHTAC bilayers in the L_β phase. At 348 K, k_{ex} in the DHTAC solution is ~20 times smaller than in the $NiCl_2$ solution, which indicates that muoniated radicals are far removed from the Ni^{2+} ions. This means that in the L_α phase PEA is located within the DHTAC bilayers.

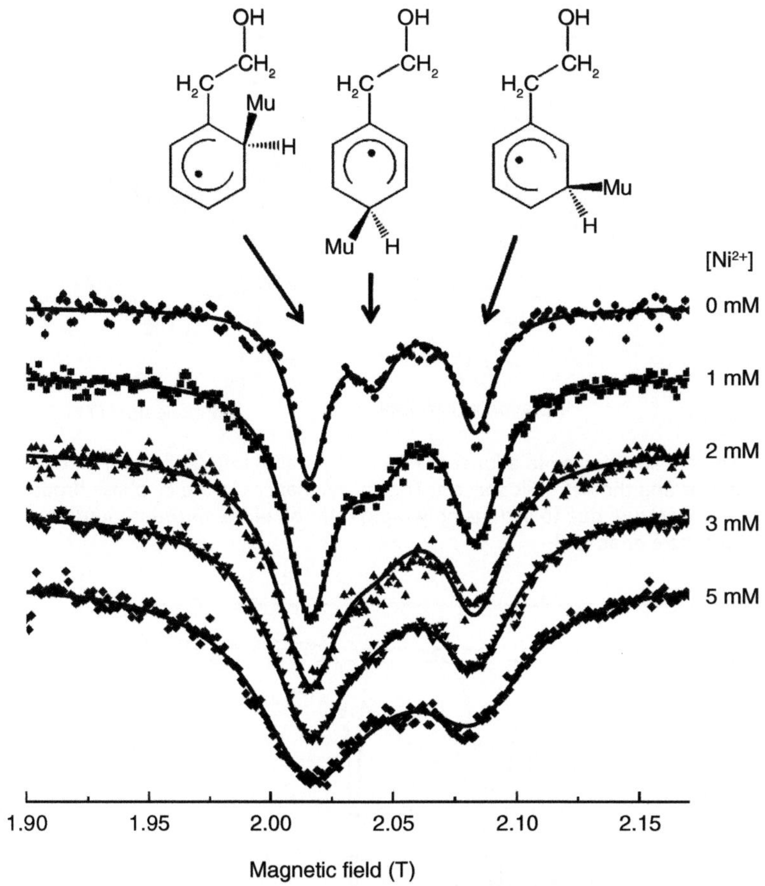

Figure 9.14 ALC-μSR spectra of the Mu adducts of 2-phenylethanol in aqueous $NiCl_2$ solutions at 298 K. The Δ_0 resonances of the three isomers broaden because of the spin-exchange reaction with paramagnetic Ni^{2+} [30].

ALC-μSR has also been used to study the location of PEA in an aqueous dispersion of the nonionic surfactant $C_{12}E_4$ (tetra(ethylene glycol) n-dodecyl ether) [29]. $C_{12}E_4$ forms bilayers above ~290 K (L_α phase) and micelles below it (L_1 phase). The Δ_0 resonance fields indicate that muoniated radicals reside in a non-aqueous environment. The spectra from the L_α phase are similar to those of DHTAC, again suggesting that the radicals undergo rapid uniaxial rotation about their long axis and fluctuations around their preferred orientation. There is a dramatic change in the spectra in the L_1 phase. The Δ_1 resonances are much broader and the width increases with increasing temperature (Figure 9.15). This is due to slow isotropic rotation of the micelles, with rotational correlation times (τ_{Rot}) between a few tens of ns and 1 μs, and τ_{Rot} decreasing with increasing temperature (Figure 9.16).

Surfactant-based bilayers can also be used as model biomembranes, which act as selectively permeable barriers in living things. Biomembranes are composed of a phospholipid bilayer containing integral and peripheral proteins that can

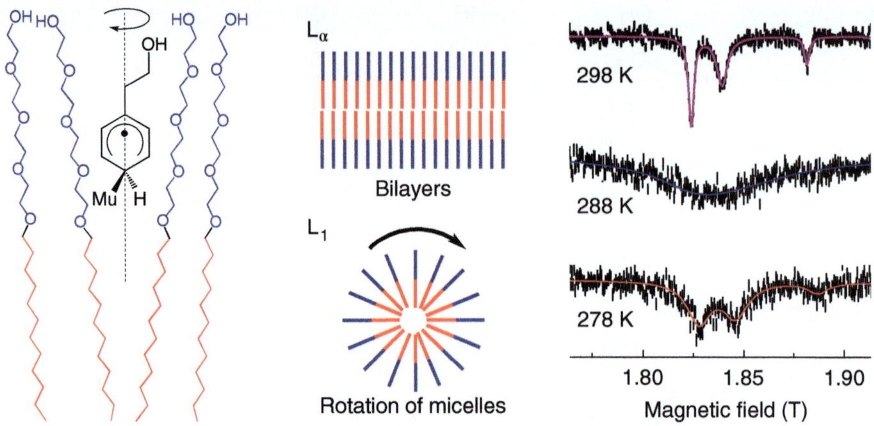

Figure 9.15 Diagram showing a Mu adduct of PEA incorporated into $C_{12}E_4$, schematics of the L_α and L_1 phases and the ALC-μSR spectra. The Δ_1 resonances in the L_1 phase broaden with increasing temperature due to increasingly rapid rotation of the micelles. Source: Adapted from McKenzie et al. [29].

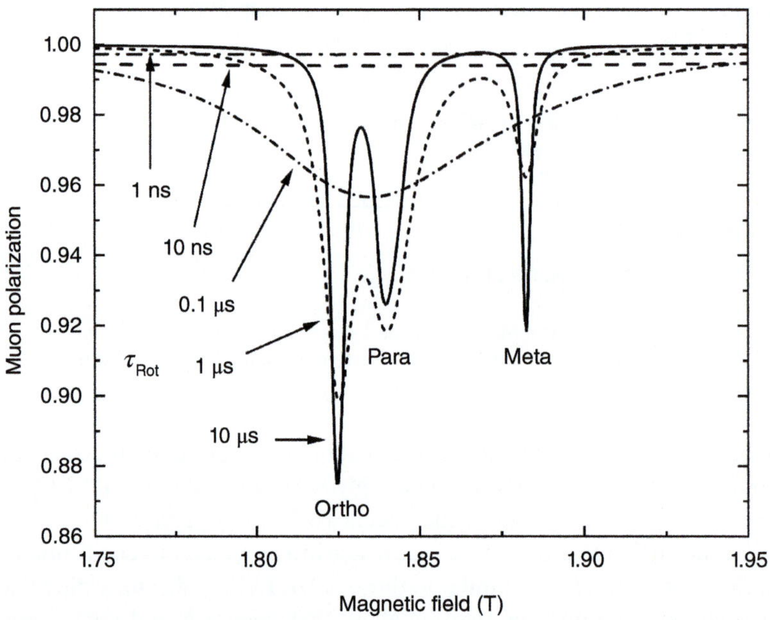

Figure 9.16 Simulated Δ_1 resonances of the Mu adducts of PEA as a function of rotational correlation time (τ_{Rot}).

form pores or channels that allow the selective transport of ions and the transportation of nutrients and waste products. Cell membranes are highly dynamic, with frequencies of motion ranging from 10^0 to 10^{12} Hz. Membrane dynamics on small molecular-length scales strongly affect functional aspects, such as diffusion and parallel or perpendicular transport through a bilayer, and it is important to obtain information about dynamics using experimental techniques sensitive to different frequencies. Muon spin spectroscopy is sensitive to fluctuations in the range ~10^5–10^9 Hz, thus bridging the gap between NMR [31] and neutron scattering [32]. μSR does not use fluorescent probes or nitroxide spin labels, which can be significant perturbations [33]. Except for the missing phosphate group, the DHTAC surfactant molecules are analogues of the typical phospholipid molecules which are the main constituents of cell membranes of living organisms. DHTAC bilayers are therefore a simple model system of a cell membrane. Muon spin spectroscopy could be used to determine whether a drug molecule resides mostly in the aqueous medium of a cell fluid or whether it intercalates into the cell membrane, and could therefore be used for studying some basic drug formulation problems [34].

9.4 Polymers

9.4.1 Dynamics in Non-conjugated Polymers

Understanding the dynamic processes in polymers is of fundamental interest and important for many technological applications [35]. The majority of polymers form an amorphous glass when cooled from a high-temperature viscoelastic 'rubbery' state, but the glass transition is not well understood. Various structural relaxation processes, which involve different types of motion of the polymer, determine the precise viscoelastic behaviour of a polymer glass [36]. Muoniated radicals can be introduced as probes in non-conjugated polymers by the reaction of Mu with isolated unsaturated bonds or phenyl groups. LF-μSR can be used to study various dynamical processes in non-conducting polymers, as motion of the polymer causes muon spin relaxation.

As an example, consider μSR studies of atactic polystyrene, which is an ideal glass former. There are several relaxation processes in polystyrene; the α-relaxation results from molecular motions which involve relatively large segments of the polymer backbone chain, while the secondary (β, γ, δ) relaxation modes involve much more localized molecular motions [36]. The frequencies of these modes range over many orders of magnitude (Figure 9.17). Mu is expected to add to each of the ortho, meta and para carbons of the styrene rings. In this section we will show that the ensuing radicals can be used to determine the glass transition temperature, T_g, and the activation energy of the γ-relaxation mode, which involves flips of phenyl rings. The dynamics and T_g depend on the chain length of the polymer, which is defined by the number of monomers N. Comparisons will be made between polystyrene with average N values of 12 and 1870.

LF-μSR spectra of polystyrene (Figure 9.18) are well described by the following model:

$$P(t) = P_R \exp\left[-\left(t/T_1^{\mu}\right)^{\kappa}\right] + P_{NR}, \tag{9.13}$$

where P_R is the initial polarization of the relaxing fraction, which is due to the muoniated radicals, P_{NR} is the non-relaxing polarization, which is due to diamagnetic

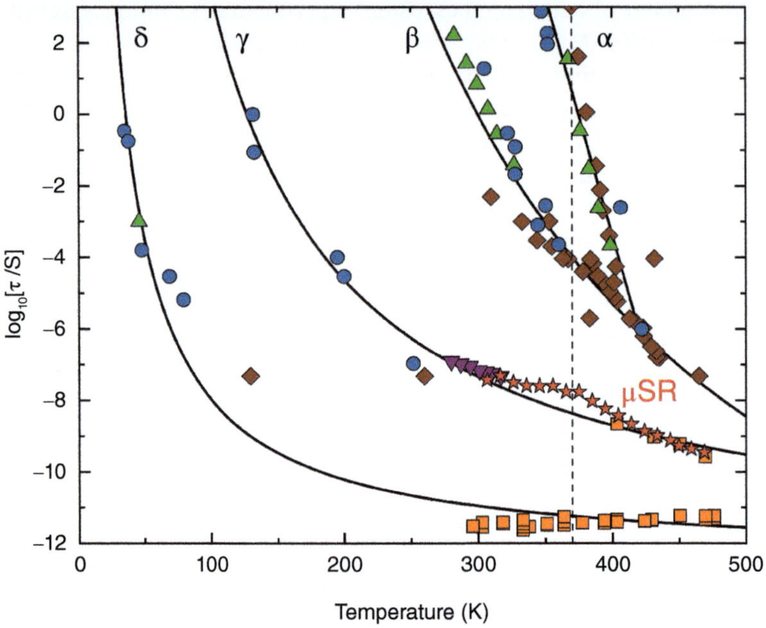

Figure 9.17 Relaxation map of bulk, high-molecular-weight polystyrene showing correlation times (τ) obtained from different analytical techniques as a function of temperature: mechanical (blue circles); dielectric (green up triangles); NMR (brown diamonds); neutron scattering (orange squares); βNMR [37] (purple down triangles); μSR [38] (red stars). The dotted vertical line marks the bulk glass transition temperature.

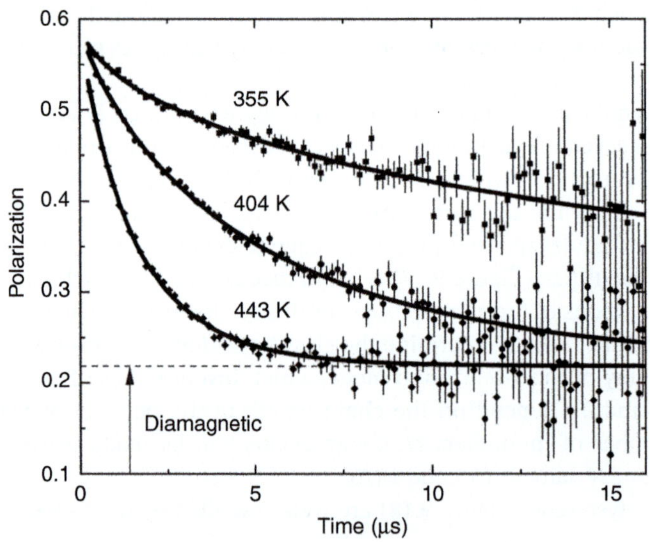

Figure 9.18 LF-μSR spectra for polystyrene composed of $N = 1870$ monomers in a magnetic field of 85 G at 355, 404 and 443 K. The solid lines are fits to Eq. (9.13). The diamagnetic polarization (P_{NR}) was determined from TF-μSR measurements. Source: Adapted with permission of IOP Publishing from a figure by McKenzie et al. [38].

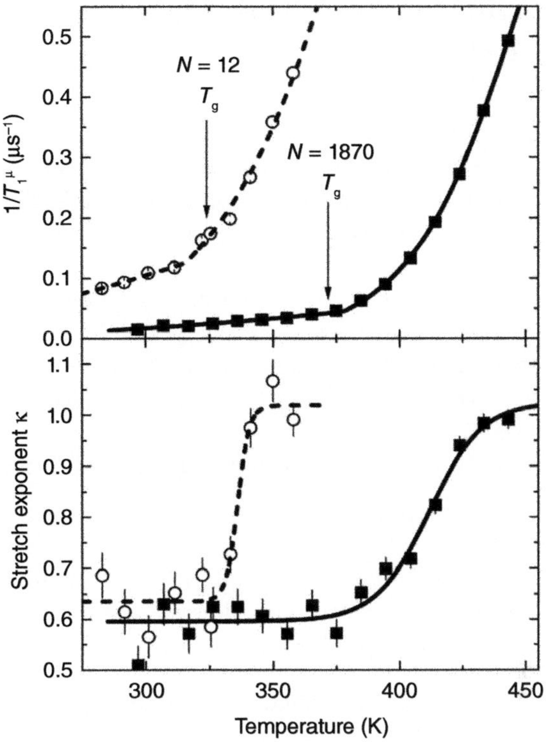

Figure 9.19 Temperature dependence of $1/T_1^\mu$ and the stretch exponent κ for polystyrene samples with $N = 12$ and 1870 in a longitudinal magnetic field of 85 G. Source: McKenzie et al. [38].

muons and is determined from TF-μSR measurements, $1/T_1^\mu$ is the muon spin-lattice relaxation rate and κ is the stretch exponent (we use κ rather than the more common β for the stretch exponent to avoid confusion with the β-relaxation process of polystyrene). The stretched exponential behaviour is due to there being a distribution of local environments, each with a different spin-lattice relaxation rate. The fitted values of $1/T_1^\mu$ and κ as a function of temperature at 85 G are shown in Figure 9.19.

The average muon spin-lattice relaxation time, $T_1^{\mu,\mathrm{avg}}$, is given by

$$T_1^{\mu,\mathrm{avg}} = \frac{T_1^\mu}{\kappa}\Gamma\left(\frac{1}{\kappa}\right), \tag{9.14}$$

where Γ is the Gamma function.

The spin relaxation mechanism that is expected to dominate at 85 G is the indirect process, where electron spin flips of the radical cause muon spin flips due to the isotropic hyperfine coupling. This has the form:

$$\frac{1}{T_1^{\mu,\mathrm{avg}}} = \left[\frac{1}{1 + (B/B_0)^2}\right]\frac{1}{T_1^{\mathrm{e}}}, \tag{9.15}$$

where B_0 is the 'contact hyperfine field' of the muon at the electron ($B_0 = A_\mu/\gamma_\mathrm{e}$), which is 178 G for the Mu adducts of polystyrene. There are several processes that can contribute to electron spin relaxation of a radical in a solid [39], but the mechanism that dominates in polystyrene is the modulation of the anisotropic hyperfine interaction by thermally activated dynamics, which can be modelled by the BPP equation [40]:

$$\frac{1}{T_1^{\mathrm{e}}} = K\left[\frac{\tau}{1+\omega_{\mathrm{e}}^2\tau^2} + \frac{4\tau}{1+4\omega_{\mathrm{e}}^2\tau^2}\right], \tag{9.16}$$

where K depends on the relaxation mechanism, ω_{e} is the electron Larmor frequency and τ is the correlation time for the motion causing the spin relaxation. Consider the long-chain ($N = 1870$) sample where $T_{\mathrm{g}} \sim 370$ K. The very small correlation time of the δ-relaxation (~10^{-11} s at T_{g}) results in negligible electron spin-lattice relaxation over the muon lifetime. The large correlation times for the α- and β-relaxation modes (~100 and ~10^{-4} s, respectively, at T_{g}) also result in very slow electron spin-lattice relaxation. The correlation time of the γ-relaxation process is of the appropriate magnitude near T_{g} (~10^{-7} s) to induce spin relaxation that is measurable by μSR.

In the slow fluctuation limit ($\omega_{\mathrm{e}}^2\tau^2 \gg 1$) Eq. (9.16) reduces to

$$\frac{1}{T_1^{\mathrm{e}}} = \frac{2K}{\omega_{\mathrm{e}}^2\tau}. \tag{9.17}$$

Combining Eqs. (9.15) and (9.17), averaging over all local environments, and assuming the correlation time of the γ-relaxation process has an Arrhenius temperature dependence gives

$$\ln\left(T_1^{\mu,\mathrm{avg}}\right) = \ln\left[\omega_{\mathrm{e}}^2\tau_0\left(1+(B/B_0)^2\right)\right] + \frac{E_{\mathrm{A}}}{(k_{\mathrm{B}}T)}. \tag{9.18}$$

An Arrhenius plot of the LF-μSR data is shown in Figure 9.20. There is a change in slope at 376 K in the long-chain ($N = 1870$) sample and 316 K in the short-chain

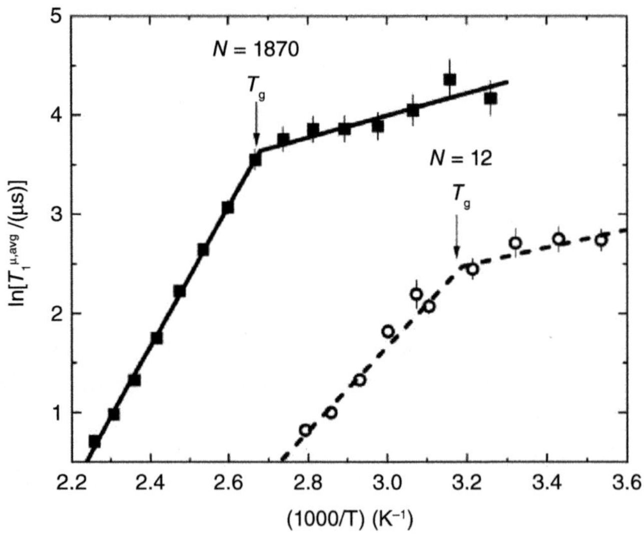

Figure 9.20 Arrhenius plots of $\ln\left(T_1^{\mu,\mathrm{avg}}\right)$ versus inverse temperature for polystyrene samples with $N = 12$ and 1870. The lines above T_{g} were obtained from fits using Eq. (9.18). Source: McKenzie et al. [38].

($N = 12$) sample, which corresponds to T_g. The activation energy above T_g was found to be 0.609(9) eV for the $N = 1870$ sample and 0.40(4) eV for the $N = 12$ sample; the lower activation energy in the short-chain polymer is due to the lower amount of entanglement exerting fewer constraints on the γ-relaxation motion.

The temperature dependence of the stretch exponent provides information about the local environment of the spin probe. A stretched exponential with $\kappa < 1$ is used when there is a broad distribution of correlation times. NMR experiments on bulk polystyrene have shown that this is the case for the backbone motion (α-relaxation) around T_g [41]. This dynamic heterogeneity is a common feature of glassy materials [42]. The stretch exponent in the $N = 1870$ sample is ~0.6 below T_g and becomes approximately 1.0 at ~60 K above T_g. A similar change in the stretch exponent is observed in the $N = 12$ sample above T_g. The distribution of local environments at and below T_g is due to the motion of the polystyrene backbone (specifically the α-relaxation) being so slow that it appears to be static to the muon. κ approaches 1 when the correlation time for the backbone motion approaches the μs timescale (Figure 9.17). The motion of the polymer chains on the timescale of the muon lifetime results in all muoniated spin probes experiencing the same local environment, leading to exponential spin-lattice relaxation.

9.4.2 Electron Conduction in Conjugated Polymers

Most polymers are insulators, but conjugated polymers can be made to conduct electricity by doping with electron-donating or electron-accepting substances. These materials have a wide range of applications due to their useful mechanical, optical and electronic properties. A conjugated polymer typically has a long chain of alternating single and double bonds so that the $2p_z$ orbitals of each atom interact, resulting in the formation of bands of π and π^* orbitals. The states in the π band are occupied, but the π^* states are empty and form the conduction band. Typical π– π^* gaps in undoped conjugated polymers are 1–2 eV, which is comparable to that of inorganic semiconductors.

Defects in the polymer chain are believed to be responsible for electron conduction. Doping increases the concentration of defects and can result in the conductivity increasing by many orders of magnitude. Muonium addition to a material with conjugated double bonds can create paramagnetic defects and can be considered as doping at an infinitely dilute level. In conducting polymers, there is strong electron–lattice coupling, and the moving electrons create a large distortion that polarizes the lattice. For an alkene chain this is manifest as a change in bond lengths. The combination of electron and local lattice distortion is known as a *polaron* (Figure 9.21). In general, depending on the strength of the coupling and the symmetry of the lattice, there is a variety of quasi-particles resulting from electron–lattice coupling, the most famous of which (and very typical for one-dimensional systems) is the *soliton* [43]. Conventional magnetic resonance techniques such as ESR can be used to study polaron motion in doped conducting polymers, but μSR can also be used to study polaron/soliton motion in undoped polymers (for which there is insignificant carrier density to give an ESR signal).

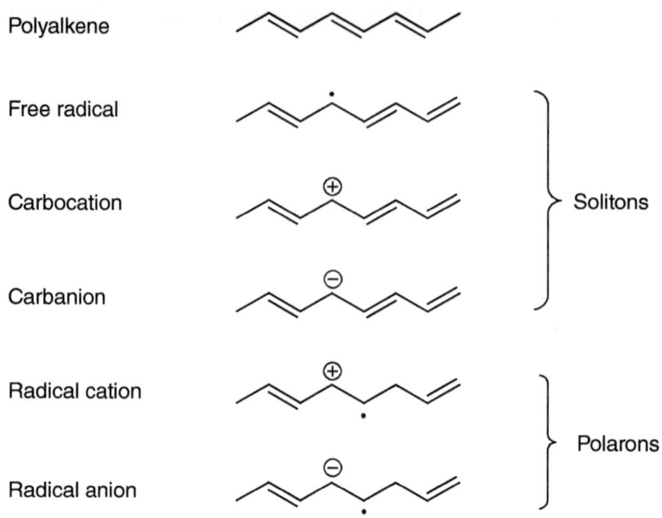

Figure 9.21 Solitons and polarons formed by local defects in polyalkene chains.

The spin dynamics provides very interesting insights into the microscopic electronic conduction processes. The mobile electronic excitations are believed to take the form of solitons and polarons, but the analysis of μSR spectra is the same regardless of the identity of the defect.

After its creation by Mu addition to the polymer chain, the defect can move away from its initial site and diffuse up and down the chain, but it cannot cross the point at which the muon has added, which therefore acts as a barrier. The defect diffuses up and down the polymer chain at the on-chain diffusion rate, $D_{\parallel}$, with less frequent hops to adjacent chains at the inter-chain diffusion rate, $D_{\perp}$ (Figure 9.22). Diffusion of the defect away from the site where the muon is covalently bound effectively switches off the muon hyperfine interaction, but every time the defect returns the hyperfine coupling is turned back on. This results in a fluctuating magnetic field and leads to relaxation of the muon spin. Measurement of the magnetic-field dependence of this relaxation yields the spectral density function associated with the random

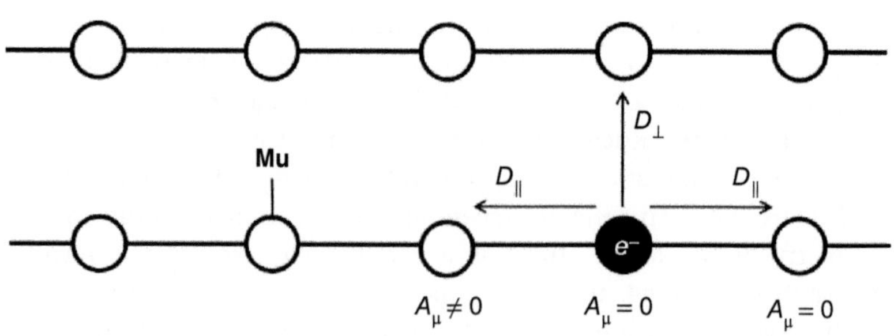

Figure 9.22 Defect diffusion model based on that of Risch and Kehr extended to include interchain diffusion.

walk of the unpaired spin, and can be used to infer the rates and dimensionality of the diffusion [44].

Risch and Kehr developed a direct stochastic theory of muon spin relaxation in conducting polymers based on the one-dimensional diffusion of defects along the polyacetylene chains and the intermittent hyperfine interaction [45]. The general form of the muon spin relaxation function in the Risch–Kehr (RK) model is

$$P_z(t) = P_z(0)e^{\Gamma t}\mathrm{erfc}(\sqrt{\Gamma t}), \tag{9.19}$$

for $\lambda_e t_{max} \gg 1$. λ_e is the electron spin flip rate, t_{max} is the experimental timescale and erfc is the complementary error function. The relaxation rate parameter, Γ, is given by

$$\Gamma = \lambda_e \left(1 + 2D_{\|} \frac{\sqrt{2\omega_e \lambda_e}}{\omega_0^2}\right)^{-2}, \tag{9.20}$$

where ω_e is the electron Larmor frequency, $\omega_0 = 2\pi A_\mu^{iso}$ is the isotropic hyperfine coupling constant when the unpaired electron is localized on the neighbouring site to the muon and $D_{\|}$ is the intrachain diffusion rate. In the fast-diffusion limit the relaxation parameter is inversely proportional to the magnetic field and independent of λ_e, with

$$\Gamma = \frac{\omega_0^4}{2\omega_e D_{\|}^2}, \tag{9.21}$$

which allows for a straightforward determination of $D_{\|}$ provided A_μ is known. For a given relaxation rate $D_{\|}$ is proportional to $A_\mu{}^2$, which means that any uncertainty in the value of A_μ can have a large impact on the calculated $D_{\|}$ value. It is generally assumed that the electron can be fully localized on the neighbouring carbon; however, another possibility is that the defect hops between finite conjugated segments along the chain. In such a case the spin density would be delocalized over the finite conjugated segment adjacent to the muon and could therefore have a much lower A_μ value. Neither of these options have been experimentally validated, and this remains an open question.

Interchain diffusion results in a deviation of Γ from the inverse field-dependence at low magnetic fields [41]. The RK model does not include interchain diffusion, so an empirical low-field cut-off is introduced of the form:

$$\Gamma = \Gamma_0/(1 + \omega_e/D_{\perp}), \tag{9.22}$$

where $D_{\perp}$ is the inter-chain diffusion rate and is typically much lower than $D_{\|}$.

LF-μSR has been used to study the conjugated polymer polyaniline (PANI) [46]. The electronic properties of PANI depend on the doping (protonation or oxidation) level. The emeraldine base form (PANI:EB), whose structure is shown in Figure 9.23a, is neutral and is regarded as the most useful form of polyaniline, due to its high stability at room temperature and the fact that, upon doping with acid, the resulting emeraldine salt form of polyaniline is highly electrically conducting. PANI:EB has a non-degenerate ground state and polarons are believed

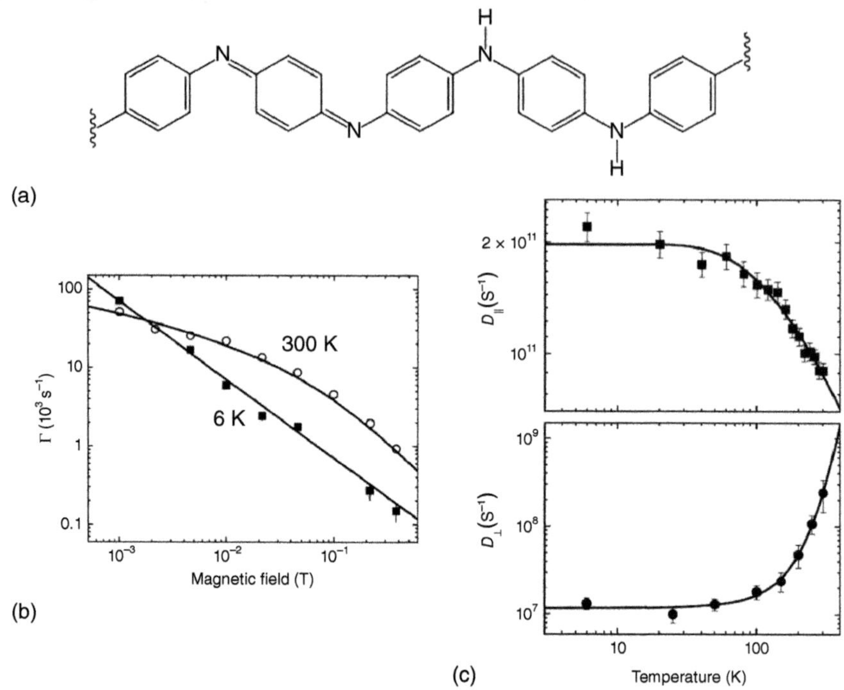

Figure 9.23 (a) The structure of polyaniline:emeraldine base (PANI:EB). (b) Magnetic field dependence of the RK relaxation parameter observed in PANI:EB at 6 and at 300 K. (c) The temperature dependence of the intrachain $D_{\parallel}$ and interchain $D_{\perp}$ diffusion rates in PANI:EB. Source: Pratt et al. [46].

to be the dominant charge carriers. LF-μSR spectra of PANI:EB were obtained as a function of magnetic field and temperature. The relaxation parameter Γ was determined from fits of Eq. (9.19) (Figure 9.23b). Eqs. (9.20) and (9.22) were used to determine $D_{\parallel}$ and $D_{\perp}$ as a function of temperature (Figure 9.23c). The relatively slow inter-chain diffusion is a thermal process while the fast intra-chain diffusion, which decreases slightly with increasing temperature, is ascribed to phonons in the material.

9.5 Organic Materials

9.5.1 Magnetic Ordering in Organic Materials

Unpaired electron spins are an essential component of any magnetic material. Magnetism is traditionally associated with transition metals, such as Fe, Co or Ni, or metal oxides such as Fe_3O_4, where the unpaired electron spins that contribute to the net magnetic moment reside only on metal atoms, in d- or f-orbitals. Stable organic radicals, such as p-nitrophenyl nitronyl nitroxide (p-NPNN), where the unpaired

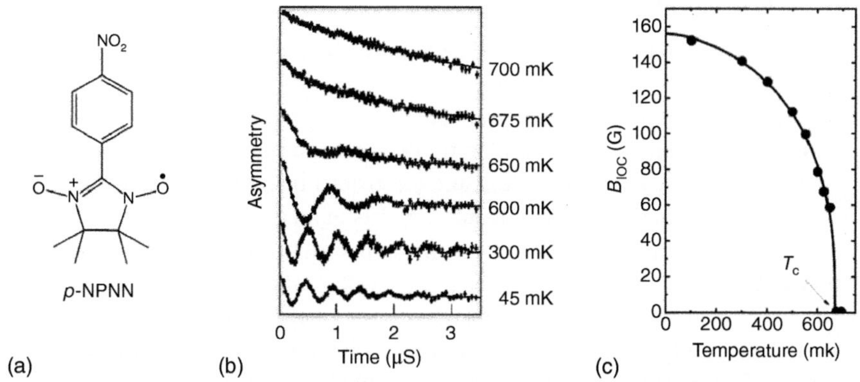

Figure 9.24 (a) The structure of p-nitrophenyl nitronyl nitroxide (p-NPNN). (b) ZF-μSR spectra showing the time evolution of the muon polarization in a polycrystalline mosaic of p-NPNN. (c) The local magnetic field as a function of temperature. Ordering occurs at 0.67 K. Source: (b) Adapted from a figure by Blundell et al. [48] with permission of Springer Nature. (c) Blundell et al. [48].

electron spin is on the N and O atoms, can also order magnetically [47]. The structure of p-NPNN is shown in Figure 9.24a.

Even though the molecules contain unsaturated bonds, only diamagnetic signals are observed. In zero applied magnetic field the muon polarization can still evolve with time because the surrounding spins (protons or electrons) give rise to a local field, $|B_{\mathrm{loc}}|$. The amplitude of the oscillations depends on the angle between the local field and the initial direction of the muon's spin, θ:

$$P_z(t) = \cos^2\theta + \sin^2\theta \cos(2\pi\gamma_\mu |B_{\mathrm{loc}}| t). \tag{9.23}$$

In a polycrystalline material one has to average over all values of θ: $\langle\cos^2\theta\rangle = \frac{1}{3}$ and $\langle\sin^2\theta\rangle = \frac{2}{3}$, giving

$$P_z(t) = \frac{1}{3} + \frac{2}{3}\cos(2\pi\gamma_\mu |B_{\mathrm{loc}}| t). \tag{9.24}$$

Examples of zero-field muon spin relaxation (ZF-μSR) spectra of p-NPNN are shown in Figure 9.24b [48]. Above the critical temperature the muon spin polarization, which is proportional to the asymmetry, decreases monotonically. There are no oscillations, which indicates that there is no long-range order, just a distribution of magnetic fields with different orientations. When the sample is cooled below the critical temperature, T_c (670 mK for p-NPNN), oscillations are observed, consistent with the magnetic order. The magnitude of the local field increases as the temperature is lowered. Such measurements can be used to determine the temperature dependence of the magnetic order parameter, which is proportional to the local magnetic field (Figure 9.24c). The solid line in Figure 9.24c shows a fit of $B(T) \propto [1-(T/T_c)^\alpha]^\beta$ with critical exponents $\alpha = 1.7$ and $\beta = 0.36$. The critical exponents provide information about the dimensionality and range of the magnetic interactions.

9.5.2 Localized Defects in Organic Semiconductors

Organic semiconductors are being considered for use in a vast array of electronic applications and there is considerable research in this area in order to better understand the behaviour at the molecular level. There have been several μSR studies of the organic semiconductor tris[8-hydroxyquinoline]aluminium (Alq_3), which is widely used as a luminescent and electron-transporting material in organic light-emitting diodes [49]. This molecule consists of an aluminium atom with three bidentate 8-hydroxyquinolate ligands in an octahedral geometry. In the crystal there is good π–π orbital overlap between ligands of different molecules along the c axis, and this results in highly one-dimensional motion of charge carriers.

The initial LF-μSR study of Alq_3 claimed to have measured the intrinsic hopping rate of electrons to be $1.4\times10^{12}\,s^{-1}$ at 290 K, based on the magnetic-field dependence of the Risch–Kehr relaxation rate [50]. The result of this μSR experiment generated considerable interest as it is very difficult to extract intrinsic transport properties from bulk transport measurements, which are typically dominated by extrinsic properties such as inter-grain hopping. Unfortunately, this conclusion must be incorrect, because the spin relaxation is due to a muoniated radical with localized spin density. The initial study implicitly assumed that the muon is a passive probe and that the unpaired electron from Mu goes into the lowest unoccupied molecular orbital (LUMO) of Alq_3, in essence allowing the material to be doped at an incredibly small level. DFT calculations indicate that the electronic structure of the radical (the Mu adduct of Alq_3) is considerably different from that of Alq_3 itself [51]. The highest occupied molecular orbitals (HOMO) of the radical lie in the HOMO–LUMO gap of Alq_3. Electron transfer in either direction between the radical and a neighbouring Alq_3 molecule would be very unfavourable (ranging from 4.09 to 5.68 eV). This is shown schematically in Figure 9.25.

ALC-μSR spectroscopy has been used to confirm that the electron is localized on the same molecule to which the muon is attached [53]. Such localized electron spin density should give rise to Δ_1 and methylene-proton Δ_0 resonances between 0.8 and 1.9 T. Monte Carlo simulations performed to model the effect of electron hopping on the ALC-μSR spectra found that the Δ_1 resonance broadens substantially for hop rates of 10^6 to $10^7\,s^{-1}$ and disappears when the hop rate is comparable to the isotropic muon hfc. A new resonance forms at 1/10th the magnetic field of the original resonance when the hop rate is much larger than A_μ; this is simply due to the unpaired spin density being equally distributed over the 10 molecules used in the simulation (Figure 9.26). Hopping at $10^{12}\,s^{-1}$ in a system containing a larger number of molecules would result in negligible unpaired spin density on the muoniated molecule and hence no ALC resonance. It is unreasonable to assume that the unpaired electron is hopping between limited numbers of Alq_3 molecules, because once it hops away from the muoniated radical it will continue to further Alq_3 molecules rather than return to the molecule bearing the muon.

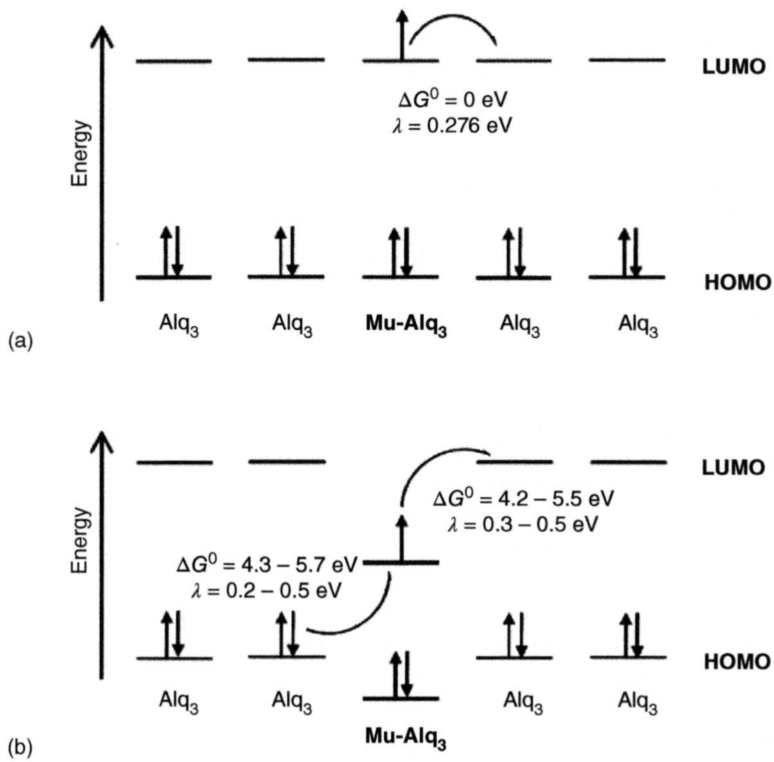

Figure 9.25 (a) The muon as a passive observer. The Gibbs energy for electron transfer, ΔG^0, is 0, and λ is the reorganization energy for electron transport between Alq_3 molecules [52]. (b) The muon as a perturbation; the energy levels of the muoniated radical are different from those of the parent compound. ΔG^0 and λ were calculated for electron transfer to and from the muoniated molecule [51]. The singly-occupied molecular orbital lies between the HOMO and LUMO of Alq_3. The large ΔG^0 and λ values indicate that the unpaired electron is localized on the molecule bearing the muon.

ALC-μSR experiments on Alq_3 have revealed multiple resonances between 0.8 and 1.7 T [53, 54], consistent with Mu addition at different sites of the Alq_3 molecule. At 300 K two resonances are prominent (Figure 9.27). The observation of Δ_1 resonances conclusively shows that the unpaired electron is localized on the molecule bearing Mu.

ALC resonances have also been observed from 6,13-bis(triisopropylsilylethynyl) pentacene (TIPS-pentacene) and other organic semiconductors based on polyaromatic hydrocarbons [55] so it is likely that spin localization occurs in these systems as well. This is not to say that the Mu adducts of organic semiconductors could not provide information about organic semiconductors. The hyperfine coupling constants and spin relaxation of the muoniated probe could potentially provide information about the electronic structure and charge conduction, but further investigation is required.

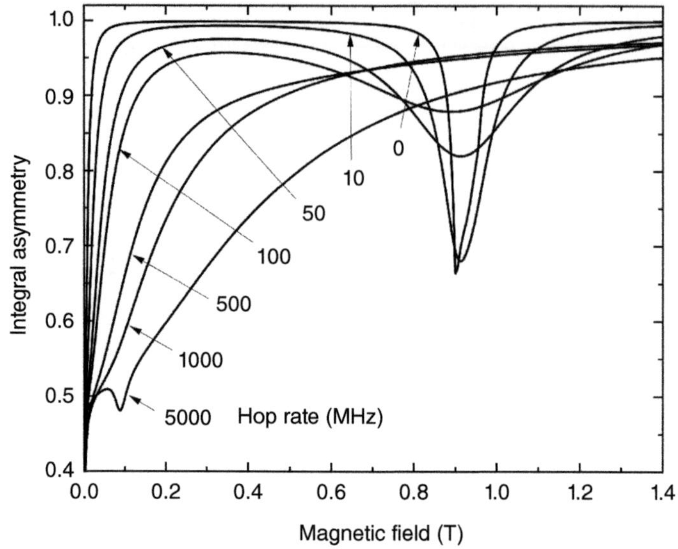

Figure 9.26 Simulated ALC-μSR spectra for electron hopping between 10 molecules of which only one has a non-zero muon hfc (isotropic muon hfc $A_\mu = 250$ MHz and dipolar muon hfc $D_\mu = 10$ MHz). The curves correspond to different electron hopping rates between adjacent molecules. Source: Adapted from a figure by McKenzie et al. [53] with the permission of AIP Publishing.

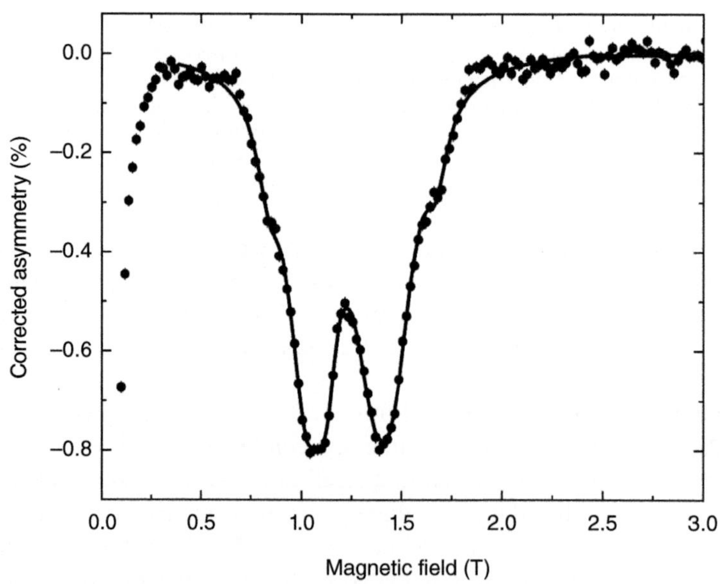

Figure 9.27 Background-subtracted ALC-μSR spectrum obtained from Alq_3 powder at 298 K. Source: Adapted from a figure by McKenzie et al. [53] with the permission of AIP Publishing.

9.6 Biological Systems

Biological systems have not been extensively studied by muon spin spectroscopy. This is most likely due to the complicated structures and that one cannot control where the muon ends up; i.e. one cannot label specific parts of a large biomolecule. This is a disadvantage of μSR compared with site-directed spin labelling and ESR. The relative yields of radicals formed by Mu addition are determined by the relative rates of reaction at the various addition sites. Given the large number of potential sites with similar reactivity, a plethora of muoniated radicals should be formed in most biological materials. Distributing the muon spin polarization among so many radicals can result in signals too small to be detected.

9.6.1 μSR of DNA and Its Constituents

DNA is the cornerstone of life, as it contains the information used in the growth, development, functioning and reproduction of all known living organisms. DNA has a double helix structure that is composed of two strands of nucleotides. Each nucleotide is made up of one of four nucleobases (adenine (A), guanine (G), cytosine (C) and thymine (T)), a sugar called deoxyribose and a phosphate group. Ionizing radiation can damage DNA and cause drastic effects such as genetic mutations or the death of a living entity [56]. The hydrogen atom is an important intermediate in the radiolysis of water [57] and H atoms have been detected by ESR in samples of UV- and γ-irradiated DNA [58–60] where it is believed that they are produced by the decomposition of radical ions. In vivo, DNA molecules are subject to radiation damage from attack by radicals such as $^{\bullet}$OH, e_{aq}^{-} and H$^{\bullet}$ formed from the radiolysis of water. Hydrogen atom adducts of nucleobases are important transient radical intermediates in the complex process of DNA radiation damage.

Multiple muoniated radicals can be formed by Mu addition to adenine, cytosine, guanine and thymine. The naming scheme used in the following discussion is as follows: the first letter refers to the nucleobase; the next term refers to the addition site (with the numbering used in Figure 9.28) and the last term refers to whether the radical is produced by H or Mu addition. For example, the radical C-C5-Mu is produced by Mu addition to C5 of cytosine. ALC-μSR spectra of the muoniated nucleobases are shown in Figure 9.29 [61]. Assignment of the resonances is based on two considerations: (i) the observed muoniated radicals are likely to correspond to the radicals with the lowest energies [62]; (ii) matching the muon hfc with the relevant proton hfc of the H-atom adduct, after accounting for known isotope effects. The proton hfcs are known from the ESR literature and computation [61, 63].

The ALC-μSR spectra of the purine bases are relatively simple. There are two resonances in the spectrum of adenine at 298 K, one at 1.52 T and a weaker one at 1.73 T. Both are Δ_1 resonances, and they correspond to muon hfc values of 414.3 and 471.4 MHz, respectively. These are due to the A-C8-Mu and A-C2-Mu radicals with relative yields of ~70 and 30%. There is a single Δ_1 resonance at 1.47 T in the spectrum of guanine, corresponding to an A_μ value of 400.8 MHz. This is due to the G-C8-Mu radical; the amino group at C2 blocks addition at that site.

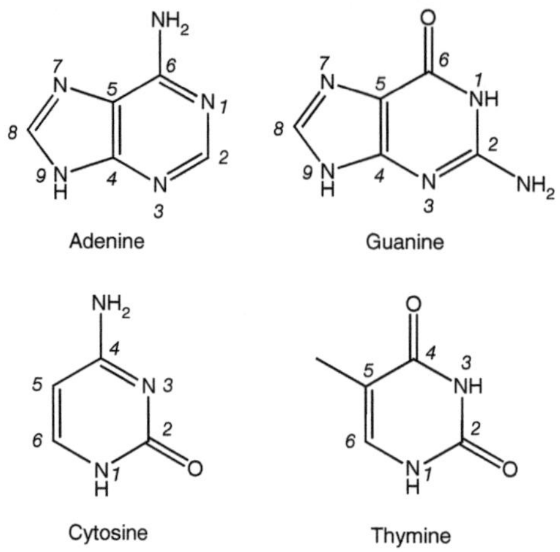

Figure 9.28 Structures of the purine nucleobases adenine and guanine, and the pyrimidine nucleobases cytosine and thymine.

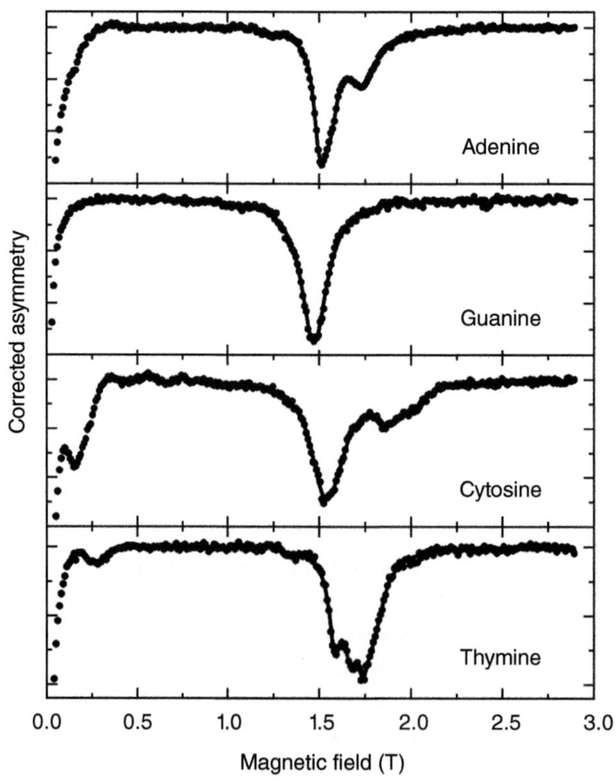

Figure 9.29 ALC-μSR spectra of the nucleic acid bases adenine and guanine (purine derivatives) and cytosine and thymine (pyrimidine derivatives) [61].

The ALC-μSR spectra of the pyrimidine bases are more complicated. The spectrum of cytosine has a large Δ_1 resonance at 1.55 T (A_μ = 420.9 MHz) which is assigned to the C-C5-Mu radical. The weaker resonances at 1.87 and 2.01 T are the Δ_1 and the proton Δ_0 resonance of the C-C6-Mu radical, with A_μ = 510.2 MHz and A_p = 133.6 MHz. The relative yields of C-C5-Mu and C-C6-Mu are about 80 and 20%. There is also a small Δ_1 resonance at 0.16 T ($|A|$ = 42 MHz) that is due to the C-N3-Mu radical. There are several overlapping resonance in the spectrum of thymine. The resonances at 1.58 and 1.74 T are the Δ_1 and the methylene proton Δ_0 resonances of the T-C6-Mu radical, with A_μ = 431.4 MHz and A_p = 106.5 MHz. The resonance at 1.67 T is a Δ_1 corresponding to A_μ = 454.8 MHz and is due to the T-C5-Mu radical. Mu adds to both the secondary C6 and tertiary C5 in thymine with approximately equal yields, which is surprising given the methyl group at C5. There is also a Δ_1 resonance at 0.27 T ($|A|$ = 74 MHz) that is due to the T-O4-Mu radical.

The μSR measurements made so far on the nucleic acid bases have established that there are multiple resonances in the same field range, making it difficult to separate the contributions from different radicals in a mixture of bases. Additional resonances have been found in the ALC-μSR spectra of a double-stranded DNA sample, indicating that the behaviour of implanted muons in DNA is more complex than that of the individual bases [63]. We conclude that muon spin spectroscopy has the potential to make an impact in the study of radiation damage in biological materials, but thorough study of simpler systems is a prerequisite.

9.6.2 Muoniated Radicals Formed from Proteins

Proteins are large biomolecules consisting of one or more long chains of amino acid residues. Proteins perform a vast array of functions within organisms, including catalysing metabolic reactions, DNA replication, responding to stimuli, providing structure to cells and organisms and transporting molecules from one location to another. Proteins differ from one another primarily in their sequence of amino acids, which is dictated by the nucleotide sequence of their genes, and which usually results in protein folding into a specific three-dimensional structure that determines its activity.

There are 20 amino acids commonly found in proteins, of which 4 have side chains with aromatic groups (histidine, phenylalanine, tryptophan, tyrosine). There is a carbonyl group in each peptide bond (amide bond), which is a covalent chemical bond linking two consecutive amino acid monomers along a peptide or protein chain. Predicting what radicals are formed requires some chemical intuition, but one can make reasonable guesses by considering a few simple rules. The rate constant for Mu addition to C=C bonds is approximately 10^{10} M^{-1} s^{-1}, to aromatic rings it is ~10^9 M^{-1} s^{-1} and to unconjugated carbonyl groups it is ~10^7 M^{-1} s^{-1}. Thus, when C=C bonds or aromatic groups are present, Mu addition is at these sites. The only exception is when the C=O bond is conjugated with the aromatic ring [64]. Thus, Mu addition to the carbonyl group is observed in acetophenone but not in benzylacetone, where the benzene ring and carbonyl group are separated by two $-CH_2-$ groups. Given the relative reactivities, Mu addition should mainly produce radicals at unsaturated side chains. As discussed in Section 3.3, there is a marked preference for addition at the least substituted end of a double bond. Thus, the likely muoniated radicals that could form in proteins are:

(1) At histidine: two radicals from addition to the secondary carbons of the imidazole ring. This is supported by the results of experiments on 1-methylimidazole, where the only radicals observed were formed by Mu addition to the secondary carbon atoms of the ring [65].
(2) At phenylalanine: three cyclohexadienyl-type radicals from addition at the phenyl ring.
(3) At tryptophan: five radicals from Mu addition to the indole ring.
(4) At tyrosine: two cyclohexadienyl-type radicals from Mu addition to the hydroxyphenyl ring.

The structures of the possible muoniated radicals are shown in Figure 9.30.

The considerable challenges faced by μSR studies of biomolecules can be illustrated by the case of cytochrome C. This is a small hemeprotein (mass ~12 kDa) found loosely associated with the inner membrane of mitochondria. Its primary structure consists of a chain of 104 amino acids (in humans) along with a heme C unit. In human cytochrome C there are 3 histidines, 3 phenylalanines, 1 tryptophan and 4 tyrosines, giving 12 different types of muoniated radical that would likely form by Mu addition to the protein chain, and these could generate up to 28 radical signals if the hyperfine parameters depend significantly on the amino acid sequence. There could be slight differences in A_μ for the cyclohexadienyl radicals formed by Mu addition to phenylalanine, for example, depending on where they are located in the protein. This could broaden ALC resonances, although it has not been studied experimentally. In addition, for aqueous solutions more than half of the muon polarization is found in the diamagnetic fraction. Accordingly, the detection of multiple radical signals would be exceedingly difficult.

The heme unit of cytochrome C is involved in electron transfer processes that occur within the cell, during which the oxidation state of the iron changes. The iron is always in a low-spin state; Fe^{3+} has one unpaired electron and is paramagnetic whereas Fe^{2+} is diamagnetic. In principle muoniated radicals could form by Mu addition to the secondary carbons of the heme ring with diamagnetic Fe^{2}, or Mu^{+} addition to the heme ring with paramagnetic Fe^{3+}. However, implantation of muons in paramagnetic compounds has been found to produce only diamagnetic signals ($P_D \sim 1$) [66].

Attempts have been made to study the e-transfer process in biomolecules using LF-μSR, but this has been unsuccessful as there are contributions to the spin relaxation from all muoniated species, from the diamagnetic muons as well as all the paramagnetic species. Given the variety of muoniated radicals that can form, with each having different hyperfine parameters and field dependence for the spin-lattice relaxation rate, it is not possible to extract meaningful information from the spectra reported so far. In particular, a single Risch–Kehr relaxation function is not appropriate to analyse the LF-μSR spectra of cytochrome C [67]. Numerous things are

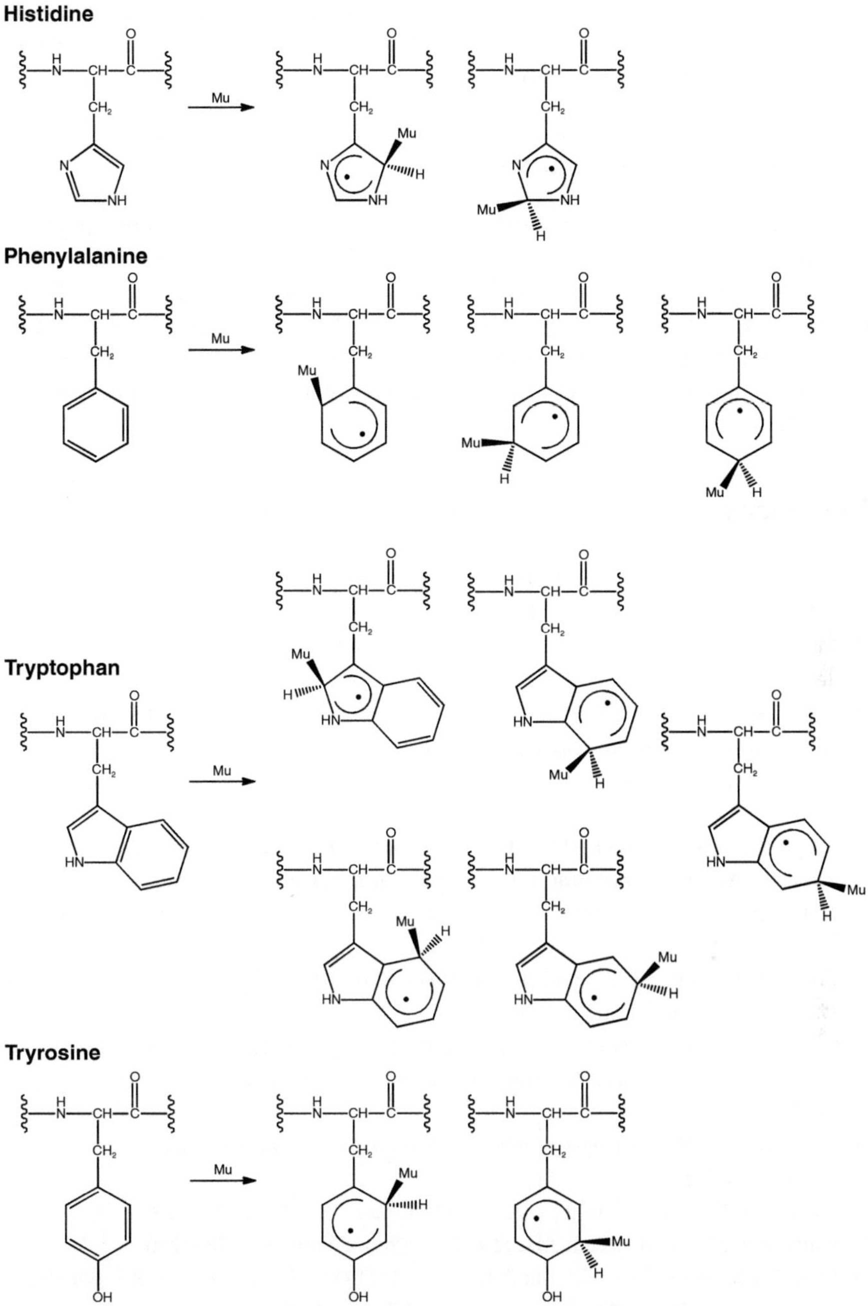

Figure 9.30 Structures of amino acids with unsaturated side groups (histidine, phenylalanine, tryptophan and tyrosine) and the corresponding potential Mu adducts.

happening simultaneously, and it is not possible to deconvolute them. Fitting with any relaxation function will give qualitatively similar results: the relaxation rate will decrease with increasing magnetic field.

9.7 Concluding Thoughts

Muon spin spectroscopy can be used to study radicals in complex systems, and variations in the hyperfine coupling or the spin relaxation can provide information about the microscopic properties of materials. However, the information one can glean is only as good as one's knowledge of the muoniated spin probe and how it interacts with its environment. Muon spin spectroscopy is a powerful technique for studying soft matter and organic materials, and it also has potential for studying biological materials.

References

1 Hamley, I.W. (2007). *Introduction to Soft Matter: Synthetic and Biological Self-Assembling Materials*. Wiley.
2 Jones, R.A.L. (2002). *Soft Condensed Matter*. Oxford University Press.
3 de Gennes, P.-G. (2005). *Soft Matter* 1: 16–16.
4 Garcia Sakai, V., Alba-Simionesco, C., and Hsin Chen, S. (ed.) (2012). *Dynamics of Soft Matter: Neutron Applications*. Springer US.
5 Gabrys, B.J. (ed.) (1995). *Applications of Neutron Scattering to Soft Condensed Matter*. CRC Press.
6 Higgins, J.S. and Benoit, H.C. (1997). *Polymers and Neutron Scattering, Oxford Series on Neutron Scattering in Condensed Matter*. Clarendon Press.
7 de Gennes, P.-G. and Prost, J. (1993). *The Physics of Liquid Crystals*, 2ed. Oxford University Press.
8 Collings, P.J. and Hird, M. (1997). *Introduction to Liquid Crystals: Chemistry and Physics*. Taylor and Francis Ltd.
9 Chandrasekhar, S. (1992). *Liquid Crystals*, 2ed. Cambridge University Press.
10 Kneppe, H., Reiffenrath, V., and Schneider, F. (1982). *Chem. Phys. Lett.* 87: 59–62.
11 Dong, R.Y. (1997). *Nuclear Magnetic Resonance of Liquid Crystals*. Springer-Verlag.
12 Burnell, E.E. and de Lange, C.A. (1998). *Chem. Rev.* 98: 2359–2388.
13 Polnaszek, C.F. and Freed, J.H. (1975). *J. Phys. Chem.* 79: 2283–2306.
14 Rao, K.V.S., Polnaszek, C.F., and Freed, J.H. (1977). *J. Phys. Chem.* 81: 449–456.
15 Lin, W.-J. and Freed, J.H. (1979). *J. Phys. Chem.* 83: 379–401.
16 Meirovitch, E., Igner, D., Igner, E. et al. (1982). *J. Chem. Phys.* 77: 3915–3938.
17 Bordignon, E. (2018). *EPR Spectroscopy: Fundamentals and Methods*, vol. ch. 14 (ed. D. Goldfarb and S. Stoll), 277–301. Wiley.
18 McKenzie, I., Scheuermann, R., Sedlak, K., and Stoykov, A. (2011). *J. Phys. Chem. B* 115: 9360–9368.

19 Yu, D., Percival, P.W., Brodovitch, J.-C. et al. (1990). *Chem. Phys.* 142: 229–236.
20 Vujŏsevìc, D., Dilger, H., McKenzie, I. et al. (2007). *J. Phys. Chem. A* 111: 199–208.
21 Israelachvili, J., Sjösten, J., Eriksson, L.E.G. et al. (1975). *Biochim. Biophys. Acta* 382: 125–141.
22 McKenzie, I., Dilger, H., Stoykov, A., and Scheuermann, R. (2009). *J. Phys. Chem. B* 113: 10135–10142.
23 Cox, S.F.J. and Sivia, D.S. (1994). *Hyperfine Interact.* 87: 971–976.
24 Lovett, B.W., Blundell, S.J., Stießberger, J.S. et al. (2001). *Phys. Rev. B* 63: 054204.
25 Scheuermann, R., Tucker, I.M., Creeth, A.M. et al. (2002). *Phys. Chem. Chem. Phys.* 4: 1510–1512.
26 Scheuermann, R., Tucker, I.M., Dilger, H. et al. (2004). *Langmuir* 20: 2652–2659.
27 Martyniak, A., Scheuermann, R., Dilger, H. et al. (2006). *Physica B* 374–375: 328–331.
28 Martyniak, A., Dilger, H., Scheuermann, R. et al. (2006). *Phys. Chem. Chem. Phys.* 8: 4723–4740.
29 McKenzie, I., Scheuermann, R., and Tucker, I. (2017). *Phys. Chem. Chem. Phys.* 19: 9551–9557.
30 Dilger, H., Martyniak, A., Scheuermann, R. et al. (2006). *Physica B* 374–375: 317–320.
31 Molugu, T.R., Lee, S., and Brown, M.F. (2017). *Chem. Rev.* 117: 12087–12132.
32 M.C. Rheinstätder. In: V. Garcia Sakai, C. Alba-Simionesco and S.H. Chen, editors, *Dynamics of Soft Matter: Neutron Applications*, Chapter 10, Springer Verlag (2012).
33 Veatch, S.L., Leung, S.S.W., Hancock, R.E.W., and Thewalt, J.L. (2007). *J. Phys. Chem. B* 111: 502–504.
34 Roduner, E. (2003). *Physica B* 326: 19–24.
35 Berthier, L. and Ediger, M.D. (2016). *Phys. Today* 69: 40–46.
36 Hutchinson, J.M. (1997). *The Physics of Glassy Polymers*, Chapter 3, 2ee, 85–154. Springer Science + Business Media.
37 McKenzie, I., Chai, Y., Cortie, D.L. et al. (2018). *Soft Matter* 14: 7324–7334.
38 McKenzie, I., Cordoni-Jordan, D., Cannon, J., and Cottrell, S.P. (2020). *J. Phys. Condens. Matter* 33: 065102.
39 Eaton, S.S. and Eaton, G.R. (2018). *EPR Spectroscopy: Fundamentals and Methods* (ed. D. Goldfarb and S. Stoll) Chapter 9. Wiley.
40 Bloembergen, N., Purcell, E.M., and Pound, R.V. (1948). *Phys. Rev.* 73: 679–712.
41 Pschorn, U., Rössler, E., Sillescu, H. et al. (1991). *Macromolecules* 24: 398–402.
42 Schnauss, W., Fujara, F., Hartmann, K., and Sillescu, H. (1990). *Chem. Phys. Lett.* 166: 381–384.
43 Roth, S. and Carroll, D. (2015). *One-Dimensional Metals: Conjugated Polymers, Organic Crystals, Carbon Nanotubes and Graphene*, 3ed. Wiley.
44 Pratt, F.L. (2004). *J. Phys. Condens. Matter* 16: S4779–S4796.
45 Risch, R. and Kehr, K.W. (1992). *Phys. Rev. B* 46: 5246–5257.
46 Pratt, F.L., Blundell, S.J., Hayes, W. et al. (1997). *Phys. Rev. Lett.* 79: 2855–2858.
47 Miller, J.S. (2011). *Chem. Soc. Rev.* 40: 3266–3296.

48 Blundell, S.J., Pattenden, P.A., Pratt, F.L. et al. (1997). *Hyperfine Interact.* 104: 251–256.

49 Tang, C.W. and Van Slyke, S.A. (1987). *Appl. Phys. Lett.* 51: 913–915.

50 Drew, A.J., Pratt, F.L., Hoppler, J. et al. (2008). *Phys. Rev. Lett.* 100: 116601.

51 McKenzie, I. (2010). *J. Phys. Chem. A* 114: 12759–12763.

52 Lin, B.C., Cheng, C.P., You, Z.-Q., and Hsu, C.-P. (2005). *J. Am. Chem. Soc.* 127: 66–67.

53 McKenzie, I., Cannon, J., Cordoni-Jordan, D. et al. (2022). *J. Chem. Phys.* 157: 064702.

54 Nuccio, L., Willis, M., Schulz, L. et al. (2013). *Phys. Rev. Lett.* 110: 216602.

55 Schulz, L., Willis, M., Nuccio, L. et al. (2011). *Phys. Rev. B* 84: 085209.

56 von Sonntag, C. (2006). *Free-Radical-Induced DNA Damage and its Repair: A Chemical Perspective.* Springer-Verlag.

57 Draganić, I.G. and Draganić, Z.D. (1971). *The Radiation Chemistry of Water.* Academic Press, Inc.

58 Heller, H.C. and Cole, T. (1965). *Proc. Natl. Acad. Sci. U. S. A.* 54: 1486–1490.

59 Müller, A. and Dertinger, H. (1968). *Z. Naturforsch. B* 23: 83–89.

60 Herak, J.N. and Galogaza, V. (1969). *Proc. Natl. Acad. Sci. U. S. A.* 64: 8–12.

61 McKenzie, I. (2019). *J. Phys. Chem. B* 123: 4540–4549.

62 Goli, M. and Jalili, S. (2018). *Int. J. Quantum Chem.* 118: e25758.

63 Hubbard, P.L., Oganesyan, V.S., Sulaimanov, N. et al. (2004). *J. Phys. Chem. A* 108: 9302–9309.

64 Stride, J.A., Jayasooriya, U.A., and Reid, I.D. (2002). *J. Phys. Chem. A* 106: 244–250.

65 Rhodes, C.J., Morris, H., and Reid, I.D. (1995). *J. Chem. Soc. Perkin Trans.* 2: 2107–2114.

66 Kubo, M.K. (2001). *Anal. Sci.* 17: i653–i656.

67 Nagamine, K. (2001). *Hyperfine Interact.* 136–137: 759–762.

10

Future Developments and Outlook

First invented over 75 years ago, NMR and ESR have become indispensable tools in chemistry and materials science. Indeed, the applications of NMR are widespread and it has even become ubiquitous in modern medicine, albeit renamed MRI (magnetic resonance imaging). Can we foresee a similar future for muon spin spectroscopy? Will it remain the province of specialist researchers, or can we envisage technical developments that enable applications in mainstream science?

We begin by considering the special aspects of the muon and the advantages and limitations of muon spin spectroscopy. Many of these have already been discussed earlier in this book; the purpose of this chapter is to examine potential extrapolations of muon spectroscopy, including non-resonant techniques, to prompt plausible predictions of future directions.

10.1 Light Mass and Isotope Effects

As regards chemistry, the preeminent property of the muon is its light mass. This results in a large de Broglie wavelength for Mu, three times greater than H at a given temperature. The extended probability distribution has consequences for the effective size of the atom and its diffusion rate in condensed matter [1], but most importantly on the penetration of potential barriers, i.e. quantum tunnelling. The 'world record of a kinetic isotope effect' [2] is reported as Mu transfer from the muoniated cyclohexadienyl radical to the dimethylbutadiene molecule. Kinetic isotope effects also arise from zero-point vibrational energy differences along the reaction path, as clearly demonstrated by fundamental studies of muonium kinetics in the gas phase (Chapter 4).

Zero-point vibrational isotope effects are also manifest in aspects of structure, as evident from bond elongation and conformational effects in muoniated radicals (Chapter 6). Perhaps the most remarkable effect is for the minimum-energy structures of the Br–L–Br system, from unsymmetrical van der Waals complexes for L = H (and heavier atoms) to the symmetrical vibrationally bonded BrMuBr

Muon Spin Spectroscopy: Methods and Applications in Chemistry and Materials Science, First Edition.
Donald G. Fleming, Iain McKenzie, and Paul W. Percival.

radical (Section 4.8). Interestingly, a switch from an unsymmetrical to a symmetrical structure has also been discussed for the 'muonium bond' in water [3] and in muoniated malonaldehyde [4]. In such cases the increased vibrational energy in the muoniated system can exceed the potential barrier in the usual double-potential hydrogen-bonded system.

10.2 Muon Spin Spectroscopy: Advantages and Limitations

With their radioactive nature and spin polarization, beam muons have a unique signature, permitting their use in situations where the target of an investigation would otherwise be lost in a sea of hydrogen atoms. The use of a muoniated label is particularly valuable for studying complex systems, such as soft matter (Chapter 9). In some circumstances Mu can even be used as a passive spectator in kinetic studies, for example radical cyclization and ring fission reactions [5], where there are minimal isotope effects because the muon is remote from the reaction site.

Another advantage provided by muon spin polarization is sensitivity. Quantitative comparison of ESR and μSR is difficult due to the large number of dissimilar factors. However, taking typical sensitivity factors for X-band ESR [6] a 10 μM solution of a nitroxide sample ($\sim 10^{13}$ spins) should give a clearly detectable spectrum (signal-to-noise ratio of 10). In contrast, a similar quality spectrum of a muoniated radical requires as few as 10^7 events. Note that this million-fold factor in sensitivity refers to *accumulated* events for μSR and ignores the possibility of signal averaging over a longer time period for ESR of stable radicals.

In addition to the sensitivity accorded by muon spin polarization, the short muon lifetime and the particle counting techniques used in muon spin spectroscopy mean that only a few muons are present in the sample at any one time. Thus, there is negligible probability of interactions between muoniated species. In contrast, ESR linewidths are broadened by spin exchange between radicals, and secondary reactions complicate kinetics and potentially lead to multiple radical products.

Perhaps the most important feature of μSR that distinguishes it from other forms of magnetic resonance is the ability to inject muons into any sort of sample. This includes extreme environments, such as gases at high temperatures and pressures, hydrothermal systems and supercritical water. Reaction of muonium with keto and enol forms of acetone was briefly mentioned in Section 3.3, but several more muoniated radicals have been studied in hot, pressurized water [7]. Furthermore, muonium decay kinetics has been studied in supercritical water (Section 5.3), under conditions similar to those of a water-cooled nuclear power reactor.

Other technologically important problems difficult to study by conventional magnetic resonance techniques include catalytic reactions on surfaces and in zeolites and microporous materials (Chapter 8). Free radicals have been investigated by both ESR and μSR in clathrate hydrates (Section 8.3), but studying them under high pressure (ocean-floor conditions) will only be possible with high-momentum muons.

10.3 New Methodologies

10.3.1 Transient Targets

The short lifetime of the muon means that μSR is well suited to study short-lived species, such as the transient free radicals described in Chapter 6. However, we can also conceive of experiments in which a muon or muonium reacts with a short-lived target molecule, such as a photo-excited or photogenerated molecule. This procedure has already been described for Mu reacting with the vibrationally excited H_2 molecule (Section 4.6) and could obviously be extended to other excited molecules. In fact other proof-of-principle photo-μSR experiments have already been carried out [8, 9], although some of the effects are so weak that their statistical significance has been questioned [10]. Future, more intense muon sources should allow better overlap of muon- and photo-irradiated sample volumes, and other applications will ensue. For example, hydrated electrons can be produced photolytically [11]. A study of Mu interacting with controlled amounts of e^-_{aq} would provide a valuable test of the missing fraction mechanism proposed in the spur model of muonium formation (Section 3.2).

Matrix isolation is a general strategy used in other fields to stabilize otherwise short-lived species. The difficulty as regards μSR would be to ensure encounter of the muon or muonium atom with the dilute reactant; fast diffusion in liquids and gases avoids this problem. A different, potentially valuable strategy could be to use a flow system to transport freshly generated transients through the muon-irradiated region of the sample assembly. Transient generation might be by chemical, electrochemical or other means, not just photochemical.

If a measurement is made on a *working* system, it is called an *operando* experiment. The feasibility of μSR measurements during charge and discharge cycles of an electrochemical cell has already been demonstrated [12]. Improved muon beams should lead to more experiments of this sort, with potential applications ranging from the study of the unstable intermediates that cause knock in internal combustion engines to radical spin traps in living cells.

10.3.2 Optical Spectroscopy

Another area ripe for future development involves coupling muon spin spectroscopy with optical spectroscopy, as was done in the chemiluminescence study of electronic transitions in the NeMu* molecule [13]. Although the short muon lifetime broadens optical transitions, making precision molecular spectroscopy impractical, there is significant potential for using μSR to probe optically pumped systems. An example is the proposal to monitor the μSR signal of a radical while a near-IR laser frequency is swept to excite a vibrational transition [14]. Since hyperfine frequencies are typically highly sensitive to vibrational state it should be possible to detect a vibrational mode in this manner. As discussed in Section 6.2.1, stretching vibrations involving Mu are about three times higher than an equivalent bond containing H, so this sort of measurement would allow exploration of the upper, more anharmonic region of the vibrational potential. A proof-of-principle experiment has already been carried out, albeit with a broad-band light source (a xenon lamp) [15].

10.3.3 Spin Manipulation

Pump-probe experiments are not limited to light, and radio-frequency (or microwave) excitation is a potentially productive area for development. Early demonstrations of RF-μSR were aimed at an alternative means of detecting resonance (Section 2.9) for application in situations where TF-μSR is not feasible, for example dilute solutions [16], or where the time structure of pulsed muon beams prevents detection of high precession frequencies (Section 2.3). However, modern NMR (and to a lesser extent ESR) makes wide use of RF pulses for spin manipulation, the textbook example being the spin-echo procedure to recover coherent spin polarization otherwise lost through magnetic field inhomogeneity. RF inversion pulses and the consequent muon spin inversion have already been demonstrated [17], and it seems inevitable that more advanced pulse sequences will find application in muon spin resonance. Of particular interest to μSR is the use of RF to 'decouple' spins that contribute to the local nuclear dipolar field. Not only is this of practical advantage in reducing muon spin relaxation, by irradiating at the appropriate nuclear resonant frequency it is possible to identify the particular nuclei responsible for the local field at a muon site [18]. Future applications of RF decoupling may be used to confirm nuclei involved in hyperfine interactions in muoniated radicals.

10.4 Going Beyond Muon Spin Spectroscopy

In searching for clues to the future of muon spin spectroscopy we should also consider muon applications outside that defined field. Two examples follow.

10.4.1 Muon Tomography

The most straightforward application of muons is to tomography. In its simplest form, muon transmission radiography uses the attenuation of cosmic radiation to map cavities in large objects, such as glaciers, pyramids and volcanos. Cosmic ray muons reaching the earth's surface have energies of a few GeV and can easily penetrate the thick walls of these objects. Technical advances such as muon scattering tomography have led to a proliferation of uses in recent years, as described in various review articles [19-22]. Recent applications range from mineral exploration (muon *geotomography*) to imaging nuclear waste containers and even nuclear reactors themselves [23].

Of course, tomography with cosmic radiation is slow and only practical for very large objects, since the flux of muons reaching the Earth's surface is only of order 1 muon per square centimetre per minute. At the other end of the scale is the transmission muon microscopy proposed for the ultra-slow muon source at

J-PARC [24]. This will require acceleration of the low-energy muons to provide a high-intensity penetrating microbeam, either with a muon cyclotron [25] or a linac based on radio-frequency technology [26]

When the necessary 'pencil' beams have become available, the combination of tomographic methods with existing μSR techniques should enable a variety of novel applications, such as the visualization of 3D magnetic fields to nanometre resolution. The potential for depth-resolved studies already exists at the low-energy muon beam at PSI, and technical developments will soon allow detailed studies of surfaces and interfaces, with applications ranging from biological samples [27] to battery materials [28].

10.4.2 Muon-Induced X-ray Emission

By using negative muons in a scanning microscope it will be possible to map the 3D distribution of elements, even individual isotopes, in a sample, by using the X-rays emitted as the muon cascades down to the muonic 1s atomic orbital (Section 3.4). Elemental analysis can provide valuable information in a broad range of scientific fields, such as geochemical analysis, metallurgy and archaeology. Muon Induced X-Ray Emission (MIXE) spectroscopy (also sometimes referred to as muon X-ray fluorescence or μXRF) has significant advantages over many other techniques used in elemental analysis, particularly the ability to control the depth of penetration of the sample. It has already been successfully demonstrated for the non-destructive testing of cultural heritage objects [29–31].

10.5 The Outlook for Muon Science

Over the past decade the existing muon laboratories have gradually become more user-friendly. Consequently, there is significantly less requirement for technical expertise from outside topical experts who want to apply muon spin spectroscopy in their particular fields of research. Even so, the infrastructure required for muon production, and the associated culture of accelerator laboratories, still dictate long-term planning: experiment proposals and beam-time scheduling. This conflicts with the expectations of many potential users, for example those wanting an immediate test to characterize new materials. Moreover, the existing muon facilities are greatly oversubscribed. The emergence of new muon facilities should ameliorate this problem. MuSIC at the Research Center for Nuclear Physics (RCNP) in Osaka, Japan, is the newest operational μSR facility and provides a continuous muon beam [32]. A new pulsed μSR facility called MELODY (Muon station for sciEnce, technoLOgy and inDustrY) is under construction at the China Spallation Neutron Source (CSNS) at Dongguan in Guangdong province of the People's Republic of China [33]. User operations are expected to begin in 2029. A μSR facility for the RAON heavy ion accelerator complex in the Sindong area near Daejeon-city, South Korea, is being planned [34], and another is being discussed for the Spallation Neutron Source (SNS) of Oak Ridge National Laboratory [35].

New facilities will be useful in improving access to μSR, but the main conclusion of the US Muon Workshop 2021 was [36]: *The primary need for enabling new science with μSR is access to muon beams of higher flux.* What can one do with a higher muon flux? Increased counting rates lead to better statistics and the ability to study weak signals. Alternatively, beam profiles can be reduced for microscopic work. These improvements are not simply extrapolations from current studies; having more muons enables more difficult problems, for example crossed-beam experiments which couple muon spectroscopy with another technique, such as photo-excitation. Secondly, shorter measurement times facilitate *operando* experiments, i.e. measurements on a working system, such as an operating battery [12]. Flux however is not enough. The facilities need to have appropriate spectrometers for chemistry studies, and this is not always the case. Spectrometers for ALC-μSR are currently only available at TRIUMF and ISIS.

Despite the encouraging prospects of higher muon fluxes, the fundamental limitation of muon spectroscopy is the major infrastructure requirement of a particle accelerator. Imagine the opportunities which a portable muon generator would bring! A compact muon source would require a novel design, but this is not out of the question. Indeed, a conceptual sketch of a TeV muon generator shows that it could fit on the deck of an aircraft carrier [37]. A promising new direction is to use laser plasma acceleration to generate muons [38].

10.6 Conclusions

In this book we have demonstrated that muon spin spectroscopy is a powerful tool that can provide information about chemical systems that is not available from more conventional experiments. μSR can be used for studying the kinetics of Mu reactions, the structure and dynamics of radicals in a range of environments, and spin probes in complex materials. It is important to include a word of caution, as μSR is not appropriate for all systems and great care must be taken to consider the strengths and weaknesses of μSR before beginning a research project. However, there are many areas that are amenable to study by μSR that have not yet been examined, so there are fantastic opportunities for the interested researcher.

References

1 Ceriotti, M., Fang, W., Kusalik, P.G. et al. (2016). *Chem. Rev.* 116: 7529–7550.
2 Roduner, E. (2005). *Isotope Effects In Chemistry and Biology* (ed. A. Cohen and H.-H. Limbach), 433–450. CRC Press.
3 Brocklehurst, B. and Cook, D.B. (1987). *Chem. Phys. Lett.* 142: 329–333.
4 Goli, M. and Shahbazian, S. (2016). *Chem. Eur. J.* 22: 2525–2531.
5 Burkhard, P., Roduner, E., Hochmann, J., and Fischer, H. (1984). *J. Phys. Chem.* 88: 773–777.

6 Eaton, G.R., Eaton, S.S., Barr, D.P., and Weber, R.T. (2010). *Quantitative EPR*. Springer.
7 Percival, P.W., Brodovitch, J.-C., Ghandi, K. et al. (2005). *J. Am. Chem. Soc.* 127: 13714–13719.
8 Ghandi, K., Clark, I.P., Lord, J.S., and Cottrell, S.P. (2007). *Phys. Chem. Chem. Phys.* 9: 353–359.
9 Wang, K., Murahari, P., Yokoyama, K. et al. (2017). *Nat. Mater.* 16: 467–473.
10 Scheuermann, R. and McKenzie, I. (2022). *Nat. Mater.* 21: 1108–1109.
11 Naumann, R., Kerzig, C., and Goez, M. (2017). *Chem. Sci.* 8: 7510–7520.
12 Ohishi, K., Igarashi, D., Tatara, R. et al. (2022). *ACS Appl. Energy Mater.* 5: 12538–12544.
13 Baer, S., Fleming, D.G., Sloan, J.J. et al. (1994). *J. Chem. Phys.* 101: 1202–1218.
14 Cox, S.F.J. (1995). *Philos. Trans. R. Soc. London, Ser. A* 350: 171–187.
15 Jayasooriya, U.A., Pratt, F.L., Aston, G.M. et al. (2004). *ChemPhysChem* 5: 257–261.
16 Percival, P.W., Addison-Jones, B., Brodovitch, J.-C., and Sun-Mack, S. (1996). *Appl. Magn. Reson.* 11: 315–323.
17 Clayden, N.J., Cottrell, S.P., and McKenzie, I. (2012). *J. Magn. Reson.* 214: 144–150.
18 Clayden, N.J. and Cottrell, S.P. (2006). *Phys. Chem. Chem. Phys.* 8: 3094–3098.
19 Kaiser, R. (2019). *Phil. Trans. R. Soc. London Ser. A* 377: 20180049.
20 Bonechi, L., D'Alessandro, R., and Giammanco, A. (2020). *Rev. Phys.* 5: 100038.
21 Lechmann, A., Mair, D., Ariga, A. et al. (2021). *Earth Sci. Rev.* 222: 103842.
22 Miyadera, H. and Morris, C.L. (2022). *Appl. Opt.* 61: C154–C161.
23 Procureur, S., Attié, D., Gallego, L. et al. (2023). *Sci. Adv.* 9: eabq8431.
24 Miyake, Y., Nagatani, Y., Yamazaki, T., and Strasser, P. (2018). *Microscopy* 67: i3–i3.
25 Yamazaki, T. and Nagatani, Y. (2020). *International Conference on Cyclotrons and their Applications*, 208–211. Cape Town, South Africa: JACoW Publishing.
26 Bae, S., Choi, H., Choi, S. et al. (2018). *Phys. Rev. Accel. Beams* 21: 050101.
27 Sugawara, Y. (2019). *JPS Conf. Proc.* 25: 011008.
28 McClelland, I., Johnston, B., Baker, P.J. et al. (2020). *Annu. Rev. Mater. Res.* 50: 371–393.
29 Ninomiya, K., Kubo, M.K., Nagatomo, T. et al. (2015). *Anal. Chem.* 87: 4597–4600.
30 Hillier, A.D., Paul, D.M., and Ishida, K. (2016). *Microchem. J.* 125: 203–207.
31 Biswas, S., Gerchow, L., Luetkens, H. et al. (2022). *Appl. Sci.* 12: 2541.
32 Tomono, D., Fukuda, M., Hatanaka, K. et al. (2018). *PoS*, NuFact2017, 111. https://doi.org/10.22323/1.295.0111.
33 Bao, Y., Chen, J., Chen, C. et al. (2023). *J. Phys. Conf. Ser.* 2462: 012034.
34 Jeong, J.Y., Kim, J.C., Kim, Y. et al. (2021). *Nucl. Eng. Technol.* 53: 2909–2917.
35 Williams, T.J., MacDougall, G.J., Riemer, B.W. et al. (2023). *Rev. Sci. Instrum.* 94: 033908.
36 Louca, D., MacDougall, G.J., and Williams, T.J. (2022). *Neutron News* 33: 8–21.

37 Nagamine, K. (2003). *Introductory Muon Science*, Chap. 10. Cambridge University Press.

38 Calvin, L., Tomassini, P., Doria, D. et al. (2023). *Front. Phys.* 11: https://doi.org/10.3389/fphy.2023.1177486.

Appendix A

Derivation of Muon Polarization Expressions

A.1 Polarization in a μSR Experiment

The quantity of interest in a μSR experiment is how the ensemble polarization varies with time. The observable muon polarization along a direction u is the sum of contributions $P_u^X(t)$ from all species X containing muons,

$$P_u(t) = \sum_X P_u^X(t). \tag{A.1}$$

The polarization as a function of time in the arbitrary u direction for a species 'X' that is formed at $t = 0$ with a probability h_X, which varies between 0 and 1, and decays with a pseudo-first-order rate constant λ is given by [1]:

$$P_u^X(t) = \frac{4h_X e^{-\lambda t}}{N} \sum_m \sum_n \left| \langle m | \hat{I}_u^\mu | n \rangle \right|^2 e^{2\pi i(\nu_n - \nu_m)t}, \tag{A.2}$$

N is the dimension of the Pauli spin operator matrix, which is equal to the number of spin states in the system. $N = \Pi_k(2I_k + 1)$ where the product is over the electron, muon and nuclear spins. N is therefore 2 for a muon, 4 for muonium, 8 for a muoniated radical containing a single proton and 12 for a muoniated radical containing a single ^{14}N spin, for example.

$|m\rangle$ and $|n\rangle$ are eigenstates with energies of $h\nu_m$ and $h\nu_n$, respectively. The eigenstates are obtained by solving the relevant spin Hamiltonian. We will consider diamagnetic muons, muonium and muoniated radicals in longitudinal and transverse magnetic fields.

A.2 Diamagnetic Muons in a Longitudinal Field

Muons in a diamagnetic environment have the following spin Hamiltonian when the magnetic field is aligned along the axis of polarization, chosen to be the z-axis by convention:

$$\hat{H}/h = -\nu_\mu \hat{I}_z^\mu, \tag{A.3}$$

ν_μ is the muon Larmor frequency and equals $\gamma_\mu B$ where γ_μ is the muon gyromagnetic ratio and has a value of 135.54 kHz/G. $\hat{I}_z^\mu$ is the z-component of the muon

Muon Spin Spectroscopy: Methods and Applications in Chemistry and Materials Science, First Edition.
Donald G. Fleming, Iain McKenzie, and Paul W. Percival.

spin operator. The μ superscript means that it operates only on the muon spin. The general form for the $\hat{I}_z$ spin operator written in a basis set of $|\alpha\rangle$ and $|\beta\rangle$ is:

$$\hat{I}_z = \frac{1}{2}\begin{pmatrix} 1 & 0 \\ 0 & -1 \end{pmatrix}. \tag{A.4}$$

There are two eigenstates for this spin Hamiltonian,

$$|\alpha\rangle = \begin{pmatrix} 1 \\ 0 \end{pmatrix}, \quad |\beta\rangle = \begin{pmatrix} 0 \\ 1 \end{pmatrix}, \tag{A.5}$$

which correspond to the muon spin up and down, respectively. These are orthonormal in that $\langle\alpha\,|\,\alpha\rangle| = 1$, $\langle\beta\,|\,\beta\rangle| = 1$ and $\langle\alpha\,|\,\beta\rangle = \langle\beta\,|\,\alpha\rangle = 0$.

Another way these can be specified is in terms of the I and m_I values, with $|\alpha\rangle = |\frac{1}{2}, \frac{1}{2}\rangle$ and $|\beta\rangle = |\frac{1}{2}, -\frac{1}{2}\rangle$. The value returned by $\hat{I}_z$ is the m_I value, which is +1/2 for $|\alpha\rangle$ and −1/2 for $|\beta\rangle$.

$$\hat{I}_z|\alpha\rangle = \frac{1}{2}|\alpha\rangle, \quad \hat{I}_z|\beta\rangle = -\frac{1}{2}|\beta\rangle. \tag{A.6}$$

In cases where there are multiple spins, we will use superscripts to specify which nucleus is being referred to. In the case of the positive muon, where γ_μ is positive, the α state is lower in energy with an energy of $-\frac{1}{2}h\nu_\mu$. The energy of the β state is $+\frac{1}{2}h\nu_\mu$ and the difference between the two states is $h\nu_\mu$.

The polarization for a diamagnetic muon in a longitudinal field is obtained by setting $u = z$ in Eq. (A.1) and using the spin states in Eq. (A.4). This gives:

$$P_z^{\mathrm{D}}(t) = 2h_{\mathrm{D}}\mathrm{e}^{-\lambda t}\left\{\left|\langle\alpha|\hat{I}_z^{\mu}|\alpha\rangle\right|^2 + \left|\langle\alpha|\hat{I}_z^{\mu}|\beta\rangle\right|^2\mathrm{e}^{2\pi i\nu_\mu t} + \left|\langle\beta|\hat{I}_z^{\mu}|\alpha\rangle\right|^2\mathrm{e}^{2\pi i\nu_\mu t} + \left|\langle\beta|\hat{I}_z^{\mu}|\beta\rangle\right|^2\right\}. \tag{A.7}$$

The $\hat{I}_z$ operator is diagonal and only connects states with the same spin, $\langle\alpha|\hat{I}_z|\alpha\rangle = \langle\beta|\hat{I}_z|\beta\rangle = \frac{1}{2}$ and $\langle\alpha|\hat{I}_z|\beta\rangle = \langle\beta|\hat{I}_z|\alpha\rangle = 0$, so Eq. (A.7) reduces to

$$P_z^{\mathrm{D}}(t) = h_{\mathrm{D}}\mathrm{e}^{-\lambda t}. \tag{A.8}$$

The polarization does not evolve with time except through decay of the state.

A.3 Diamagnetic Muons in a Transverse Field

The expression for the spin polarization in a transverse magnetic field is the same as for the longitudinal-field case except that it is the $\hat{I}_x^{\mu}$ operator, which is the x-component of the muon spin operator, that is involved.

$$P_x^{\mathrm{D}}(t) = 2h_{\mathrm{D}}\mathrm{e}^{-\lambda t}\left\{\left|\langle\alpha|\hat{I}_x^{\mu}|\alpha\rangle\right|^2 + \left|\langle\alpha|\hat{I}_x^{\mu}|\beta\rangle\right|^2\mathrm{e}^{2\pi i\nu_\mu t} + \left|\langle\beta|\hat{I}_x^{\mu}|\alpha\rangle\right|^2\mathrm{e}^{2\pi i\nu_\mu t} + \left|\langle\beta|\hat{I}_x^{\mu}|\beta\rangle\right|^2\right\} \tag{A.9}$$

The general form of the $\hat{I}_x$ operator is given by:

$$\hat{I}_x = \frac{1}{2}\begin{pmatrix} 0 & 1 \\ 1 & 0 \end{pmatrix} \tag{A.10}$$

$$\hat{I}_x|\alpha\rangle = \frac{1}{2}|\beta\rangle, \quad \hat{I}_x|\beta\rangle = -\frac{1}{2}|\alpha\rangle \tag{A.11}$$

The matrix elements of the $\hat{I}_x$ operator are non-zero only for the $|\langle\alpha|\hat{I}_x|\beta\rangle|^2$ and $|\langle\beta|\hat{I}_x|\alpha\rangle|^2$ terms, which have a value of 1/4. Using Euler's identity and taking the real part of (A.7) we find that:

$$P_x^{\mathrm{D}}(t) = h_{\mathrm{D}}\mathrm{e}^{-\lambda t}\cos(2\pi\nu_\mu t). \tag{A.12}$$

The muon polarization precesses at the Larmor frequency, $2\pi\nu_\mu$, matching the more intuitive classical picture.

A.4 Muonium in a Transverse Field

The spin Hamiltonian for isotropic muonium in a magnetic field in the z direction is given by:

$$\hat{H}/h = \nu_{\mathrm{e}}\hat{S}_z^{\mathrm{e}} - \nu_\mu\hat{I}_z^\mu + A_\mu\hat{S}_z\hat{I}_z^\mu + \frac{1}{2}A_\mu\left[\hat{S}_+\hat{I}_-^\mu + \hat{S}_-\hat{I}_+^\mu\right], \tag{A.13}$$

where ν_{e} is the electron Larmor frequency, A_μ is the muon hyperfine coupling constant, $\hat{S}_z$ is the z component of the electron spin operator, $\hat{S}_+$ and $\hat{S}_-$ are the electron-raising and -lowering operators, and $\hat{I}_+^\mu$ and $\hat{I}_-^\mu$ are the muon-raising and -lowering operators. The raising and lowering operators only connect states with the opposite spin and are only non-zero in the following cases: $\langle\alpha|\hat{I}_+|\beta\rangle = \langle\beta|\hat{I}_-|\alpha\rangle = 1$. ν_{e} is considerably larger than ν_μ because the electron gyromagnetic ratio ($\gamma_{\mathrm{e}} = 2.80249$ MHz/G) is much larger than that of the muon.

The four eigenstates of this spin system can be written as linear combinations of the product Zeeman states. In zero field they are grouped into a triplet of degenerate states, $|\alpha^{\mathrm{e}}\alpha^\mu\rangle$, $|\beta^{\mathrm{e}}\beta^\mu\rangle$ and $\frac{1}{\sqrt{2}}(|\alpha^{\mathrm{e}}\beta^\mu\rangle + |\beta^{\mathrm{e}}\alpha^\mu\rangle)$, and a lower energy singlet, $\frac{1}{\sqrt{2}}(|\alpha^{\mathrm{e}}\beta^\mu\rangle - |\beta^{\mathrm{e}}\alpha^\mu\rangle)$. The degeneracy of the triplet states is broken by applying a magnetic field. The eigenstates and their energies are shown in Table A.1.

The evolution of the polarization for Mu can be calculated using eq. (A.1) with $u = x$ and the eigenstates listed in Table A.1. As with diamagnetic muons, the matrix elements of the $\hat{I}_x$ operator are non-zero only for the off-diagonal terms.

$$P_x^{\mathrm{Mu}}(t) = 2h_{\mathrm{Mu}}\mathrm{e}^{-\lambda t}\left\{\left|\langle 1|\hat{I}_x^\mu|2\rangle\right|^2\mathrm{e}^{2\pi i\nu_{12}t} + \left|\langle 2|\hat{I}_x^\mu|3\rangle\right|^2\mathrm{e}^{2\pi i\nu_{23}t} + \left|\langle 1|\hat{I}_x^\mu|4\rangle\right|^2\mathrm{e}^{2\pi i\nu_{14}t} + \left|\langle 3|\hat{I}_x^\mu|4\rangle\right|^2\mathrm{e}^{2\pi i\nu_{43}t}\right\}. \tag{A.14}$$

Table A.1 Eigenvectors and energies for muonium.

$\|1\rangle = \|\alpha^{\mathrm{e}}\alpha^\mu\rangle$	$E_1/h = \frac{1}{4}A_\mu + \frac{1}{2}(\nu_{\mathrm{e}} - \nu_\mu)$
$\|2\rangle = c\|\alpha^{\mathrm{e}}\beta^\mu\rangle + s\|\beta^{\mathrm{e}}\alpha^\mu\rangle$	$E_2/h = -\frac{1}{4}A_\mu + \frac{1}{2}A_\mu\left[1 + x_{\mathrm{B}}^2\right]^{1/2}$
$\|3\rangle = \|\beta^{\mathrm{e}}\beta^\mu\rangle$	$E_3/h = \frac{1}{4}A_\mu - \frac{1}{2}(\nu_{\mathrm{e}} - \nu_\mu)$
$\|4\rangle = -s\|\alpha^{\mathrm{e}}\beta^\mu\rangle + c\|\beta^{\mathrm{e}}\alpha^\mu\rangle$	$E_4/h = -\frac{1}{4}A_\mu - \frac{1}{2}A_\mu\left[1 + x_{\mathrm{B}}^2\right]^{1/2}$

$c^2 = \frac{1}{2}\left[1 + \left(x_{\mathrm{B}}/\left[1 + x_{\mathrm{B}}^2\right]^{1/2}\right)\right]$; $s^2 = \frac{1}{2}\left[1 - \left(x_{\mathrm{B}}/\left[1 + x_{\mathrm{B}}^2\right]^{1/2}\right)\right]$;
$x_{\mathrm{B}} = (\nu_{\mathrm{e}} + \nu_\mu)/A_\mu$; $c = s = 1/\sqrt{2}$ at $B = 0$; $c \to 1$ and $s \to 0$ when $\nu_{\mathrm{e}} + \nu_\mu \gg A_\mu$.

Table A.2 Precession frequencies for muonium.

Frequency	Amplitude
$\nu_{12} = \frac{1}{2}(\nu_e - \nu_\mu) + \frac{1}{2}A_\mu - \frac{1}{2}A_\mu\left[1 + x_B^2\right]^{1/2}$	$\propto c^2$
$\nu_{23} = \frac{1}{2}(\nu_e - \nu_\mu) - \frac{1}{2}A_\mu + \frac{1}{2}A_\mu\left[1 + x_B^2\right]^{1/2}$	$\propto s^2$
$\nu_{14} = \frac{1}{2}(\nu_e - \nu_\mu) + \frac{1}{2}A_\mu + \frac{1}{2}A_\mu\left[1 + x_B^2\right]^{1/2}$	$\propto s$
$\nu_{43} = \frac{1}{2}(\nu_e - \nu_\mu) - \frac{1}{2}A_\mu - \frac{1}{2}A_\mu\left[1 + x_B^2\right]^{1/2}$	$\propto c$

$x_B = (\nu_e + \nu_\mu)/A_\mu$.

The factor of two comes from the degenerate terms. The non-zero matrix elements are:

$$\langle 1|\hat{I}_x^\mu|2\rangle = -\frac{1}{2}c, \tag{A.15}$$

$$\langle 2|\hat{I}_x^\mu|3\rangle = -\frac{1}{2}s, \tag{A.16}$$

$$\langle 1|\hat{I}_x^\mu|4\rangle = \frac{1}{2}s, \tag{A.17}$$

$$\langle 3|\hat{I}_x^\mu|4\rangle = \frac{1}{2}c. \tag{A.18}$$

Using Euler's identity and taking the real part one finds:

$$P_x^{\mathrm{Mu}}(t) = \frac{1}{2}h_{\mathrm{Mu}}\mathrm{e}^{-\lambda t}[c^2\cos(2\pi\nu_{12}t) + s^2\cos(2\pi\nu_{23}t) + s^2\cos(2\pi\nu_{14}t) + c^2\cos(2\pi\nu_{43}t)]. \tag{A.19}$$

There are four precession frequencies (Table A.2) whose amplitudes are proportional to the coefficients c^2 and s^2. The formulas for c and s are given in Table A.1. All four frequencies have equal amplitude at zero field. The amplitude of the ν_{23} and ν_{14} frequencies decreases with increasing field and disappears when $\nu_e + \nu_\mu \gg A_\mu$. Transitions between the $|1\rangle$ and $|3\rangle$ states and the $|2\rangle$ and $|4\rangle$ states are forbidden in transverse field due to the selection rule that $\Delta M = \pm 1$, where M is the total spin quantum number.

In zero field the intra-triplet transitions (ν_{12} and ν_{23}) are 0, while the ν_{14} and ν_{43} transitions are degenerate and equal A_μ, which is 4463 MHz for Mu in vacuum. In low transverse fields, where $\nu_e + \nu_\mu \ll A_\mu$, the ν_{12} and ν_{23} transitions are degenerate with a frequency of $\frac{1}{2}(\nu_e - \nu_\mu)$.

A.5 Muonium in a Longitudinal Field

The selection rule in longitudinal field is $\Delta M = 0$. The polarization for Mu in a longitudinal field as a function of time is given by:

$$P_z^{\mathrm{Mu}}(t) = h_{\mathrm{Mu}}\mathrm{e}^{-\lambda t}\left\{\left|\langle 1|\hat{I}_z^\mu|1\rangle\right|^2 + \left|\langle 2|\hat{I}_z^\mu|2\rangle\right|^2 + \left|\langle 3|\hat{I}_z^\mu|3\rangle\right|^2 + \left|\langle 4|\hat{I}_z^\mu|4\rangle\right|^2 + 2\left|\langle 2|\hat{I}_z^\mu|4\rangle\right|^2\mathrm{e}^{2\pi i\nu_{24}t}\right\}. \tag{A.20}$$

The factor of 2 in front of the $\langle 2|\hat{I}_z^{\mu}|4\rangle$ term is to account for the degenerate $\langle 4|\hat{I}_z^{\mu}|2\rangle$ transition. The relevant matrix elements are:

$$\langle 1|\hat{I}_z^{\mu}|1\rangle = \frac{1}{2}, \tag{A.21}$$

$$\langle 2|\hat{I}_z^{\mu}|2\rangle = -\frac{1}{2}c^2 + \frac{1}{2}s^2 = -\frac{x_B}{2\sqrt{1+x_B^2}}, \tag{A.22}$$

$$\langle 2|\hat{I}_z^{\mu}|4\rangle = cs = \frac{1}{2}\frac{1}{\sqrt{1+x_B^2}}, \tag{A.23}$$

$$\langle 3|\hat{I}_z^{\mu}|3\rangle = -\frac{1}{2}, \tag{A.24}$$

$$\langle 4|\hat{I}_z^{\mu}|4\rangle = -\frac{1}{2}s^2 + \frac{1}{2}c^2 = \frac{x_B}{2\sqrt{1+x_B^2}}. \tag{A.25}$$

Combining Eqs. (A.20) through (A.25), using Euler's identity, and taking the real part one finds:

$$P_z^{\mathrm{Mu}}(t) = h_{\mathrm{Mu}} e^{-\lambda t}\left\{\frac{1}{2} + \frac{1}{2}\left(\frac{x_B^2}{1+x_B^2}\right) + \frac{1}{2}\left(\frac{1}{1+x_B^2}\right)\cos(2\pi\nu_{24}t)\right\}, \tag{A.26}$$

ν_{24} is a high frequency and not resolved for typical spectrometers, so the final term of Eq. (A.25) averages to zero. This gives:

$$P_z^{\mathrm{Mu}}(t) = h_{\mathrm{Mu}} e^{-\lambda t}\left\{\frac{1}{2} + \frac{1}{2}\left(\frac{x_B^2}{1+x_B^2}\right)\right\}. \tag{A.27}$$

A.6 Muoniated Radicals in Isotropic Environments

We will examine a muoniated radical with one unpaired electron, one muon and one proton. This can be used to derive all relevant expressions. General expressions are given by Roduner elsewhere [1, 2]. The spin Hamiltonian of this muoniated radical in an isotropic environment is given by:

$$\begin{aligned}\hat{H}/h &= \nu_e\hat{S}_z - \nu_\mu\hat{I}_z^{\mu} - \nu_p\hat{I}_z^{p} + A_\mu\hat{S}_z\hat{I}_z^{\mu} + A_p\hat{S}_z\hat{I}_z^{p}\\ &\quad + \frac{1}{2}A_\mu\left[\hat{S}_+\hat{I}_-^{\mu} + \hat{S}_-\hat{I}_+^{\mu}\right] + \frac{1}{2}A_p\left[\hat{S}_+\hat{I}_-^{p} + \hat{S}_-\hat{I}_+^{p}\right],\end{aligned} \tag{A.28}$$

where ν_p is the proton Larmor frequency, A_p is the hfc of the proton, $\hat{I}_+^{p}$ and $\hat{I}_-^{p}$ are the proton-raising and -lowering operators, respectively. We will deal with muoniated radicals in high magnetic fields where $\nu_e \gg A_\mu$ and A_p. The eigenfunctions can be written as a product of basis functions:

$$|\chi^i\rangle = |\chi^e\rangle|\chi^{\mu}\rangle\Pi_k|\chi^k\rangle. \tag{A.29}$$

The eigenvectors are linear combinations of $|\chi^i\rangle$.

$$|n\rangle = \sum_i c_{ni}|\chi^i\rangle. \tag{A.30}$$

Table A.3 Eigenvectors and energies for a muoniated radical (e–μ–p) in high magnetic fields.

$\|1a\rangle = \|\alpha^e\alpha^\mu\alpha^p\rangle$	$E_{1a}/h = \frac{1}{4}A_\mu + \frac{1}{2}(\nu_e - \nu_\mu - \nu_p) + \frac{1}{4}A_p$
$\|1b\rangle = \|\alpha^e\alpha^\mu\beta^p\rangle$	$E_{1b}/h = \frac{1}{4}A_\mu + \frac{1}{2}(\nu_e - \nu_\mu + \nu_p) - \frac{1}{4}A_p$
$\|2a\rangle = c\|\alpha^e\beta^\mu\alpha^p\rangle + s\|\beta^e\alpha^\mu\alpha^p\rangle$	$E_{2a}/h = -\frac{1}{4}A_\mu + \frac{1}{2}A_\mu\left[1+x_B^2\right] - \frac{1}{2}\nu_p + \frac{1}{4}A_p(c^2 - s^2)$
$\|2b\rangle = c\|\alpha^e\beta^\mu\beta^p\rangle + s\|\beta^e\alpha^\mu\beta^p\rangle$	$E_{2b}/h = -\frac{1}{4}A_\mu + \frac{1}{2}A_\mu\left[1+x_B^2\right] + \frac{1}{2}\nu_p - \frac{1}{4}A_p(c^2 - s^2)$
$\|3a\rangle = \|\beta^e\beta^\mu\alpha^p\rangle$	$E_{3a}/h = \frac{1}{4}A_\mu - \frac{1}{2}(\nu_e - \nu_\mu + \nu_p) - \frac{1}{4}A_p$
$\|3b\rangle = \|\beta^e\beta^\mu\beta^p\rangle$	$E_{3b}/h = \frac{1}{4}A_\mu - \frac{1}{2}(\nu_e - \nu_\mu - \nu_p) + \frac{1}{4}A_p$
$\|4a\rangle = -s\|\alpha^e\beta^\mu\alpha^p\rangle + c\|\beta^e\alpha^\mu\alpha^p\rangle$	$E_{4a}/h = -\frac{1}{4}A_\mu - \frac{1}{2}A_\mu\left[1+x_B^2\right] - \frac{1}{2}\nu_p - \frac{1}{4}A_p(c^2 - s^2)$
$\|4b\rangle = -s\|\alpha^e\beta^\mu\beta^p\rangle + c\|\beta^e\alpha^\mu\beta^p\rangle$	$E_{4b}/h = -\frac{1}{4}A_\mu - \frac{1}{2}A_\mu\left[1+x_B^2\right] + \frac{1}{2}\nu_p + \frac{1}{4}A_p(c^2 - s^2)$

$c^2 = \frac{1}{2}\left[1 + \left(x_B/[1+x_B^2]^{1/2}\right)\right]; s^2 = \frac{1}{2}\left[1 - \left(x_B/[1+x_B^2]^{1/2}\right)\right]; x_B = (\nu_e + \nu_\mu)/A_\mu.$

The proton Zeeman and hyperfine terms are treated as perturbations.

$$\hat{H}/h = \hat{H}_0/h - \nu_p\hat{I}_z^p + A_p\hat{S}_z\hat{I}_z^p + \frac{1}{2}A_p\left[\hat{S}_+\hat{I}_-^p + \hat{S}_-\hat{I}_+^p\right], \tag{A.31}$$

where $\hat{H}_0/h$ is the spin Hamiltonian for muonium (A.13). The first-order eigenvectors and energy levels for the spin states of the muoniated radical are listed in Table A.3.

A.7 Muoniated Radicals in High Transverse Fields

The polarization as a function of time in the x direction for a muoniated radical R that is formed at $t = 0$ with a probability h_R, is calculated using (A.2) with $u = x$. The $\hat{I}_x$ operator only mixes states with opposite muon spins but the same electron and nuclear spins. This results in the selection rule in transverse field of:

$$\Delta m_\mu = \pm 1, \Delta m_e = 0, \Delta m_p = 0. \tag{A.32}$$

The polarization for a muoniated radical in a transverse field as a function of time is given by:

$$P_x^R(t) = \frac{1}{2}h_R e^{-\lambda t}\left\{\begin{array}{l} 2\left|\langle 1a|\hat{I}_x^\mu|2a\rangle\right|^2 e^{2\pi i\nu_{12}t} + 2\left|\langle 1b|\hat{I}_x^\mu|2b\rangle\right|^2 e^{2\pi i\nu_{12}t} + 2\left|\langle 1a|\hat{I}_x^\mu|4a\rangle\right|^2 e^{2\pi i\nu_{14}t} \\ + 2\left|\langle 1b|\hat{I}_x^\mu|4b\rangle\right|^2 e^{2\pi i\nu_{14}t} + 2\left|\langle 2a|\hat{I}_x^\mu|3a\rangle\right|^2 e^{2\pi i\nu_{23}t} + 2\left|\langle 2b|\hat{I}_x^\mu|3b\rangle\right|^2 e^{2\pi i\nu_{23}t} \\ + 2\left|\langle 4a|\hat{I}_x^\mu|3a\rangle\right|^2 e^{2\pi i\nu_{43}t} + 2\left|\langle 4b|\hat{I}_x^\mu|3b\rangle\right|^2 e^{2\pi i\nu_{43}t} \end{array}\right\}. \tag{A.33}$$

The factor of two comes from the degenerate terms. Using the eigenvectors in Table A.3 to evaluate the matrix elements in the preceding equation, applying Euler's identity, and taking the real part gives:

$$P_x^R(t) = \frac{1}{2}h_R e^{-\lambda t}[c^2\cos(2\pi\nu_{12}t) + s^2\cos(2\pi\nu_{14}t) + s^2\cos(2\pi\nu_{23}t) + c^2\cos(2\pi\nu_{43}t)]. \tag{A.34}$$

Table A.4 Transition frequencies for muoniated radicals.

Frequency	Amplitude
$\nu_{12} = -\frac{1}{2}(\nu_e - \nu_\mu) - \frac{1}{2}A_\mu + \frac{1}{2}A_\mu\left[1 + x_B^2\right]^{1/2} - \frac{1}{2}A_p s^2$	$\propto c^2$
$\nu_{14} = -\frac{1}{2}(\nu_e - \nu_\mu) - \frac{1}{2}A_\mu - \frac{1}{2}A_\mu\left[1 + x_B^2\right]^{1/2} - \frac{1}{2}A_p c^2$	$\propto s^2$
$\nu_{23} = -\frac{1}{2}(\nu_e - \nu_\mu) + \frac{1}{2}A_\mu - \frac{1}{2}A_\mu\left[1 + x_B^2\right]^{1/2} - \frac{1}{2}A_p c^2$	$\propto s^2$
$\nu_{43} = -\frac{1}{2}(\nu_e - \nu_\mu) + \frac{1}{2}A_\mu + \frac{1}{2}A_\mu\left[1 + x_B^2\right]^{1/2} - \frac{1}{2}A_p s^2$	$\propto c^2$

$c^2 = \frac{1}{2}\left[1 + \left(x_B/\left[1 + x_B^2\right]^{1/2}\right)\right]; s^2 = \frac{1}{2}\left[1 - \left(x_B/\left[1 + x_B^2\right]^{1/2}\right)\right]; x_B = (\nu_e + \nu_\mu)/A_\mu.$

There are four precession frequencies, and these are listed in Table A.4.

In large magnetic fields, where $(\nu_e + \nu_\mu) \gg A_\mu$, $\left[1 + x_B^2\right]^{1/2} \to (\nu_e + \nu_\mu)/A_\mu$, $s^2 \to 0$ and $c^2 \to 1$. In this high-field limit ν_{23} and ν_{14} are on the order of the electron Larmor frequency, which is too fast to measure. The values of ν_{12} and ν_{43} in this limit are:

$$\nu_{12} = \nu_\mu - \frac{1}{2}A\mu, \tag{A.35}$$

$$\nu_{43} = \nu_\mu + \frac{1}{2}A\mu. \tag{A.36}$$

There are two radical frequencies equally spaced about the diamagnetic muon precession frequency.

A.8 Muoniated Radicals in Longitudinal Fields Outside Level Crossings

The polarization for a muoniated radical in a longitudinal field as a function of time is given by:

$$P_z^R(t) = \frac{1}{2}h_R e^{-\lambda t}\left\{\begin{array}{l} \left|\langle 1a|\hat{I}_z^\mu|1a\rangle\right|^2 + \left|\langle 1b|\hat{I}_z^\mu|1b\rangle\right|^2 + \left|\langle 2a|\hat{I}_z^\mu|2a\rangle\right|^2 + \left|\langle 2b|\hat{I}_z^\mu|2b\rangle\right|^2 \\ + \left|\langle 3a|\hat{I}_z^\mu|3a\rangle\right|^2 + \left|\langle 3b|\hat{I}_z^\mu|3b\rangle\right|^2 + \left|\langle 4a|\hat{I}_z^\mu|4a\rangle\right|^2 + \left|\langle 4b|\hat{I}_z^\mu|4b\rangle\right|^2 \\ + 2\left|\langle 2a|\hat{I}_z^\mu|4a\rangle\right|^2 e^{2\pi i\nu_{24}t} + 2\left|\langle 2b|\hat{I}_z^\mu|4b\rangle\right|^2 e^{2\pi i\nu_{24}t} \end{array}\right\}. \tag{A.37}$$

The factor of two comes from the degenerate terms. Using the eigenvectors in Table A.3 to evaluate the matrix elements in the preceding equation, applying Euler's identity, and taking the real part gives:

$$P_z^R(t) = h_R e^{-\lambda t}\left\{\frac{1}{2} + \frac{1}{2}\left(\frac{x_B^2}{1 + x_B^2}\right) + \frac{1}{2}\left(\frac{1}{1 + x_B^2}\right)\cos(2\pi\nu_{24}t)\right\}. \tag{A.38}$$

As with Mu, the ν_{24} frequency is very high and not resolved for typical spectrometers, so the final term of Eq. (A.38) averages to zero. This gives:

$$P_z^{\mathrm{R}}(t) = h_{\mathrm{R}} \mathrm{e}^{-\lambda t} \left\{ \frac{1}{2} + \frac{1}{2} \left(\frac{x_{\mathrm{B}}^2}{1 + x_{\mathrm{B}}^2} \right) \right\}. \tag{A.39}$$

This is analogous to Eq. (A.27). The muon spin repolarizes in large longitudinal fields, and this can be used to determine A_μ. It should be noted that this expression was derived using high-field expressions for the spin states of the muoniated radical. The hyperfine coupling to the protons results in the polarization at low longitudinal fields being lower than ½.

In the high-field limit (i.e. $x_{\mathrm{B}} \gg 1$) the preceding equation simplifies to give:

$$P_z^{\mathrm{R}}(t) = h_{\mathrm{R}} \mathrm{e}^{-\lambda t}. \tag{A.40}$$

This means there is no evolution of the muon spin in high longitudinal field except through decay of the state.

A.9 Avoided Level-Crossing Muon Spin Resonance in Isotropic Environments

We are considering level crossings in high magnetic field where the spin states (Table A.3) are reduced to the product of the Zeeman terms (Table A.5). Examining the energies of the spin states in Table A.5 shows that for $A_\mu > A_{\mathrm{p}}$, there is a crossing of $|1\mathrm{b}\rangle = |\alpha^{\mathrm{e}}\alpha^{\mu}\beta^{\mathrm{p}}\rangle$ and $|2\mathrm{a}\rangle = |\alpha^{\mathrm{e}}\beta^{\mu}\alpha^{\mathrm{p}}\rangle$ (i.e. $E_{1\mathrm{b}}/h = E_{2\mathrm{a}}/h$) at the field:

$$B_{\mathrm{res}}^{\Delta_0} = \frac{A_\mu - A_{\mathrm{p}}}{2(\gamma_\mu - \gamma_{\mathrm{p}})}. \tag{A.41}$$

These spin states differ in the orientation of the muon and proton spin. In higher order, crossing is avoided by coupling of the two states to a common energetically distant state ($|4a\rangle = |\beta^{\mathrm{e}}\alpha^{\mu}\alpha^{\mathrm{p}}\rangle$, which has the opposite electron spin) via non-secular terms of the spin Hamiltonian (A.28). Arranging the matrix as above results in a block diagonal. It is sufficient to determine the sub-determinant:

$$\det \begin{pmatrix} E_{4\mathrm{a}} - E & A_{\mathrm{p}}/2 & A_\mu/2 \\ A_{\mathrm{p}}/2 & E_{1\mathrm{b}} - E & 0 \\ A_\mu/2 & 0 & E_{2\mathrm{a}} - E \end{pmatrix} = 0. \tag{A.42}$$

The solution to Eq. (A.42) gives the energy of the three spin states as a function of applied magnetic field. The Δ_0 resonance field is the applied magnetic field when the energy gap between $|1b\rangle$ and $|2a\rangle$ is a minimum. The Δ_0 resonance field is

$$B_{\mathrm{res}}^{\Delta_0} = \frac{A_\mu - A_{\mathrm{p}}}{2(\gamma_\mu - \gamma_{\mathrm{p}})} - \frac{A_\mu + A_{\mathrm{p}}}{2\gamma_{\mathrm{e}}}. \tag{A.43}$$

This is very similar to (A.41) with the second term being a small correction due to γ_{e} in the denominator. The energy gap between $|1b\rangle$ and $|2a\rangle$, which is the ALC

Table A.5 Eigenvectors and energies for a muoniated radical (e−μ−p) in high magnetic fields.

$\lvert 1a\rangle = \lvert\alpha^e\alpha^\mu\alpha^p\rangle$	$E_{1a}/h = \frac{1}{2}(\nu_e - \nu_\mu - \nu_p) + \frac{1}{4}A_\mu + \frac{1}{4}A_p$
$\lvert 1b\rangle = \lvert\alpha^e\alpha^\mu\beta^p\rangle$	$E_{1b}/h = \frac{1}{2}(\nu_e - \nu_\mu + \nu_p) + \frac{1}{4}A_\mu - \frac{1}{4}A_p$
$\lvert 2a\rangle = \lvert\alpha^e\beta^\mu\alpha^p\rangle$	$E_{2a}/h = \frac{1}{2}(\nu_e + \nu_\mu - \nu_p) - \frac{1}{4}A_\mu + \frac{1}{4}A_p$
$\lvert 2b\rangle = \lvert\alpha^e\beta^\mu\beta^p\rangle$	$E_{2b}/h = \frac{1}{2}(\nu_e + \nu_\mu + \nu_p) - \frac{1}{4}A_\mu - \frac{1}{4}A_p$
$\lvert 3a\rangle = \lvert\beta^e\beta^\mu\alpha^p\rangle$	$E_{3a}/h = \frac{1}{2}(-\nu_e + \nu_\mu - \nu_p) + \frac{1}{4}A_\mu - \frac{1}{4}A_p$
$\lvert 3b\rangle = \lvert\beta^e\beta^\mu\beta^p\rangle$	$E_{3b}/h = \frac{1}{2}(-\nu_e + \nu_\mu + \nu_p) + \frac{1}{4}A_\mu + \frac{1}{4}A_p$
$\lvert 4a\rangle = \lvert\beta^e\alpha^\mu\alpha^p\rangle$	$E_{4a}/h = \frac{1}{2}(-\nu_e - \nu_\mu - \nu_p) - \frac{1}{4}A_\mu - \frac{1}{4}A_p$
$\lvert 4b\rangle = \lvert\beta^e\alpha^\mu\beta^p\rangle$	$E_{4b}/h = \frac{1}{2}(-\nu_e - \nu_\mu + \nu_p) - \frac{1}{4}A_\mu + \frac{1}{4}A_p$

	$\lvert\alpha^e\alpha^\mu\alpha^p\rangle$ $\lvert 1a\rangle$	$\lvert\beta^e\alpha^\mu\alpha^p\rangle$ $\lvert 4a\rangle$	$\lvert\alpha^e\alpha^\mu\beta^p\rangle$ $\lvert 1b\rangle$	$\lvert\alpha^e\beta^\mu\alpha^p\rangle$ $\lvert 2a\rangle$	$\lvert\beta^e\alpha^\mu\beta^p\rangle$ $\lvert 4b\rangle$	$\lvert\beta^e\beta^\mu\alpha^p\rangle$ $\lvert 3a\rangle$	$\lvert\alpha^e\beta^\mu\beta^p\rangle$ $\lvert 2b\rangle$	$\lvert\beta^e\beta^\mu\beta^p\rangle$ $\lvert 3b\rangle$
$\langle 1a\rvert$	$(E_{1a}-E)$	0	0	0	0	0	0	0
$\langle 4a\rvert$	0	$(E_{4a}-E)$	$A_p/2$	$A_\mu/2$	0	0	0	0
$\langle 1b\rvert$	0	$A_p/2$	$(E_{1b}-E)$	0	0	0	0	0
$\langle 2a\rvert$	0	$A_\mu/2$	0	$(E_{2a}-E)$	0	0	0	0
$\langle 4b\rvert$	0	0	0	0	$(E_{4b}-E)$	0	$A_\mu/2$	0
$\langle 3a\rvert$	0	0	0	0	0	$(E_{3a}-E)$	$A_p/2$	0
$\langle 2b\rvert$	0	0	0	0	$A_\mu/2$	$A_p/2$	$(E_{2b}-E)$	0
$\langle 3b\rvert$	0	0	0	0	0	0	0	$(E_{3b}-E)$

transition frequency, as a function of applied magnetic field is

$$\nu^{\Delta_0}(B) = \sqrt{\left(2B(\gamma_\mu - \gamma_p) - (A_\mu - A_p) + \frac{A_\mu^2 - A_p^2}{2B\gamma_e}\right)^2 + \left(\frac{A_\mu A_p}{B\gamma_e}\right)^2}. \tag{A.44}$$

On resonance, the ALC transition frequency becomes

$$\nu_{res}^{\Delta_0} \approx \frac{A_\mu A_p}{B_{res}^{\Delta_0}\gamma_e}. \tag{A.45}$$

The time-integrated polarization as a function of applied magnetic field is

$$\overline{P}_z^{\mathrm{R}}(B) = h_{\mathrm{R}} - \frac{(2h_{\mathrm{R}}/N)\left(\nu_{\mathrm{res}}^{\Delta_0}\right)^2}{(k_1/2\pi)^2 + \left(\nu_{\mathrm{res}}^{\Delta_0}\right)^2 + \left[(\gamma_\mu - \gamma_{\mathrm{p}})\left(B - B_{\mathrm{res}}^{\Delta_0}\right)\right]^2}, \tag{A.46}$$

where k_1 is the rate of any process that takes the radical out of resonance, such as the muon radioactive decay or chemical reactions. This is a Lorentzian with an amplitude given by

$$\overline{P}_z^{\mathrm{R}}\left(B_{\mathrm{res}}^{\Delta_0}\right) \propto \left(\frac{2h_{\mathrm{R}}}{N}\right) \frac{\left(\nu_{\mathrm{res}}^{\Delta_0}\right)^2}{(k_1/2\pi)^2 + \left(\nu_{\mathrm{res}}^{\Delta_0}\right)^2}. \tag{A.47}$$

If either the muon or proton hfc is small, ν_{res} will be small, and this will lead to the amplitude of the Δ_0 resonance being small. The resonances will also be small if there is hyperfine coupling to many nuclei with $I > 0$, i.e. N is large. The full width at half max (FWHM) of the Δ_0 resonance, $\Delta B_{1/2}^{\Delta_0}$, is given by

$$\Delta B_{1/2}^{\Delta_0} = \frac{2\left(\nu_{\mathrm{res}}^{\Delta_0}\right)^2}{(\gamma_\mu - \gamma_{\mathrm{p}})} \sqrt{1 + \frac{(k_1/2\pi)^2}{\left(\nu_{\mathrm{res}}^{\Delta_0}\right)^2}}. \tag{A.48}$$

A.10 Avoided Level-Crossing Muon Spin Resonance in Anisotropic Environments

For our discussion of radicals in anisotropic environments, we will consider a muoniated radical with no other nuclei with spin $I > 0$. The spin states for this radical are the same as Mu except with a smaller muon hfc. We will also assume that this radical has an axially symmetric hyperfine tensor with principal components $D_\mu^{\parallel}$ and $D_\mu^{\perp}$. The dipolar tensor is traceless, so

$$D_\mu^{\parallel} = -2D_\mu^{\perp}. \tag{A.49}$$

The muon hfc as a function of the angle θ, which is the angle between the symmetry axis and the magnetic field, is given by

$$A_\mu(\theta) = A_\mu + \frac{1}{2} D_\mu^{\parallel}(3\cos^2\theta - 1). \tag{A.50}$$

The spin Hamiltonian of this radical is given by:

$$\begin{aligned}
\hat{H}/h = {} & \nu_e \hat{S}_z - \nu_\mu \hat{I}_z^\mu + \left[A_\mu + \frac{1}{2} D_\mu^{\parallel}(3\cos^2\theta - 1)\right] \hat{S}_z \hat{I}_z^\mu \\
& + \frac{1}{2}\left[A_\mu - \frac{1}{4} D_\mu^{\parallel}(3\cos^2\theta - 1)\right]\left[\hat{S}_+ \hat{I}_-^\mu + \hat{S}_- \hat{I}_+^\mu\right] \\
& + \frac{3}{4} D_\mu^{\parallel} \sin\theta\cos\theta \left[\hat{S}_z \hat{I}_+^\mu + \hat{S}_+ \hat{I}_z^\mu\right] + \frac{3}{4} D_\mu^{\parallel} \sin\theta\cos\theta \left[\hat{S}_z \hat{I}_-^\mu + \hat{S}_- \hat{I}_z^\mu\right] \\
& + \frac{3}{8} D_\mu^{\parallel} \sin^2\theta \left[\hat{S}_+ \hat{I}_+^\mu\right] + \frac{3}{8} D_\mu^{\parallel} \sin^2\theta \left[\hat{S}_- \hat{I}_-^\mu\right],
\end{aligned} \tag{A.51}$$

Table A.6 Eigenvectors and energies for muonium or a muoniated radical with no other spins with $I > 0$ in the high-field limit.

$\vert 1\rangle = \vert \alpha^{e}\alpha^{\mu}\rangle$	$E_1/h = \frac{1}{4}A_{\mu}(\theta) + \frac{1}{2}(\nu_e - \nu_{\mu})$
$\vert 2\rangle = \vert \alpha^{e}\beta^{\mu}\rangle$	$E_2/h = -\frac{1}{4}A_{\mu}(\theta) + \frac{1}{2}(\nu_e + \nu_{\mu})$
$\vert 3\rangle = \vert \beta^{e}\beta^{\mu}\rangle$	$E_3/h = \frac{1}{4}A_{\mu}(\theta) - \frac{1}{2}(\nu_e - \nu_{\mu})$
$\vert 4\rangle = \vert \beta^{e}\alpha^{\mu}\rangle$	$E_4/h = -\frac{1}{4}A_{\mu}(\theta) - \frac{1}{2}(\nu_e + \nu_{\mu})$

The eigenstates and energies of Mu are shown in Table A.1. The corresponding high-field limiting values are given in Table A.6. These also apply to a muoniated radical with no other spins with $I > 0$.

It is easy to show that in first order, using the expressions in Table A.6, the $|1\rangle$ and $|2\rangle$ spin states and the $|3\rangle$ and $|4\rangle$ spin states cross at

$$B_{\text{res}}^{\Delta_1}(\theta) = \frac{A_{\mu}(\theta)}{2\gamma_{\mu}}. \tag{A.52}$$

The $\hat{S}_z\hat{I}_+^{\mu}$, $\hat{S}_+\hat{I}_z^{\mu}$, $\hat{S}_z\hat{I}_-^{\mu}$ and $\hat{S}_-\hat{I}_z^{\mu}$ terms in (A.51) lead to new off-diagonal elements that are not present in the isotropic system. The transition can be called a muon spin flip with $\Delta M = \pm 1$. We will ignore the $\Delta M = 0 \left(\hat{S}_+\hat{I}_-^{\mu} + \hat{S}_-\hat{I}_+^{\mu}\right)$ and $\Delta M = 2\left(\hat{S}_+\hat{I}_+^{\mu}, \hat{S}_-\hat{I}_-^{\mu}\right)$ terms in (A.51) to simplify the derivation. The resonance field is obtained by diagonalizing two sub-determinants:

$$\begin{array}{c|cc|cc}
 & |\alpha^{e}\alpha^{\mu}\rangle & |\alpha^{e}\beta^{\mu}\rangle & |\beta^{e}\beta^{\mu}\rangle & |\beta^{e}\alpha^{\mu}\rangle \\
 & |1\rangle & |2\rangle & |3\rangle & |4\rangle \\
\hline
\langle 1| & (E_1 - E) & \frac{3}{8}D_{\mu}^{\parallel}\sin\theta\cos\theta & & \\
\langle 2| & \frac{3}{8}D_{\mu}^{\parallel}\sin\theta\cos\theta & (E_2 - E) & & \\
\hline
\langle 3| & & & (E_3 - E) & -\frac{3}{8}D_{\mu}^{\parallel}\sin\theta\cos\theta \\
\langle 4| & & & -\frac{3}{8}D_{\mu}^{\parallel}\sin\theta\cos\theta & (E_4 - E)
\end{array}$$

The Δ_1 resonance field is

$$B_{\text{res}}^{\Delta_1} = \frac{A_{\mu}(\theta)}{2\gamma_{\mu}} - \frac{A_{\mu}(\theta)}{2\gamma_{e}}. \tag{A.53}$$

This is very similar to (A.52) with the second term being a small correction. On resonance, the ALC transition frequency becomes

$$\nu_{\text{res}}^{\Delta_1} \approx \frac{3}{4}D_{\mu}^{\parallel}\sin\theta\cos\theta. \tag{A.54}$$

The time-integrated polarization as a function of applied magnetic field is

$$\overline{P}_z^{\mathrm{R}}(B,\theta) = h_{\mathrm{R}} - \frac{(2h_{\mathrm{R}}/N)\left(\nu_{\mathrm{res}}^{\Delta_1}\right)^2}{(k_1/2\pi)^2 + \left(\nu_{\mathrm{res}}^{\Delta_1}\right)^2 + \left[\gamma_\mu\left(B - B_{\mathrm{res}}^{\Delta_0}\right)\right]^2}. \tag{A.55}$$

The resonance depends on θ due to the θ-dependence of $\nu_{\mathrm{res}}^{\Delta_1}$. The Δ_1 resonance is a Lorentzian for a given θ. For powders, the signal has to be integrated over all orientations. The amplitude of the resonance for a particular θ is given by

$$\overline{P}_z^{\mathrm{R}}\left(B_{\mathrm{res}}^{\Delta_1},\theta\right) \propto \left(\frac{2h_{\mathrm{R}}}{N}\right)\frac{\left(\nu_{\mathrm{res}}^{\Delta_1}\right)^2}{(k_1/2\pi)^2 + \left(\nu_{\mathrm{res}}^{\Delta_1}\right)^2}. \tag{A.56}$$

The Δ_1 resonance disappears when the muon dipolar coupling is zero or when θ is a multiple of $\pi/2$ as $\nu_{\mathrm{res}}^{\Delta_1} \to 0$. The FWHM of the Δ_1 resonance as a function of θ is

$$\Delta B_{1/2}^{\Delta_1} = \frac{2}{\gamma_\mu}\sqrt{(k_1/2\pi)^2 + \left(\nu_{\mathrm{res}}^{\Delta_1}\right)^2}. \tag{A.57}$$

References

1 Roduner, E. (1988). *The Positive Muon as a Probe in Free Radical Chemistry*, Lecture Notes in Chemistry, vol. 49. Springer-Verlag.

2 Roduner, E. (1999). Aspects of muon chemistry, Chap. 8. In: *Muon Science: Muons in Physics, Chemistry, and Materials* (ed. S.L. Lee), 173–209. https://doi.org/10.1201/9780203746196.

Appendix B

Muonium Rate Constants for Reactions in Solution

Table B.1 Rate constants for selected abstraction reactions in aqueous solution at 298 K.

Solute	Reaction	k_{Mu} (M^{-1} s^{-1})	References	k_H (M^{-1} s^{-1})	References	Isotope effect k_{Mu}/k_H
Methanol	$Mu + CH_3OH \rightarrow MuH + \dot{C}H_2OH$	3×10^4	[1]	$(2.84 \pm 0.07) \times 10^6$	[2]	$\sim 10^{-2}$
Ethanol	$Mu + CH_3CH_2OH \rightarrow MuH + CH_3\dot{C}HOH$	$<3 \times 10^5$	[3]	$(2.04 \pm 0.06) \times 10^7$	[4, 5]	<0.015
Ethanol-d_5	—	—	—	$(2.77 \pm 0.26) \times 10^6$	[4, 5]	—
2-Propanol	$Mu + (CH_3)_2C(H)OH \rightarrow MuH + (CH_3)_2\dot{C}OH$	$\sim 7 \times 10^5$	[3]	$(1.01 \pm 0.07) \times 10^8$	[4, 5]	~0.007
2-Propanol-d_7	—	—	—	$(1.56 \pm 0.07) \times 10^7$	[4, 5]	—
2-Butanol	$Mu + CH_3CH(OH)CH_2CH_3 \rightarrow MuH + CH_3\dot{C}(OH)CH_2CH_3$	1.1×10^6	[3]	1×10^8	[6]	$\sim 10^{-2}$
Formate ion	$Mu + HCO_2^- \rightarrow MuH + CO_2^{-\cdot}$	$(5.6 \pm 0.8) \times 10^6$	[7]	$(1.91 \pm 0.08) \times 10^8$	[7]	0.029
d-Formate ion	$Mu + DCO_2^- \rightarrow MuD + CO_2^{-\cdot}$	$(3.4 \pm 0.3) \times 10^5$	[7]	$(3.5 \pm 0.3) \times 10^7$	[7]	$\sim 10^{-2}$

Muon Spin Spectroscopy: Methods and Applications in Chemistry and Materials Science, First Edition.
Donald G. Fleming, Iain McKenzie, and Paul W. Percival.

Table B.2 Rate constants for Mu/H addition to alkenes in solution.

Compound and solvent	Reaction	k_{Mu} ($M^{-1}\ s^{-1}$)	References	k_H ($M^{-1}\ s^{-1}$)	References	k_{Mu}/k_H
Ethene in water	$Mu + H_2C{=}CH_2 \rightarrow MuH_2C{-}\dot{C}H_2$	1.4×10^{10} at 295 K	[8]	3×10^9 at 288 K and pH ~ 1	[9]	4.7
Ethene in methane		2.1×10^{10} at 110 K	[8]	—	—	—
Ethene in 2-methylbutane		2.6×10^{10} at 295 K	[8]	—	—	—
2,3-Dimethylbuta-1,3-diene in *n*-hexane	$Mu + H_2C{=}C(CH_3)C(CH_3){=}CH_2 \rightarrow MuH_2C{-}\dot{C}(CH_3)C(CH_3){=}CH_2$	$(1.4 \pm 0.2) \times 10^{11}$ at 288 K	[10]	—	—	—
2,3-Dimethylbuta-1,3-diene in methanol		$(7.4 \pm 0.1) \times 10^{10}$ at 289 K	[10]	—	—	—
2,3-Dimethylbuta-1,3-diene in octane		$(8.8 \pm 1.1) \times 10^{10}$ at 294 K	[10]	—	—	—
Acrylamide in water	$Mu + H_2C{=}CHC({=}O)NH_2 \rightarrow MuH_2C\dot{C}HC({=}O)NH_2$	$(1.90 \pm 0.13) \times 10^{10}$	[11]	3.1×10^{10}	[12]	0.61
Acrylonitrile in water	$Mu + H_2C{=}C(H)C{=}N \rightarrow MuH_2C{-}\dot{C}(H)C{\equiv}N$	$(1.14 \pm 0.20) \times 10^{10}$	[11]	2.2×10^9	[13]	5.2
Acrylic acid in water	$Mu + H_2C{=}CHC({=}O)OH \rightarrow MuH_2C\dot{C}HC({=}O)OH$	$(1.55 \pm 0.25) \times 10^{10}$	[11]	3.3×10^9	[13]	4.7
Methylmethacrylate in water	$Mu + H_2C{=}C(CH_3)C({=}O)OCH_3 \rightarrow MuH_2C\dot{C}(CH_3)C({=}O)OCH_3$	$(9.5 \pm 1.9) \times 10^9$	[11]	7.2×10^8	[13]	13.2

Table B.3 Rate constants for Mu/H addition to aromatic compounds in solution at 298 K.

Compound and solvent	Reaction	k_{Mu} ($M^{-1}\ s^{-1}$)	References	k_H ($M^{-1}\ s^{-1}$)	References	k_{Mu}/k_H
Benzene in water	$Mu + C_6H_6 \rightarrow C_6H_6Mu$ (muoniated cyclohexadienyl radical)	$(1.9 \pm 0.4) \times 10^9$	[14]	9.1×10^8	[6]	3.3
Benzene – neat		$(8.9 \pm 0.6) \times 10^9$	[15]	—	—	—
Benzene in *n*-hexane		$(3.7 \pm 0.8) \times 10^9$	[14]	—	—	—
Benzene in isopentane		4.51×10^9	[16]	—	—	—
Benzene in methanol		$(8.5 \pm 1.3) \times 10^9$	[14]	—	—	—
Phenol in water	$Mu + C_6H_5OH \rightarrow MuC_6H_5OH$ (*ortho*, *meta*, *para* isomers distributed in ratio 42 : 42 : 16 [17])	$(7.2 \pm 0.4) \times 10^9$	[18]	1.7×10^9	[6]	4.1
Phenol in *n*-hexane		$(7 \pm 1) \times \times 10^9$	[1]	—	—	—
Phenol in methanol		$(3.3 \pm 1.0) \times 10^{10}$	[19]	—	—	—
p-Nitrophenol in water	$Mu + O_2NC_6H_4OH \rightarrow O_2NC_6H_4(Mu)OH$ isomers	$(8.0 \pm 0.7) \times 10^9$	[18]	—	—	—
Nitrobenzene in *n*-hexane	$Mu + C_6H_5NO_2 \rightarrow C_6H_5(Mu)NO_2$ isomers	$(1.3 \pm 0.15) \times 10^{10}$	[20]	—	—	—
Nitrobenzene in water		1.1×10^{10}	[21]	1.6×10^9	[6]	6.9
Aniline in *n*-hexane	$Mu + C_6H_5NH_2 \rightarrow MuC_6H_5(Mu)NH_2$ (*ortho*, *meta*, *para* isomers distributed in ratio 48 : 30 : 22 [17])	$(6.7 \pm 0.8) \times 10^9$	[14]	—	—	—
Aniline in water		8.8×10^9	[21]	2.4×10^9	[6]	3.7
Hexafluorobenzene in *n*-hexane	$Mu + C_6F_6 \rightarrow C_6F_6Mu$ muoniated hexafluoro-cyclohexadienyl radical	$(6 \pm 4) \times 10^7$	[14]	—	—	—
Pyridine in water	$Mu + C_5H_5N \rightarrow C_5H_5NMu$ muoniated azacyclohexadienyl isomers + N-addition (*ortho*, *meta*, *para* and *N* isomers distributed in ratio 28 : 30 : 0 : 42 [22])	$(5.8 \pm 0.4) \times 10^9$ at pH ~7	[23]	6×10^8	[24]	9.7
Naphthalene in water	Mu + Naphthalene → Naphthalene Mu-adduct isomers	$(1.34 \pm 0.25) \times 10^9$	[19]	3.4×10^9	[25]	0.4
Naphthalene in *n*-hexane		$(1.3 \pm 0.2) \times 10^{11}$	[19]	—	—	—

Table B.4 Rate constants for Mu/H addition to heteroatom unsaturated bonds in solution.

Compound and solvent	Reaction	k_{Mu} (M^{-1} s^{-1})	References	k_H (M^{-1} s^{-1})	References	k_{Mu}/k_H
Acetone in water	$Mu + (CH_3)_2C{=}O \rightarrow (CH_3)_2\dot{C}{-}OMu$	8.7×10^7	[3]	1.97×10^6 (abstraction)	[26]	
Acetone in *n*-heptane		$(1.05 \pm 0.2) \times 10^8$	[27]	—	—	—
Acetaldehyde in water	$Mu + CH_3CHO \rightarrow CH_3\dot{C}HOMu$	$(1.75 \pm 0.20) \times 10^8$	[27]	3.1×10^7	[9]	5.6
2-Butanone in water	$Mu + CH_3C({=}O)CH_2CH_3 \rightarrow CH_3\dot{C}(OMu)CH_2CH_3$	$(9.9 \pm 1.0) \times 10^7$	[27]	$(3.82 \pm 0.32) \times 10^7$	[26]	2.6
3-Pentanone in water	$Mu + CH_3CH_2C({=}O)CH_2CH_3 \rightarrow CH_3CH_2\dot{C}(OMu)CH_2CH_3$	$(8.4 \pm 1.5) \times 10^7$	[27]	—	—	—
Cyclohexanone in water	$Mu + c\text{-}C_5H_{10}C{=}O \rightarrow c\text{-}C_5H_{10}\dot{C}{-}OMu$	$(1.17 \pm 0.20) \times 10^8$	[27]	—	—	—
Acetonitrile in water	$Mu + CH_3C{\equiv}N \rightarrow CH_3\dot{C}{=}NMu$	5.1×10^7	[28]	3.6×10^6	[29]	14

References

1 Ito, Y., Ng, B.W., Jean, Y.C., and Walker, D.C. (1980). *Can. J. Chem.* 58: 2395–2401.
2 Mezyk, S.P. and Bartels, D.M. (1994). *J. Phys. Chem.* 98: 10578–10583.
3 Percival, P.W., Roduner, E., and Fischer, H. (1979). *Adv. Chem. Ser.* 175: 335.
4 Mezyk, S.P. and Bartels, D.M. (1997). *J. Phys. Chem. A* 101: 1329–1333.
5 Lossack, A.M., Roduner, E., and Bartels, D.M. (1998). *J. Phys. Chem. A* 102: 7462–7469.
6 Buxton, G.V., Greenstock, C.L., Helman, W.P., and Ross, A.B. (1988). *J. Phys. Chem. Ref. Data* 17: 513–886.
7 Lossack, A.M., Roduner, E., and Bartels, D.M. (2001). *Phys. Chem. Chem. Phys.* 3: 2031–2037.
8 Jean, Y.C., Ganti, R.L., Cheng, K.L. et al. (1986). *Hyperfine Interact.* 32: 813–817.
9 Neta, P., Fessenden, R.W., and Schuler, R.H. (1971). *J. Phys. Chem.* 75: 1654–1666.
10 Roduner, E., Louwrier, P.W.F., Brinkman, G.A. et al. (1990). *Ber. Bunsen. Phys. Chem.* 94: 1224–1230.
11 Stadlbauer, J.M., Ng, B.W., Walker, D.C. et al. (1981). *Can. J. Chem.* 59: 3261–3266.
12 Chambers, K.W., Collinson, E., and Dainton, F.S. (1970). *Trans. Faraday Soc.* 66: 142–162.
13 Beckert, D. and Mehler, K. (1983). *Ber. Bunsen. Phys. Chem.* 87: 587–591.
14 Louwrier, P.W.F., Brinkman, G.A., and Roduner, E. (1986). *Hyperfine Interact.* 32: 831–835.
15 Roduner, E. (1984). *Hyperfine Interact.* 19: 785–791.
16 Ganti, R. and Jean, Y.C. (1986). *Hyperfine Interact.* 32: 819–823.
17 Roduner, E., Brinkman, G.A., and Louwrier, P.W.F. (1984). *Chem. Phys.* 88: 143–153.
18 Jean, Y.C., Brewer, J., Fleming, D.G. et al. (1978). *Chem. Phys. Lett.* 57: 293–297.
19 Jean, Y.C., Ng, B.W., Stadlbauer, J.M., and Walker, D.C. (1981). *J. Chem. Phys.* 75: 2879.
20 Karolczak, S., Gillis, H.A., Porter, G.B., and Walker, D.C. (2003). *Can. J. Chem.* 81: 175–178.
21 Stadlbauer, J.M., Ng, B.W., Ganti, R., and Walker, D.C. (1984). *J. Am. Chem. Soc.* 106: 3151–3153.
22 Rhodes, C.J., Morris, H., and Reid, I.D. (2001). *Magn. Reson. Chem.* 39: 438–442.
23 Wu, Z., Stadlbauer, J.M., and Walker, D.C. (1992). *J. Am. Chem. Soc.* 114: 3988–3989.
24 Neta, P. and Schuler, R.H. (1972). *J. Phys. Chem.* 76: 2673–2679.
25 Evers, E.L., Jayson, G.G., Robb, I.D., and Swallow, A.J. (1980). *J. Chem. Soc., Faraday Trans. 1 F* 76: 528–536.
26 Mezyk, S.P. and Bartels, D.M. (1994). *Can. J. Chem.* 72: 2516–2520.

27 Stadlbauer, J.M., Venkateswaran, K., and Walker, D.C. (1997). *Can. J. Chem.* 75: 74–76.

28 Stadlbauer, J.M., Ng, B.W., Jean, Y.C., and Walker, D.C. (1983). *J. Phys. Chem.* 87: 841–844.

29 Draganic, I., Draganic, Z., Petkovic, L., and Nikolic, A. (1973). *J. Am. Chem. Soc.* 95: 7193–7199.

Index

Muon Spin Spectroscopy: Methods and Applications in Chemistry and Materials Science, First Edition. Donald G. Fleming, Iain McKenzie, and Paul W. Percival.